Using Computers in Chemistry and Chemical Education

Edited by

Theresa Julia Zielinski
and
Mary L. Swift

American Chemical Society
Washington, DC

D1451535

Library of Congress Cataloging-in-Publication Data

Using computers in chemistry and chemical education / Theresa Julia
Zielinski and Mary L. Swift, editors.

    p.   cm.
    Includes bibliographical references and index.
    ISBN 0–8412–3465–5

    1. Chemistry—Data processing. 2. Chemistry—Computer-
assisted instruction. I. Zielinski, Theresa Julia, 1941–   . II. Swift,
Mary L., 1946–   .
QD39.3.E46U84   1997
542′.85—DC21

        97–14325
        CIP

Dedication

To all those faculty and students who have spent innumerable hours writing and testing software for doing and learning chemistry.

About the Editors

Theresa Julia Zielinski is a professor of chemistry in the Department of Chemistry and Physics at Niagara University. Prior to joining the Niagara faculty she held a tenured position at the College of Mount Saint Vincent in the Bronx borough of New York City, N.Y. She has taught a variety of chemistry courses for chemistry majors and nonscience majors for 35 years. She received her M.S. and Ph.D. in physical chemistry from Fordham University in the Bronx.

After 15 years of active computational chemistry research spanning the fields of quantum chemistry and molecular dynamics, she has turned her efforts to focus on the uses of computers in chemical education. The major thrust here is to develop instructional materials for the physical chemistry curriculum using commercially available software. These materials are designed to take advantage of the research in the learning styles and intellectual development stages of adults and young adults and to create more active learning environments within which students can learn more effectively and efficiently. She is currently creating a WWW site for physical chemistry instructional materials, http://www.niagara.edu/~tjz. Other current research interests include (a) elucidation of the fine structure of DNA mini-helices through molecular dynamics computer simulations and (b) structure function relationships of biomacromolecules. She has used computers in teaching and research for her entire 35-year career.

Dr. Zielinski has authored or co-authored 36 highly cited peer-reviewed research papers and in the past five years presented over 14 talks at American Chemical Society national meetings. She is a member of the Chemical Education, Computational Chemistry, Biological Chemistry, and Physical Chemistry divisions of the American Chemical Society. She currently serves on the Computers in Chemical Education and Physical Chemistry Examination committees of the American Chemical Society Division of Chemical Education.

Mary L. Swift is an associate professor of biochemistry and molecular biology in the College of Medicine at Howard University, Washington, D.C. She received her B.S. in general physical sciences (chemistry) and Ph.D. in biochemistry from the University of Maryland at College Park, returning there later to earn a second B.S. degree in computer and management information science.

For more than 20 years Dr. Swift has been involved with graduate and professional biochemical education. During this time she has designed and implemented several new courses as well as laboratory modules. Among these have been an "Introduction to Computing and Spreadsheets" module in the Biochemistry Laboratory Course; the development of a three-credit-hour course, "Use of Computer Technology in the Life Sciences"; and the introduction of computer simulations and modeling of advanced biochemical/biological problems in upper-level biology courses.

She was selected to facilitate a post-ChemConf '93 discussion session and to organize and chair the session, "What Chemists Need to Know About Computing" at the 208th Annual Meeting of the American Chemical Society. She co-chaired a two-day session at the 210th Annual Meeting of the American Chemical Society, "Computers as Innovative and Effective Classroom Partners". Dr. Swift has served as a software reviewer for the *Journal of Chemical Information and Computer Science*.

Contents

Part IV. Computational Chemistry in the Curriculum

Part V. Teaching Chemistry with Computers

Author Affiliations

Daniel E. Atkinson
Chemistry Department
University of California at Los Angeles
3123 Malcolm Avenue
Los Angeles, CA 90034

J. Phillip Bowen
Computational Center for Molecular
  Structure and Design
Department of Chemistry
The University of Georgia
Athens, GA 30602

Carol Carr
Chemistry Library
University of Pennsylvania
Philadelphia, PA 19104-6323

Martin S. Ewing
Science and Engineering Computing
  Facility
Yale University
New Haven, CT 06520-8267

Robert G. Ford
English Department
Central College
Houston Community College System
Houston, TX 77004

James B. Foresman
Chemistry Department
York College of Pennsylvania
York, PA 17405

Warren J. Hehre
Wavefunction, Inc.
18401 Von Karman Avenue, Suite 370
Irvine, CA 92715

Carol Sweeney Judd
Chemistry Department
Central College
Houston Community College System
Houston, TX 77004

Peter C. Jurs
Department of Chemistry
Pennsylvania State University
University Park, PA 16802

Steven K. Lower
Department of Chemistry
Simon Fraser University
Burnaby, BC V5A 1S6, Canada

Peter Lykos
Chemistry Department
Illinois Institute of Technology
Chicago, IL 60616

Jeffry D. Madura
Department of Chemistry
University of South Alabama
Mobile, AL 36688

Robert Megargle
Department of Chemistry
Cleveland State University
Cleveland, OH 44115

Scott H. Northrup
Department of Chemistry
Tennessee Technological University
Cookeville, TN 38505

John P. Ranck
Department of Chemistry
Elizabethtown College
Elizabethtown, PA 17022

Karen D. Rappaport
Hoechst Celanese Technical Center
86 Morris Avenue
Summit, NJ 07901-3956

Kenneth L. Ratzlaff
Instrument Design Laboratory
Department of Chemistry
The University of Kansas
Lawrence, KS 66045

Frank Rioux
Department of Chemistry
St. John's University
Collegeville, MN 56321

Arleen N. Sommerville
Carlson Library
University of Rochester
Rochester, NY 14627-0236

Mary L. Swift
Department of Biochemistry and
Molecular Biology
College of Medicine
Howard University
Washington, DC 20059-0001

Alexander Tropsha
Laboratory for Molecular Modeling
School of Pharmacy
University of North Carolina
at Chapel Hill
Chapel Hill, NC 27599-7360

F. S. Varveri
N.R.C. "Demokritos"
Institute of Physical Chemistry
P.O. Box 60228
15310 Aghia Paraskevi Attikis, Greece

Nancy Woo
Corporate Clinical and
Regulatory Systems
Pfizer Inc.
235 E. 42nd Street
Mail Stop 150-05-41
New York, NY 10017-5755

Sidney H. Young
Department of Chemistry
University of South Alabama
Mobile, AL 36688

Theresa Julia Zielinski
Chemistry Department
Niagara University
Niagara University, NY 14109

Preface

Over the past 30 years computer use in chemistry has expanded beyond the highly specialized applications of the physical and analytical chemists into an essential tool for every chemist. At first, data analysis by computer required that the chemist have knowledge of programming, and usually, access to a mainframe computer. For example, most of the early efforts in computation of molecular properties with semiempirical molecular orbital programs were done in batch processes with locally written or locally adapted software. Today the situation is very different. Chemists have local control of the computers that they use for most of their work (desktop or laptop systems) and are able to find commercially available software that fulfills all of their routine computational needs and much of their more sophisticated scientific requirements (spreadsheets, equation engines, modeling and visualization programs). Now, most chemists do not have personal control of the development of chemical software that the pioneers of chemical computing enjoyed. Nevertheless, modern chemists routinely and facilely use computers for instrument control, data reduction, molecular modeling, personal productivity, and access to on-line information resources. Control for most chemists, especially those in chemical education, now primarily finds expression in the adaptation of the features of standard software packages to the solution of chemical problems and creation of teaching scenarios used for educating chemistry students.

The adoption of computers in chemical education parallels the growth of their usage in research, albeit with a substantial time lag. An unfortunate aspect of the evolution to the current state of computer usage in chemical education is that the application of computer technology in the academic arena frequently begins (and ends) with using it as a supplement, or add-on, to traditional instructional methods. This is contrary to the practice of modern chemistry where computer technology is central to the effort, replacing older, less efficient practices and techniques.

The American Chemical Society recognizes that the current level of computer use in chemistry is pushing the need for increased computer training and literacy for postsecondary chemistry students. It is taking a strong leadership position in fostering the greater use of computers in undergraduate and graduate chemistry programs. For example, the society sponsored the on-line chemistry conferences "Faculty Rewards: Can We Implement the Scholarship of Teaching?" in the fall of 1995 and ChemConf, "The Uses of Technology in Chemistry Education" during the summer of 1993. The latter was followed by two symposia entitled "What Every Chemist Needs to Know About Computers" at the Fall 1994 ACS National Meeting in Washington, DC Some of the chapters in this book had their beginnings in these symposia.

These recent ACS-supported activities build on the early literature for the use of computers in chemical education. For example, *Computer Education of Chemists*, edited by Peter Lykos, was one of the first books devoted solely to this topic. Reflecting the state of the technology at that time (1984), most chapters stressed the need for programming skills. Within a few years other books appeared that also emphasized programming ability in BASIC (Jurs, Isenhorn, and Wilkins, *Programming for Chemists, An Introduction*), or PASCAL (Filby and Klusmann, *Turbo PASCAL for Chemists, A Problem Solving and Practical Approach*). In addition to these, chapters in on-going series such as "Reviews in Computational Chemistry" have offered guidance in the use of computers for data acquisition in the undergraduate laboratory, the analysis of data and extraction of chemical parameters from the data, computer programming, and other aspects of computer literacy.

One interesting text, *Computer Aids to Chemistry,* edited by Vernin and Chanon, made the important connection between pedagogical issues and the appropriate use of computers in the training of chemists. The leadership for using computers in chemical education is, of course, found in the *Journal of Chemical Education.* This journal continues to be a valuable resource for ideas on implementing appropriate computer exercises in the chemistry curriculum. This implementation is becoming less and less platform-dependent.

Given the widespread use of computers in the practice of chemistry and the advances in commercially available software and hardware, it is time to address the changes in curriculum and pedagogy required for the appropriate training of the next generation of chemists. The major goal of our effort in this volume is to provide chemistry faculty, graduate students, and others interested in science education with a resource for the development and implementation of effective and modern uses of computers for chemistry instruction. The contributing authors have developed chapters that are grounded in the realities of the current computing environment at a wide range of postsecondary institutions. The focus is on bringing modern computer usage into the classroom by providing the reader with a vision of what is done in research and industry and how that can be translated into real classroom experiences.

Those readers of this book who may be computer novices will have recognized the importance of this tool for the next generation of chemical scientists. For this audience we provide some support as they proceed to develop computer competencies necessary for implementing meaningful changes in the curriculum. Other readers who are comfortable using computers would move immediately to implement some of the ideas presented here or better yet will go on to create new and innovative classroom applications and curriculum approaches that depend on the more effective and insightful use of computers. All teachers would benefit as these developers of curriculum materials share their work with the wider academic community.

This book is intended to serve as a resource for chemical educators to use to expand their vision of computer use in the classroom. This is accomplished in part by presenting some of the outstanding applications and examples found in the various branches of chemical research (industrial and academic) as well as those from educational environments. Since all important topics cannot be included in one book, we chose a set of major applications that we think are needed for chemical computer literacy. These chapters present computer usage through the examination of the significant methodologies employed in chemical research: acquiring, analyzing, and visualizing data; computing molecular properties, chemical reactivity, structure, and function; visualizing molecules; and accessing the literature. It is the opinion of the editors and authors that all of the concepts explored in this book need to be introduced as early in the curriculum as practicable and used systematically throughout the curriculum. Thus, undergraduate and graduate students can gain a vision of how computers are used in the discipline today and thereby be prepared for the advances of tomorrow. This approach serves all of the groups of students that use chemistry as a starting point for careers in industry, health sciences, chemical research, etc. Of special interest is the usefulness of the material in this book for the preparation of science teachers at all levels.

We believe that the authors have presented material illustrating chemical concepts that will be necessary for students to master for the future. In documenting the efficacy and significance as well as the limitations of computers in the work of chemists and in chemical education, we are presenting readers with the tools needed to implement change in their own environment.

Acknowledgments

The editors thank Donald Rosenthal for apprising us of this opportunity and for his unfailing support of our efforts. Many at the American Chemical Society helped throughout this process—Cheryl Shanks, Barbara Pralle, Maureen Matkovich, and all others at the ACS Books Department who helped bring this project to fruition. The continued advice and support of Masayuki Shibata have been invaluable. MLS gratefully acknowledges the generosity of Jennifer S. Fajman, director of the Academic Information Technology Service at the University of Maryland, who has granted me an Internet account. MLS further acknowledges the National Science Foundation which provided partial support for this work through grant DUE #9452475 from the Division of Undergraduate Education. TJZ acknowledges that partial support for this work was provided by the National Science Foundation Division of Undergraduate Education through grant DUE #9354473.

Part I
A View of the Codependence
of Chemistry and Computers

Beyond Lecture—A Vision of the Computer in the Modern Chemistry Classroom

Theresa Julia Zielinski
Mary L. Swift

"A teacher certified in 1890 who walked into a classroom today could go to the chalkboard and begin to teach. A doctor certified in 1890 who walked into an operating room today would be bewildered. Education has not changed much in the last 100 years."

Connie Stout, *Educom Review*, Vol. 29, Nov./Dec. 1994

Introduction

In most chemistry courses, the dominant mode of instruction is the lecture in which the instructor presents material, usually based on the textbook, to the class. Since the textbook itself is a compendium of information in condensed form, the instructor in many cases is required to supplement its contents with explanations and descriptions at the blackboard or with prepared texts distributed in class. This is a quintessential example of the traditional educational model. This model, "mastery of a body of knowledge for a complete preparation for a lifelong career," led to the current governing pedagogical paradigm, in which transferring content, primarily through lecture, is paramount.[1] This focus forces the practice of teaching to be dominated by an overbearing emphasis on content that has given life to "the tyranny of the curriculum" as described by Tobias.[2,3] It has stifled creativity in the classroom, denying most students the opportunity to develop critical thinking skills. It is probably at its functioning limit given the degree of information growth in all fields. In addition, this approach denies the student "the real learning opportunity". After all, the teacher is doing the research, making decisions about what is "important", abstracting, organizing, drawing conclusions, and presenting.[4]

The older, constricted view of education has become untenable in an era marked by continuous eruptions of information. We can no longer teach in 4 years, or even

10 years, everything a chemist needs to know. There is just too much. New strategies are needed for learning and for the creation of learning environments. These strategies need to be implemented through the wise use of technology so as to streamline the curriculum and reform the pedagogical approach. Furthermore, these strategies must be more than just faster or fancier ways of transmitting information. Legend has it that the overhead projector was widely used in bowling alleys long before its extensive use in college classrooms. A corollary to this is that sightings of computers in bowling alleys can be taken as a sure sign that they will soon find a permanent and prominent place in the classrooms of chemistry teachers.[5] This says much not only about the slow pace of technological innovation in the college classroom, but also about the willingness of academics to alter the status quo of the types of activities carried on in the classroom. It is interesting that current classroom practice can include the use of computers as super storage and projection devices. In this scenario, the paradigm of teaching or "transfer of information" also remains unchanged.

This old model of education is not enough; it will not prepare the next generation of chemists for the lifetime of self-directed learning required by rapidly changing technology and exploding information resources. A new educational model must be adopted. This new model would be defined by the demands of the work environment, by what the student deserves as preparation for a career, and by the practices of teachers in the classroom. Accommodation of these desiderata raises issues regarding the need for change in the college curriculum and the role of the college teacher in the education of the next generation of chemists.

This new model of education emphasizes the acquisition of skills and abilities that transcend traditional topic boundaries and enable students to cope in the "information age." The skills to be acquired include the wide range of higher order thinking skills such as critical thinking, quantitative reasoning, and communication. These are supported by the ability to use library and information resources and the ability to work in groups toward common goals and objectives. The required paradigm shift is one that changes the focus of instruction from a delivery-dominant, teacher-centered example to a construction-dominant, student-centered example. In addition, the new model is technology-rich, process-oriented, and multidimensional. It engages the mind of the learner in an active and interactive mode, inside and outside the classroom. The emphasis is on developing the student into an independent lifelong learner. Classroom activities must set the stage for this type of learning by fostering student intellectual growth on various levels. This is not an easy task. It will require a complete rethinking of the logistics of instruction and the role of the instructor. However, accomplishment of this feat will afford the student a more seamless transition from the work in the classroom to work in the academic or industrial laboratory setting.[6]

The new educational paradigm that we support is designed to prepare students to be autonomous learners and constructors of concepts. Computers, a technology with extraordinary potential, can be tools that support teachers in achieving this objective. Students in this new system must be prepared to use technology effectively to amplify requisite skills and abilities. The critical question is: Will teachers use computers only as additions to an already full curriculum, or will they change the curriculum and learning environment, using the computer and other technological devices as essential components, and permit more effective learning of chemistry through enhanced critical thinking skills?

In this chapter we explore some of the issues associated with facilitating learning in a college setting. Three examples of alternatives to traditional instruction will be presented for three different student groups. The special needs of these groups and the advantages and disadvantages of the methods will be described. A vision of an ideal situation for teaching with computers will also be presented. It is hoped that instructors in other teaching environments will use these examples as catalysts for innovative approaches and hitherto unimagined methods for quality learning in the information age, some of which may be radical by current standards.

The Computer's Potential

Adoption of computer technology in educational endeavors frequently begins (and ends) with using it as a supplement for traditional instruction. This approach contrasts sharply with the practice of modern chemistry in which computer technology is central to the effort, not merely an adjunct. This dichotomy is the driving force behind the push for increased computer usage, increased computer training, and higher computer literacy in postsecondary chemical education. The innovative teacher has a unique opportunity to use technology not only to support the traditional goals of education—mastery of content and learning to use a tool for doing science—but also to create new pedagogical methods and philosophies of instruction that would facilitate the development of lifelong learning skills in the student, skills that are more appropriate for the modern era.

The lag in incorporating the use of computers in chemical education may be attributed to many factors. Historically, as well as currently, emphasis has been placed on obtaining hardware.[7] Indeed, the presence of state-of-the-art PCs, Macs, and UNIX workstations in a chemistry classroom represents a level of computing power that exceeds by several orders of magnitude what was available to most research scientists, graduate students, and all other chemistry students only 15 years ago. In many cases these powerful computers were purchased with little or no consideration of how they would be used in the curriculum. Teachers often find that there is insufficient suitable software and an absence of an educational philosophy that would permit them to take advantage of the technology.[7] Given such conditions it is understandable that computer usage has not been seamlessly integrated into the curriculum. This is further exacerbated in some cases by the inadequate computer skills of the faculty.

Nevertheless, the dream of using computers for instruction survives, and with the plummeting price of computer power and software, coupled with the pervasiveness of the device in all aspects of science practice, a window of necessity has opened for introducing and using computers in chemistry classrooms. This technology, more importantly, is also providing an opportunity for rethinking the current chemistry curriculum and the pedagogical model that supports it.

Technology will inevitably be a significant partner in restructuring chemical education. It can permit students to acquire two important skills needed for success in their future careers, namely, the ability to access high-quality information and the ability to construct and evaluate concepts based on that information. To achieve both of these, students need access to high-quality, content-rich, self-paced, interactive learning materials that are available on demand from any location—the laboratory, classroom, or home. The burden for producing these materials lies with the faculty as they implement a concept of high-quality education that is based on the "achievement of learning outcomes regardless of how those outcomes are achieved."[6]

The Learning Environment

Preparation of new materials will require new ways of thinking about traditional courses. First, the content of each course must be critically examined and streamlined. It will not be enough merely to place entire textbooks on line or to replace paper with electronic storage media, although both of these seem to be inevitable consequences of the electronic age. A major effort needs to be made to prioritize topics and efficiently create a learning environment that is appropriate for the 21st century. The order of topics should reflect the current state of research in the discipline. In physical chemistry, for example, the emphasis should be on laser spectroscopy and on computer applications for solving chemical problems. No student should be permitted to complete a baccalaureate degree in chemistry without a fundamental introduction to the force field equation used in molecular mechanics and molecular dynamics. The connection of computed properties from these methodologies to thermodynamics should then follow. Next, faculty must identify the modern research issues of interest to students and bind them to an appropriate use of statistical evaluation criteria. Topics that are of special interest to many students and are within their grasp are structure, function, and reactivity of molecules, especially biomolecules, and drug design. We can no longer afford to emphasize 19th-century steam-engine models for fundamental physico-chemical concepts.

The second change in thinking about traditional courses is in how the students interact with instructional material, construct concepts, and assess their own understanding of the course content. Instructors must become fluent with theories and models of student development and learning. This is necessary because the materials that teachers develop will be completely inaccessible to the student if they are written at too high a level or become tedious if written at too low a level. In either case student interest wanes, learning declines, and attitudes about science sour. In the worst case, students develop an "I can't do this stuff" mentality that precludes further learning.

Students too must change. They must become more aware of themselves as thinking individuals by developing their own metacognitive skills. In other words they must learn how to think about their own thinking. Appropriately designed curricula and classroom activities can induce the transition from passive unresponsive students to students who are actively engaged in learning and are able to question themselves and probe their own learning progress by consciously asking "Why did I do that?" or "How did I do that?"

Pedagogical Models

Three principal pedagogical models can be used to describe students as learners. The Perry model describes students as falling into a few major developmental levels: dualist, multiplist, relativist, and committed relativist.[8-12] Most students enter college as dualists and some leave college as dualists. Critical thinking is not possible at this level because this individual is dependent on authority and views knowledge as a collection of facts and information as either right or wrong. Problem solving beyond the application of algorithms is unlikely. Students want to know how to get the right answer and the one correct path to that answer. In this stage they can be very successful operating at the knowledge and application levels of Bloom's taxonomy.[13]

During college, students should move along intellectual development paths that progress from dualism, through multiplism, and on to relativism. The multiplist student appreciates the diversity of viewpoints but is not yet able to evaluate the relative merits of differing points of view. These evaluative skills develop as the student moves to the relativist level. Students who have reached higher levels of the spectrum can understand and use lower level thinking skills, and they can function at the higher taxonomic planes of synthesis and analysis.[12,13] Students early in their development need to grow into the higher levels. Growth is facilitated by challenge and support: challenges at one level higher than their own level and support at their own level.[10,11,14] For example, students at the relativist level can use and appreciate various gas equations of state. They can discuss the advantages of each and choose an appropriate one to use for a given set of circumstances in a problem or experiment. They also can appropriately choose to use any one of several equations for an examination or homework problem. The dualist, on the other hand, wants to know which one is right so that all others can be forgotten. This student will use the van der Waals equation even if the ideal gas law is sufficient for the task at hand or for calculations where the van der Waals equation is less appropriate. Dualists cannot judge the appropriateness of the tool for the task at hand.

In the next model, described by Bodner, the emphasis is on the student who constructs new levels of understanding within the framework of his or her previous experience.[15] In this model, building on the work of Piaget[16,17] and others, the driving force is the continuing search for a "fit" between observation and the mental models or constructs that explain reality. Within this scheme the instructor no longer deposits knowledge into the mind of the student. Rather, learning environments are developed in which active learners, who learn more than passive learners, construct and test understanding, organize experiences in terms of their preexisting knowledge scaffolding, and remodel this scaffolding to accommodate new experiential data. The teacher becomes the orchestrator of the learning experience. The computer and computer simulations become primary tools in the learning process described by this model. By manipulating computer-generated models the students can engage in an exploration of ideas and parameters that can efficiently and effectively help them build the desired level of mastery of a topic. This is equivalent to moving the raison d'être for laboratory exercises, hands-on exploratory learning, into the classroom.

In the final pedagogical model to be discussed, the content-specific logic method of Richard Paul, student reasoning skills are addressed.[18] The important goal in this mode of instruction is to develop in students the ability to ask meaningful questions, construct logical sequences of activities, and evaluate processes within a context or application. This requires movement away from solving artificial, algorithmic problems and toward solving complex, real problems. Essentially, the goal is to get students to figure things out—to reason things through. Here, too, the instructor shifts focus away from course content and toward student intellectual abilities through the application of the elements of reasoning and an understanding of reasoning abilities. The "why" and "how" of understanding become important questions.

All three models for higher order thinking and learning focus attention on the students as learners. In each, the students must have greater control of the learning situation and must be responsible for the outcomes. It is the students who are moving along the paths of intellectual development as they construct an understanding of complex situations and use discipline-specific rules of logic and measures of model adequacy (when and how a model can be useful as opposed to it being simply wrong or right; for example, the flat-earth model is one of the most useful scientific models). The students are at the center, must be active, and must be responsible and accountable for their own learning. The teacher, on the other hand, assumes the appropriate role of guide/mentor to the student. Using computers, cooperative learning, and other such techniques during all or part of a "lecture" facilitates the transfer to a student-centered learning environment.

A False Dichotomy

Even though understanding the developmental levels of students is an invaluable tool for teachers who are planning to move the curriculum into the 21st century, the tools we create for students to use to learn content and to enable students to learn independently must be technology-based and must be based on sound pedagogical principles. We must go beyond delivery of content. We must use imaginative and realistic computer-based activities and must create new ways of looking at student needs, abilities, assessment, and evaluation. An objection raised by some is that the students will not get the essential content when so much of the classroom time is spent on process, assessment, or cooperative learning activities. This objection is without merit. It represents a false dichotomy. Classroom practices that foster active learning and critical thinking are also those that promote content acquisition.[12]

Just In Time Learning

A model that might be useful in educating future chemists is an adaptation of "just in time" industrial techniques. In this approach the students learn in a computer-rich environment where material is not presented in a linear traditional format but, rather, is accessed as needed in a weblike hyperlinked network of information that extends into other courses in the curriculum. The metaphor for this type of instruction is the World Wide Web (WWW). To implement this we need to develop high-quality, content-rich, self-paced, interactive learning materials that are available on demand from various locations on campus and from the home. The materials should be modular, contain hot links, include multimedia presentations, provide assessment feedback, and, most important, go beyond the page-turner model of computer use in education. Furthermore, the hot links should allow the students to call into play any relevant software package so that they can immediately launch an interactive lesson at the instant they are considering a topic of study. Although this is the ideal situation, it is still, for the most part, only on the drawing boards. Many teaching faculties do not have WWW access, and the majority of teaching software packages are standalone units. In preparation for the advent of high-level WWW access and the arrival of integrated exploratory learning materials, it is necessary to use the methods at hand to model what may be possible in the not-too-distant future. Possible scenarios for students who are computer novices or intermediate users follow.

In the 1995 Classroom

Three examples of how computer technology may be used in education follow. Two of these depend on the students having or being able to develop skills in the use and application of commercially available software to solve scientific problems. The other example takes advantage of technology to create an intellectually stimulating and realistic learning environment in which the students do not have to be "computer literate". Each aims to move the student to higher skill levels within the respective discipline of study.

Skill Levels of Incoming Students

Students arrive at postsecondary educational institutions with widely different exposures to computing, varying from none to extensive. This situation occurs for a number of reasons, and because we (as college educators) are forced to deal with these students, it might be instructive to examine how computers are used in grades K–12.

Despite the fact that the United States spends more to educate students in grades K–12 than any other industrialized nation except Switzerland,[19] a significant number of students arrive at postsecondary institutions with little or no facility in using the most important instrument of our era, the computer. Although a detailed discussion is beyond the scope of this chapter, a brief look at the use of computers in grades K–12 is beneficial to the understanding of what college teachers must deal with in trying to incorporate computers into their curriculum. In 1993 it was estimated that even though there was one computer for every 18 students in grades K–12, more than 3,500 schools in the United States had no computers.[19,20] Since education is locally controlled in the United States, there is no uniformity in computer training. Only a few states have established computer literacy requirements or set teacher certification criteria regarding computer expertise.[21] Thus, in those schools with computers the usage varies greatly. Some of the computers are in closets[7,19]; elsewhere they are heavily used, playing a major role in the curriculum.[19,21] The result is that students' opportunities to use computer technology are "accidents" of their being in certain schools. Some schools have integrated computers highly into the curriculum, whereas others have only begun to consider computer usage.

Overall, it is reported that in middle and high schools there are three times more classes in computer education than in computer-enriched mathematics or science courses.[20] In these computer education classes the learning is centered on keyboarding, use of applications (word processing), or "general computer literacy". Disappointingly, of the mathematics teachers who reported using computers as learning tools, only 3% used graphing programs more than five times during the school year, and only 1% used spreadsheets over that time frame. The situation is no better in the science classroom. Only 1% of these used computer-interfaced instruments on more than five occasions in the school year. Even more discouragingly, word processing was the most common computer application taught in mathematics and sciences classes.[21]

Given this situation it is not surprising that a significant number of students, particularly those from disadvantaged backgrounds, enter college without having had the opportunities of many of their contemporaries to learn how to use computers. Furthermore, they are unaware that the computer is much more than a word processor to the active scientist. Thoughtful use of the computer would place at their disposal a powerful tool to find information, model complex phenomena, and solve difficult problems.

One approach to helping such students is to offer short "how to" courses. These are typically sessions of 3 to 6 hours on how to use a popular word processor or spreadsheet and are taught by academic computing services. Unfortunately, the design of such courses ignores the fact that new users need time to develop their skills through trial and error, exploration, and even algorithmic instruction before they can become even modestly proficient with the programs.[20] Moreover, in this setting, the use of a tool is taught as the goal. This leaves the novice user to wonder how to apply these new skills (or if they can be applied) to coursework problems. A vision of how the same tools can be applied to research is even further beyond their grasp.

Courses That Promote Learning

A Course for the Beginner

The best way to teach students to use computers is to incorporate computer use early and frequently into all courses. However, for current students who are traversing a traditional curriculum bereft of such opportunities, other avenues for learning the importance of computing to the profession must be developed. At Howard University beginning graduate students, many of whom have had little computing experience, may elect the course "Computer Applications in the Life Sciences". The major goals of this course are software skills development and "computer literacy" set in the practice of science. Thus, computers are used in the course not only to teach computer applications but also to permit the students to get practice in solving scientific problems or overcoming difficulties in a self-sufficient manner. Every attempt is made to have the course experience reflect as accurately as possible the real world of the scientist by getting the student to function as a professional who recognizes and identifies a problem and then develops the expertise with a tool or method to solve the problem. In achieving success the student builds the confidence to confront new situations.

The exercises and assignments are designed to mimic the way scientists deal with most research problems. The larger problem is broken down into manageable steps, each of which must be addressed in turn. Operationally, for the class, a series of exercises and assignments are charted such that a "continuing" problem is developed. Each assignment must be completed in order for the succeeding one to be workable or for it to make sense. Thus, problems are solved a little at a time, often with students thinking of new approaches as each step is completed. The students do their own "background" work in determining the appropriate statistical analyses, type of graphs, and so on needed to evaluate the data. Frequently, students' questions are met with the admonition "Go to your own source material and 'find out' an answer."

Planning and executing this course was a dynamic process. Initially, the selection of course topics was based on interviews with each of the students enrolled in the course and the consensus from those who participated in the 1993 ChemConf.[22,23] Software-specific capabilities chosen for inclusion were those supported by the Chem-Conf conferees who responded to the survey distributed by Don Rosenthal,[22] and the supporting general computing topics, shown in Table 1, were drawn from those listed by Tom O'Haver.[22,23] The latter are fundamentally important, for not only are basic concepts such as the difference between program files and user-generated files taught, but current software metaphors, such as cut and paste, are also included. Familiarity with these concepts imparts a degree of skill transferability across popular software packages and computing platforms that enables the student to function in a variety of computing environments. Another theme that ran through the semester was that software is evolving rapidly, and to remain a competent user one has to keep learning. The resultant course schedule (Table 2), however, was subjected to alteration as the semester progressed. In particular, the time devoted to certain concepts or skills was determined by the difficulties and/or interests of the students.

Most of the 3-hour class sessions consisted of a brief and usually highly interactive discussion, followed by hands-on work using various Microsoft (MS) Office applications on an MS–DOS platform. The latter is mentioned because Windows relies on the DOS file system, and this necessitates devotion of some time to the operating system. One class session, which was mostly lecture, was devoted to DOS file operations (naming, renaming, directories). The students were quickly moved to Windows by making the first assignment a presentation of a Windows applet to the

---

TABLE I   Computing and software fundamentals included in the course Computer Applications in the Life Sciences.

Computing Fundamentals
    I. Software
        A. Classification
        B. Basic functions of the operating system
        C. File system basics
            1. Directories
            2. Files
            3. (Re)naming, formats, ASCII, copying, deleting
    II. Hardware
        A. Usual components
            1. CPU, monitor, mouse, other input/output
        B. Memory
            1. RAM, ROM, secondary
    III. Navigating windows
        A. Program Manager
            1. Active window/application
            2. Launching an application
            3. Task list
        B. File manager
            1. Directory tree
            2. File types, (re)naming, copying, moving, deleting
            3. File associations
        C. Applets
            1. Notepad, Write
            2. Calendar, clock, alarm
            3. Calculator
            4. Desktop options
            5. Clipboard
            6. Paintbrush
            7. Character map

Software Fundamentals
    I. Spreadsheets
        A. Methods of data entry, formulas, cell addressing, sorting, statistical functions, formatting, graphing
    II. Word processing
        A. Text formatting (continued), cut/paste, object linking and embedding, tables, drawing, lists, equation editor

---

class in the next session. This had two benefits: It took the focus off of the instructor and placed it on the computer early in the course, and, more important, it put the burden of learning about some computing functionalities on the student as an independent learner. The students' approaches to this task set the pattern for their work habits for the rest of the semester! Even though several Windows manuals were in the laboratory, only one student used them. All of the others quickly, and on their own, found the Help menu. Several figured out how to print Help pages. This pattern was followed for the duration of the course—brief overviews of the tasks to be accomplished were given, and the students worked though the mechanics of the

TABLE 2   Computer applications in the life sciences, semester schedule.

Week	Topic
1	Computing basics—software, hardware, DOS
2	Introduction to Windows
3	More Windows
4	Introduction to spreadsheets—EXCEL
5	Data, data entry, simple calculations, data sorting
6	More complex calculations/operations
7	Graphing
8	Midterm examination
9	Introduction to word processing—MS Word
10	File types, text formatting, characters, and symbols
11	Tables, drawings, and equations
12	Dynamic data exchange/object linking and embedding
13	E-mail, telnet
14	FTP, Gopher
15	Final project

operations using Help, although one student consistently sought the authority of a printed bound manual. After 2 to 3 weeks, the students began to discuss with each other the difficulties they were encountering. Usually, these discussions took place around a computer, where some experimenting went on until the students reached a workable solution.

Measures of student growth over the semester are both anecdotal and quantitative. While working on her final project, one student, as she visualized her data in several ways, exclaimed "This is just like research!" On several occasions the students regaled each other (and me) with stories of how they helped their peers and even their research mentors with computing tips. More quantitative evaluation of their growth is reflected by excellent grades on the midterm examination and the outstanding quality of the final course projects. Over the semester the students became more willing to explore and experiment with the software, often sharing their "findings" not only with each other, but with me as well. In the end the students showed many characteristics of the "autonomous learner" (Table 3).[24]

Most students enter this course as Perry dualists. They are resistant to or uncomfortable with the independence that they are granted, seeking the instructor's approval for every action, even to the point of declining to select a file name for their generated work files. The sought-after approval (advice) is met with a question such as "Does that file name make sense to you?" or "What are the DOS file-naming rules?" As the semester progresses and the assignments become more complex, the students learn (from each other) that there is more than one way to accomplish a task. At first they are surprised by this flexibility, but later they revel in it, having moved to a higher level on the Perry model, perhaps multiplism. This progress is similar to

---

TABLE 3 Some characteristics of the autonomous learner.

Methodical	Develop individual plans for achieving goals; establish personal priorities; pay close attention to details of an ongoing project
Self aware	Decide what knowledge and skills to learn; know his or her strengths and weaknesses
Flexible	Able to achieve/abandon goals; have a tolerance for frustration; able to learn through many different modes; be persistent and responsible
Developing skills in the "learning process"	Capable of reporting what he or she has learned in a variety of ways
Developing information seeking and retrieval skills	Identify and know how to use what he or she has learned

---

that described by Bodner,[15] who notes that students construct new levels of skill (in this case) by organizing their own experiences.

Computers in Advanced Courses: A Biochemistry Experience

The goals of developing students into autonomous learners and critical thinkers may be restated in part as follows: Proper education of science students must include opportunities for the students to practice behaving as scientists. That is, students must be challenged with, or learn to ask, difficult significant scientific questions; they must be given access to the tools necessary to find a valid answer; they must be given the freedom to use the tools; and they must be required to justify their conclusions. Achieving this goal, already difficult, is further constrained by the 15-week semester and severely curtailed budgets. Consider the logistics and financial implications of adopting a hands-on set of experiments to characterize the structure–function relationship of a peptide. One aspect of such an investigation might be to determine the primary structure of the peptide using an amino acid analyzer and peptide sequencer, the capital costs of which are over $200,000. Furthermore, one set of analyses could easily cost $3,000. Because sample preparation for these instruments is at least equal parts art and skill, it is apparent that expense (and time required) will balloon with student use. Another disadvantage is that as students become immersed in the myriad manipulations needed, along with the time lag in obtaining results, they are prone to "lose sight" of the requisite scientific decision making process.

One strategy to address this dilemma is the increased use of computers. Properly designed use of this technology would permit storage of, and access to large databases, allow the creation of models that can be tested against (new) data,[25] and allow the creation of interactive simulations. De novo origination of such versatile software is an intellectually challenging, time-consuming task that would require most of us to obtain a significant proficiency with either authorware or an advanced programming language. On completion of the programming phase, the new software should be tested extensively for "bugs" as well as for achieving the desired pedagogical goals. Usually, many revisions are necessary to obtain a refined product.

Fortunately, more and more examples of software that take good advantage of computer technology are becoming available, lifting the burden of authorship. A particularly effective collection of educational software is the BioQUEST Library.[26] The library, consisting of five major components—the core collection, collection candidates, first review submissions, extended learning resources, and support materials—has been built by extensive screening and peer review of submissions from faculty at many institutions. Modules accepted for inclusion are founded on the belief that science is best learned via the practice of collaborative, open-ended, scientific inquiry. Based on constructivism,[15,27] the BioQUEST "3 P" philosophy, *Problem-posing, Problem-solving, and peer Persuasion,* embraces the precepts that science "is a product of rational thought"[15] and that the learning process is incomplete without student reflection on their own work.[15,27] Problem-posing asks the student to wrestle with the difficulties of formulating good research questions. This aids students in their scientific growth as they learn differences between well-formulated textbook problems and the posing of a problem in the laboratory. They begin to understand what makes a problem "interesting", significant, and feasible to study. Problem solving is facilitated by the tools made available in the BioQUEST Library. Software modules provide sets of tools that the student may use to examine a problem. The problems can be complex, allowing for multiple hypotheses and engaging the students in open-ended inquiry often requiring collaboration with their peers to reach an answer. In the process the student learns those attributes that are required to demonstrate that the conclusion to a scientific investigation has been reached, finding that there may not be one "right" answer. Peer persuasion affords students the opportunity for in-depth analysis of their methods and results. This process is essential for the student to gain an appreciation of the ways that scientific theories and paradigms change.[28]

In an effort to enhance the life science laboratory experiences of the students at Howard University, Dr. Muriel Poston (Biology Department) and I adopted the BioQUEST Library. Several factors drove our decision:

1. the lack of appropriate laboratory experiences in several instructional areas;
2. the need to retain students in the scientific disciplines;
3. the lack of opportunities for our students to work with computers and the students' subsequent "computer illiteracy".

The latter posed a major concern and is not unknown in the college community.[29,30] Therefore, a question we needed to explore was: Will student unfamiliarity with computers create a barrier to full utilization of the BioQUEST software? The first students to use the software were given a questionnaire designed to determine their general knowledge of computing (Table 4). Although 12 students rated themselves intermediate or above, their answers to some of the basic computer questions indicated that the students were probably all in the novice or beginner categories. As BioQUEST software is predominantly Mac-based, the students' lack of exposure to the Mac was some cause for concern. Some students offered comments on their questionnaires; for example, one self-described "beginner" stated "They [computers] should be imposed more because I have had 3 years in college and I have not had to use a computer but I know I will later in life." An intermediate user said "To keep up with the rest of the world, it [the computer] should be introduced early in one's educational career."

Subsequently, these students in the Evolution and Biochemistry Laboratory courses were assigned work with BioQUEST modules Evolve and SequenceIt, respectively.

TABLE 4 Student self-assessment of computing knowledge.

Expertise rating (Self Assessment)	
Novice	6
Beginner	10
Intermediate	11
Expert	1
Number of students owning a computer	8
Number of students using a computer for schoolwork/employment	13
Platform used	
DOS or Windows	26
Mac	7

Evolve is a simulation program for the study of selection, gene flow, and genetic drift. SequenceIt permits the user to apply a full range of common protein laboratory techniques, in any order, to an unknown protein. In both courses students were "thrown in, to sink or swim". After a brief exploratory phase, most students easily mastered the computing interface (thus eliminating our early concerns) and were using the software to solve a problem. In the case of Evolve, they set the problem for themselves. Dr. Poston and I observed animated student-to-student interaction with a high level of determination among students to solve the problem themselves.

In an exit survey, students noted that one aspect of the software was particularly effective. By being able to select alternative parameters and then view the results of that alteration in an immediately displayed easy-to-understand graph, they felt that they were able to grasp the scientific principles more easily. Dr. Poston elaborated further by commenting that these students gained a fuller appreciation of the concepts than had students in the past. No comparison can be made for the biochemistry students, as this was the first class to actually "sequence a peptide". Usually, this topic is taught with a pen-and-paper problem. Typically, all experimental methods, the order of application of these methods (without explanation), and all results are described. The student is asked only to assemble the peptide from the data given, not to work out the logic of the analysis itself. Use of the BioQUEST module SequenceIt affords students the chance to select their own analytical methodologies, interpret the results of each experiment, and then assemble the resulting fragments. Students' reactions to their use of BioQUEST software are reported in Table 5.

Our experience demonstrates that technologically naive students can quickly master the vehicle used for the BioQUEST curriculum and that this tool is effective in helping students see scientific problems as being subject to rational experimentation that gives rise to qualitative and quantitative solutions. Such experiences aid students in constructing their own understanding of the science, creating a "fit" between observation and the model.[15] The processes invoked in these learning experiences by adopting the 3 P philosophy are closely akin to those advocated by Paul.[18]

TABLE 5 Student responses after bioquest experience.

SequenceIt

"...it was set up to anticipate questions of varying degrees of difficulty, and allows you to think through each step logically."

"... it was moderately effective because one still has to determine which tests to perform in order for the experiment to be carried out."

"It did not make choices for you. If you did not have any knowledge of protein sequencing and the different reactions you would not be able to use SequenceIt."

"SequenceIt made you think about what you were doing."

Evolve

"The program left room for trial and error."

"...computer experience is not necessarily a prerequisite."

"...allows the student to teach himself with the assistance of an instructor."

"...having the ability to create problems and describe what is happening in the problem" (on what they liked).

Computers in Advanced Courses: A Physical Chemistry Experience

Even after two or three years of study, students can enter a physical chemistry course at lower stages of intellectual development. The mathematically rich environment of the subject further limits their appreciation of their own powers as learners. The dominant form of instruction, almost exclusively lecture, further exacerbates the situation. Students enter passive and remain passive in most courses. One way to change this scenario is to change the nature of the classroom and the expectations of the instructional environment.

Since 1992 the physical chemistry course at Niagara University has followed a nontraditional format; that is, formal lectures have been replaced by a modified mastery learning approach that uses in-class group work based on guided readings and computer exercises. The classes usually consist of five to ten students. Reading guides are distributed to the class for each learning unit at least one week ahead of the assigned time for discussion of that unit. Each learning unit requires at least one computer-based activity for learning a physical concept.

The reading guides and computer exercises are designed to alleviate the feeling of helplessness that students experience when faced with the task of understanding a mathematically rich discipline like physical chemistry. Each guide consists of a learning matrix and a sequence of activities that the students are required to accomplish in a particular learning unit. The learning matrix sets out the list of prerequisites that a student must bring to the learning unit, the learning objectives for the unit, questions, problems, computer exercises that will lead to completion of the learning objectives, and, finally, a list of performance skills that the student can use to demonstrate a basic, intermediate, or high level of mastery. A sample learning matrix is shown in Figure 1. Here we find the matrix elements for the exploration of real gases using the van der Waals equation, a topic that is often studied early in standard physical chemistry courses. Various performance skill levels also are presented in Table 6.

Modified Mastery Learning Chart: The van der Waals Equation/Nonideal Gas Behavior		
Prerequisites	Objectives	Tasks to complete before class
To state the SI units for mass, force, work, and distance. To convert cgs units of volume to SI units and other simple conversions.	To interpret PV diagrams involving phase changes. To read, compare, and interpret data in isothermal and isobaric graphs.	
Be able to sketch and describe figures representing Boyle's law, Charles' law, Guy-Lussac's law and explain each in concise algebraic form.		Carefully examine the objectives for this exercise and make note of the sections in the chapter that address these topics.
Be able to state Avogadro's principle and explain its significance.		Use the information in the chapter and your general chemistry text to polish the skills listed as prerequisites.
Choose correct gas constant with respect to the units in a problem.	To compute pressure/volume for real gases.	Do tutorial for software to be used for this lesson if necessary.
Compute density, molecular weight, etc. for an ideal gas.	To use the concepts of state and equation of state appropriately.	Critically read the assigned chapter from your text. Write answers to questions in the guided reading notes. The guided reading notes must be completed before coming to this class.
	To compare other equations of state to the ideal gas equation.	
	To explain the molecular basis for nonideality and its relationship to the equations of state for real gases.	Focus on the development of the van der Waals equation. Compare the van der Waals equation to the ideal gas law. Describe succinctly in writing the differences between the equations and the molecular basis for the differences.
	To explain the temperature, pressure, and volume behavior of real gases in terms of the compressibility factor. To be able to explain how the compressibility factor depends on temperature and molecular properties of a substance.	Design a strategy for plotting pressure as a function of volume using the computer software of your choice. The temperature is 200 °C for one mole of CO_2 gas. You will need at least 100 points distributed over the volume range of 6.5×10^{-5} to 5.5×10^{-3}.
Familiarity with a spreadsheet program or an equation solver such as Mathcad or Mathematica.	To increase fluency with physical chemistry software tools.	
	To describe one application of the theory of gases in the work of practicing chemists.	

FIGURE 1. Modified mastery learning chart for investigation of the van der Waals equation and nonideal gas behavior. Specific use of prerequisites, objectives, and tasks helps remove some of the mystery about learning. The use of charts of this type places responsibility on the student, who now knows exactly what is expected in any particular learning unit.

TABLE 6   Mastery demonstration levels.[a]

Basic Skills

1. Compute various state properties for gases using the ideal gas equation or other simple gas laws.
2. Compute other properties of a gas from simple gas laws.
3. Be able to relate the fitting parameters of real gas equations such as the van der Waals equation or the Redlich-Kwong equation to molecular properties of the gases. In other words the trends in the values should be related to the molecular properties of the substance.
4. Do simple calculations such as computing the reduced pressure, temperature, and volume of a gas. Also to be able to use other simple equations to compute various parameters for a gas.
5. Compute the root mean square, mean, and most probable velocity of a gas.
6. Use one recommended software package for simple computation and preparation of simple graphs.

Intermediate Skills

1. Interpret graphical representations of gas properties and extract required gas properties of parameters.
2. Use real gas equations of state in calculations of gas properties.
3. Relate compressibility curves to molecular properties of gases.
4. Describe significance of energy distribution for gases in terms of molecular behavior.
5. Use two recommended software packages for simple computation and graphics or one software package for more complex work and more complex graphics.

Advanced Level

1. Use appropriate mathematical forms to evaluate state properties of gases and to justify the choice of form with sound arguments.
2. Compute molecular properties from gas equation fitting constants.
3. Use two recommended software packages for more complex work.
4. Create appropriate graphical representations of gas properties from data or equations.
5. Use nonlinear curve fitting techniques to evaluate fitting parameters for real gases to real gas equations.

[a] Some of the intermediate and mastery level skills will need to be developed over several weeks. Computer fluency is one of these.

The importance of the mastery matrix and skills list is that they set out clear expectations for student progress toward mastery of a topic. They also help students to make connections and to start to develop a web structure for their chemical science knowledge base through a clear set of prerequisites that link their current learning activities to material learned in other courses and, through the learning objectives, tie their studies to expectations. Of particular interest to the discussion in this chapter are the last two entries in the "Tasks" column of Figure 1. In some schools the students may need to learn to use the software for physical chemistry during the course, and tutorials are necessary. Even when the students are computer literate, up to and including the use of a simple spreadsheet or equation engine, the use of warm-up exercises will bring them to the point of full functional ability in the course. Some day, all students will come prepared with these skills, and instruction can proceed to interaction with course content more quickly.

The importance of computer work for effective learning is facilitated by immediate implementation of the software in a specific exercise, whereby students can work in pairs to complete a project. This is essential to reinforce the learning of content and the development of skills necessary to examine the content. The ultimate goal is to have the students immediately see the effectiveness of interactive computer simulation of physical models as represented by the mathematical equations in a chemistry text. The role of the instructor in developing this skill is significant. The instructor must provide the potential critical thinking links for the students to discover in the computer exercise. This can be accomplished through critical questions that are answered at the end of a simulation or through carefully constructed examples that will create a learning disequilibrium via an unexpected result, causing the student to stop and think or register an "ah ha!" Finally, an exercise should be capped by a suitable display of performance skills. Sample learning measures are summarized in Table 6. The emphasis on adequate use of essential software is clearly a significant component of the mastery level performance for any chemistry student.

An example of a learning disequilibrium situation that can be created for students is represented by a computer exercise in which students are asked to calculate the pressure–volume curves for a van der Waals gas at different temperatures. An exercise of this type can substitute for the traditional class lecture on a topic and, as such, should be used in place of a traditonal lecture. In this exercise the students are asked to develop a series of curves for the gas over a wide temperature range. Typical student spreadsheet and graphic results are shown in Figures 2 and 3. Because the students have had some experience with the van der Waals equation in earlier chemistry courses, they are suprised to see the pressure dip into the negative region. Students usually express disbelief in the result by asking what mistake they made in setting up the calculation, since the computer cannot make a mistake. The vulnerable position that the students enter through an exercise like this makes them very teachable. Their interest is piqued and they are ready to explore the limits of mathematical models, seek a deeper interpretation of the phase phenomena associated with the mathematical model, and then extend the concept to physical properties in a way they may not have considered before. In the case of the van der Waals equation, the extension of the gas equation to a description of the liquid state, the interpretation of the negative loop in terms of the tensile strength of a liquid, and the further extension of this phenomenon to the problems of cavitation at the surface of propellers and movement of liquids to the tops of tall trees are intriguing applications of physical chemistry to engineering and biology.[31-33] Further extension to vapor pressure, surface tension, and the vapor pressure across curved surfaces is immediately possible. The connection to Δ G at equilibrium is easily made.[34] The overall result is the webbing of scientific connections from the very first computer exercise in a course.

The essential features of this exercise that can be ported over to other topics in physical chemistry or to topics in other courses in the chemistry curriculum are to:

1. give students clear goals and objectives for each lesson;
2. provide them with a well-designed computer exercise for each lesson to be used in the classroom under teacher supervision in a collaborative interactive mode;
3. design the exercise to be more than a page-turner activity or an animation that students merely watch or interact with in only a limited fashion;

P=	(RT/(V-b))-a/V**2		
R=	8.314		
T=	373	283	233
V	P		
0.00006			10863571
0.000065			714404.97
0.00007		11899707	-3327399
0.000075		8132913.7	-4737055
0.00008		6204410.2	-4940362
0.000085		5242592.3	-4584831
0.00009		4805111.1	-3983472
0.000095	18962466	4655391.2	-3292984
0.0001	17720803	4662164	-2592635
0.000105	16761366	4750772.2	-1921780
0.00011	15996434	4878157.7	-1298662
0.000115	15368791	5019413.4	-730240.8
0.00012	14840230	5160281.7	-217467.3
0.000125	14384705	5292845.7	241812.88
0.00013	13984127	5412993.2	651252.09
0.000135	13625723	5518897.5	1015105.5
0.00014	13300329	5610092.5	1337739
0.000145	13001275	5686905.6	1623367
0.00015	12723639	5750106.7	1875922.1
0.000155	12463737	5800691.6	2098999.7

FIGURE 2. A segment of the van der Waals equation exercise spreadsheet for CO_2. The data in this spreadsheet are used to generate Figure 3.

4. expect students in upper-division courses to create their own mathematical representations of textbook material in a spreadsheet or an equation engine environment;
5. design each exercise to permit exploration of a variety of parameter ranges and graphical representations;
6. design exercises that elicit critical decisions from the students in terms of asking the question, What would happen if I changed this or did that and what is the physical significance of the representation that is presented on the computer screen.

At every turn the students should be brought back to the empirical world and forced to relate their mathematical models to real physical phenomena and decision-making processes such as those practiced by bench chemists and others who use chemistry as the basic science discipline underpinning their specific research or work environment.

Guided reading, group discussions, and thought-provoking computer exercises are useful tools for promoting learning and intellectual growth. These activities provide support to dualist and multiplist students while challenging students to move to higher developmental levels. They provide a means for teachers to create learning

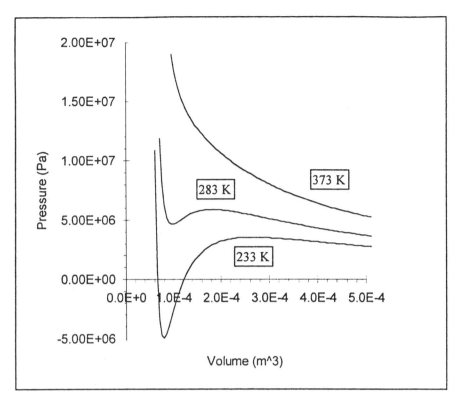

FIGURE 3. Pressure volume plot of the van der Waals equation for CO_2 at three different temperatures. At 283K the van der Waals loops would permit an estimate of the vapor pressure of liquid CO_2 at that temperature. The negative pressure loop also permits application of the equation to the liquid phase and the concept of liquid tensile strength. Students can quickly adjust the temperature in the spreadsheet to explore other properties of the van der Waals equation, including determination of the critical point.

exercises and assessment strategies that move beyond the knowledge and application levels of Bloom's taxonomy. Simultaneously, they provide students with a systematic opportunity for constructing their own knowledge and developing valuable investigative skills. The classroom *cum* computer becomes an exploratorium in which rich context-specific and interconnected learning becomes possible. The format of guided or directed inquiry supports the principles of logical critical thinking in chemistry and other disciplines that depend on a firm grasp of molecular science principles. It can support the design of materials that include a full spectrum of question types, ones that range over the entire taxonomy of learning.

Some may say that this approach is just another form of spoon-feeding students. Shouldn't students be able to listen to a lecture, read a chapter logically and systematically on their own after listening to the lecture, and then perform the required computer work as part of the normal homework load? This opinion has no merit. First, lecture itself is the paramount form of spoon-feeding, one that promotes passivity in students. Second, students in this generation don't know how to read a scientific text critically because their major experience is with didactic instruction. Reading guides provide them with a quick way to get the skills they need to learn more and learn it more effectively. Third, students are well-known for procrastinating and reserving serious study for the last days before a test. The guided inquiry method forces students to work at a steady pace. Fourth, active students learn more and retain more of what they study.[35] Fifth, the curriculum is already crowded with content. No student has the experience or time to create unaided the necessary degree of order and connectivity of content within the normal day. Traditional lecture was the old way of imposing order on content for students, a way that left them passive and unconnected. The new way holds out the promise of making them active participants in the ordering of their learning and their content under our guidance.

Three years of teaching experience using guided inquiry in physical chemistry classes clearly demonstrates that an alternative mode of instruction is as effective as lecturing for "coverage" of content, as reported by Nelson,[12] and is more effective for promoting student learning in higher education, as reported by Bonwell and Eison.[35] Indeed, content is not sacrificed when process is modified.

Conclusion

A triad of pressures on the chemistry curriculum, the expanding use of computers in the practice of chemistry, the volume of chemical information, and the skills demanded by employers all require us to reexamine not only the content we teach and how it is accessed by students but also the educational paradigm we use during instruction. The very complexity of the content forces us to teach critical thinking skills in addition to the material compiled in various texts. This necessitates the inclusion of learning skills and metacognitive skills as normal parts of every learning experience. This is best done in an interactive mode, one that is ideally represented by computer laboratory environments.

In this chapter we presented ideas and examples of ways in which an educational model based on developing life-long learning habits for a rapidly changing discipline may respond to the challenges posed by the pressures found in chemical education. The elegance of the model is that not only does it make use of a fundamental tool, the computer, in meeting these challenges, but the use of this tool itself in the practice of chemistry is one of the challenges that we are called on to meet. Furthermore, the essential connection of chemistry to a wide variety of career paths and scientific pursuits across disciplinary boundaries requires that new learning materials be closely linked to real-world situations and applied in various contexts. The effective use of the ubiquitous computer can accomplish this by providing the hammer for forging the links to information, skills development, concept construction, and mastery in chemistry both during and after passing through formal education programs. It is the educators who are called on to shape the new pedagogical paradigm that implements the best technology and pedagogical research for the learners of the future.

Acknowledgments

T. J. Zielinski acknowledges that partial support for this work was provided by the National Science Foundation Division of Undergraduate Education through grant DUE #9354473. M. L. Swift further acknowledges the National Science Foundation, which provided partial support for this work through grant DUE #9452475 from the Division of Undergraduate Education.

References

1. Twigg, Carol A. "The Changing Definition of Learning"; *Educom Rev.* **1994,** *29*(4), 23–25.
2. Tobias, Sheila; *They're Not Dumb, They're Different: Stalking the Second Tier;* Research Corporation: Tucson, AZ, 1990.
3. Tobias, Sheila; *Revitilizing Undergraduate Science: Why Some Things Work and Most Don't;* Research Corporation: Tucson, AZ, 1992.
4. Twigg, Carol A. "The Need for a National Learning Infrastructure"; *Educom Rev.* **1994,** *29*(5), 17–20.
5. Luehrmann, A.; *Science* **1989,** *245,* 15.
6. Twigg, Carol A. "Navigating the Transition"; *Educom Rev.* **1994,** 29(6), 21–24.
7. Bork, A. "Guest Editorial: Why Has the Computer Failed in Schools and Universities?"; *J. Sci. Educ. Tech.* **1995,** *4,* 97–102.
8. Finster, D. C. "Developmental Instruction, Part I. Perry's Model of Intellectual Development"; *J. Chem. Educ.* **1989,** *66,* 659–661.
9. Finster, D. C. "Developmental Instruction, Part II. Application of the Perry Model to General Chemistry"; *J. Chem. Educ.* **1991,** *66,* 659–661.
10. Perry, William G., Jr. "Forms of Intellectual Development in the College Years: A Scheme"; Holt, Rinehart, and Winston: New York, 1970.
11. Belenky, Mary F.; Clinchy, B. M.; Goldberger, N. R.; Tarule, J. M. *Women's Ways of Knowing—The Development of Self, Voice, and Mind;* Basic Books: USA, 1986.
12. Nelson, Craig E. "Skewered on the Unicorn's Horn: The Illusion of Tragic Tradeoff Between Content and Critical Thinking in the Teaching of Science", In *Enhancing Critical Thinking in the Sciences;* Crow, L. W., Ed.; Society for College Science Teachers: Washington, DC, 1989; pp 17–27.
13. Bloom, Benjamin S. *Taxonomy of Educational Objectives. The Classification of Educational Goals. Handbook 1, The Cognitive Domain;* Longman: New York, 1956.
14. Widick, Carol; Simpson, D. "Developmental Concepts in College Instruction"; In *Encouraging Development in College Students;* Parker, C. A., Ed.; University of Minnesota Press: Minneapolis, MN, 1978.
15. Bodner, George M. "Constructivism: a Theory of Knowledge"; *J. Chem. Educ.* **1986,** *63,* 873–878.
16. Herron, Dudley. "Piaget for Chemists: Explaining What 'Good' Students Cannot Understand"; *J. Chem. Educ.* **1975,** *52,* 146–150.
17. Ausubel, David P.; Novak, Joseph D.; Hanesian, Helen. *"Educational Psychology: A Cognitive View",* 2nd ed.; Holt, Rinehart and Winston: New York, 1978.
18. Paul, Richard. *Critical Thinking: What Every Person Needs to Survive in a Rapidly Changing World,* revised 2nd ed.; Binker, A. J. A., Ed.; Foundation for Critical Thinking: Santa Rosa, CA, 1992.
19. Beach, G. "Building a Foundation"; *Comm. ACM* **1993,** *36,* 13–15.
20. Beaker, H. J., "Teaching With and About Computers in Secondary Schools"; *Comm. ACM* **1993,** *36,* 69–74.
21. Litke, D. K.; Moursund, D. "Computers in Schools: Past, Present, and How We Change the Future"; *Comm. ACM* **1993,** *36,* 84–88.
22. ChemConf '93. "ChemConf '93: Applications of Technology in Teaching Chemistry, Summer 1993". Anonymous ftp: info.umd.edu, Path:/Educational Resources/Faculty Resources and Support/Chemistry Conference (CHEMCONF). LISTSERV: CHEM-CONF@umdd.umd.edu.
23. Swift, M. L., Zielinski, T. J. "What Chemists (or Chemistry Students) Need to Know about Computing"; *J. Sci. Ed. Tech.* **1995,** *4,* 171–179.
24. Candy, Philip A. *Self-Direction for Lifelong Learning;* Jossey-Bass; San Francisco, CA, 1991; pp 459–466.

25. Keeves, J. P., Ed., *An IEA Study of Sciences III: Changes in Science Education and Achievement 1970–84,* vol. 10; Pergamon; New York, 1992.

26. BioQUEST Consortium, Beloit College, Beloit WI. Available from The ePress Project, Academic Software Development Group, Computer Science Department, University of Maryland, College Park MD 20742.

27. Jungck, J. R. "Constructivism, Computer Exploratoriums, and Collaborative Learning: Constructing Scientific Knowledge"; *Teaching Educ.* **1991**, *3*, 151–170.

28. Petersen, N. S. and Jungck, J. R., "Problem-Posing, Problem-Solving, and Persuasion in Biology Education"; *Academic Computing* **1988**, *2*, 14–17.

29. Sprague, E. D., "Raw Data to Finished Report: Microcomputer Assisted"; *J. Chem. Educ.* **1993**, *70*, 997–999.

30. Earl, B. L., Emerson, D. W., Johnson, B. J., and Titus, R. L., "Teaching Practical Computer Skills to Chemistry Majors"; *J. Chem. Educ.* **1994**, *71*, 1065–1068.

31. Benson, S. W., and Gerjuoy, E., "The Tensile Strength of Liquids. I. Thermodynamic Considerations"; *J. Chem. Phys.* **1949**, *17*, 914–918.

32. Hayward, A. T. J., "New Law for Liquids:Don't Snap, Stretch!"; *New Scientist* **1970**, *45*, 196–199.

33. Vogel, S., "Nature's Pumps"; *American Scientist* **1994**, *82*, 464–471.

34. Mau, M., and McIver, J. W., "Alternative approach to Phase Transitions in the van der Waals Fluid"; *J. Chem. Ed.* **1986**, *63*, 880–810.

35. Bonwell, Charles C., and Eison, James A., "Active Learning: Creating Excitement in the Classroom"; ASHE-ERIC Higher Education Report No.1; The George Washington University, School of Education and Human Developemnt: Washington DC, 1991.

2

The Evolution of Computers in Chemistry

Peter Lykos

Introduction

In September, 1986, I helped organize a day-and-a-half symposium at the 192nd National ACS Meeting in Anaheim, CA, titled "History of Computers in Chemistry," on the occasion of the 10th anniversary of the ACS Division of Computers in Chemistry. Only selected highlights of this vast subject were addressed at that time. However, I do draw on what I learned at that symposium, appropriately augmented with events since.

Of course, the International Conference on Computers in Chemical Research and Education is part of that history. Little did we dream when we organized the first ICCCRE in 1971 that a biennial series would emerge bringing the ever-evolving story of computers in chemistry to different places in the world. The 11th ICCCRE occurred in July 1995 in Paris.

In 1978, at the 4th ICCCRE in Novosibirsk, we highlighted the advent of the personal computer, suggesting we were witnessing the dawning of a new age. IBM clearly believed that to be the case because in 1981, the IBM PC—and MS-DOS—appeared, and the acceptance of computing in that cottage industry, chemistry, has been accelerating rapidly ever since.

Hardly a day goes by without yet another announcement of a further increase in the power, and decrease in the cost, of integrated circuits on silicon chips. Computer graphics enhancements are of special significance to chemists, given that the paradigm for chemistry is the molecule. The number of commercial products, especially workstations and chemistry-oriented software, is increasing, and vendors are now targeting chemists as an important audience for computer technology. Thus, we are entering the rapidly rising part of the growth curve of computer applications to chemistry.

Yet, only within the past 10 years or so have we seen a significant interest in the chemistry of computers. Of course, that has to do with condensed-matter chemistry as well as with the design and synthesis of individual molecules that can function as computer memory units or perhaps computer function units. The 6th ICCCRE included that focus in 1982. That suggests an overlap with molecular biology, as does the emergence of the neural-net computer. Thus, chemists, especially materials chemists and biochemists, are destined to play an ever-increasing role in the design of computers.

Brief History of the Supporting Technology

The evolution of scientific computers, in contrast to data processing devices, was a direct consequence of high technology needs during World War II. The Charles Babbage Institute, an archival collection of documents and photographs related to the history of computer technology, is the source of photographs showing the first scientific computers (which were a consequence of the war effort) and derivative scientific computers, such as the 4-ft-long and 10-in.-wide arithmetic component from the ENIAC (1945) and the successively more compact components of the EDVAC (1949), ORDVAC (1952), and BRLESC (1962). The latter is included because it marked the beginning of mass production of miniaturized components—printed circuit, single vacuum tube, logical gating package. Leaping ahead to 1989 we had a Cray 1, Serial Number 14, being prepared as an exhibit at the Smithsonian Institution's Air and Space Museum. That operational machine was no longer cost effective to run!

In 1954, following the end of World War II, IBM introduced the first commercially successful electronic digital computer, the IBM 650, building on IBM's long-standing dominance of punched card-based data processing. Input/output for the 650 was via punched cards, print-out was done on a punched-card-reading accounting machine that was wire-board-programmable for formatting output. Optimistic marketing predictions were that some 200 might be sold. In fact almost 2,000 were sold before successive IBM machines, and by then other vendors' products, rendered the 650 obsolete. Although it was designed to support data processing, scientists and engineers quickly took to that machine.

With the appearance of the 650 and derivative machines, the notion of a higher level language for users emerged—FORTRAN in 1955 (and COBOL somewhat later). Although more sophisticated higher level languages have evolved, FORTRAN remains the dominant computer language of science, alternately praised and damned because of its wide acceptance in science and engineering but with its idiosyncratic constraints and its inherently linear algorithmic flow. The highly entrenched use of FORTRAN is generally viewed as the major inhibitor to the acceptance of inherently more cost-effective and increasingly available parallel hardware.

A major enhancement to computing came with the announcement in 1964 of the IBM 360 series, designed with an operating system such that all codes were upwardly compatible, and on-line secondary memory was used by a new concept in operating systems, OS 360, into which many of the functions previously performed by a human operator were incorporated. Through JCL, the job control language, the user was in command of the machine's various features. Although OS was released with bugs, was quite inefficient, and was difficult for computer center staff to use, it was nevertheless a conceptual breakthrough that influenced the design of all so-called mainframe computers.

At about the same time, experimental chemists were given a distinct enhancement to laboratory science. The Digital Equipment Company (DEC) created the PDP-8, a rack-mounted machine designed for real-time computing. Initially exploited by crystallographers, the PDP8 was soon brought into other chemists' laboratories. Tens of thousands were sold. Subsequently, computer memory costs decreased to the point where experimentalists could afford to program in a higher level language. The PDP-11 series evolved with miniaturized versions, such as the LSI-11, for laboratory applications, and larger versions, notably the PDP-11/70 (the predecessor to the VAX), supporting time sharing.

In 1968 the U.S. Department of Defense Advanced Research Projects Agency (DARPA) embarked on a major project to enhance remote sharing of major computers by DARPA contractors doing research and development. The so-called ARPANET, whereby 50-kB links provided multiple path links between major computers, was put into operation. The radically new technology divided files into packets and shipped those packets over the network, different packets perhaps traveling over different paths depending on the load on each interconnection, with reassembly of the file at the target machine. The key to the ARPANET was BBN's interface message processor (IMP), which received packets and optimally routed them to the next nodal IMP. BBN's packet-switching technology became widely used: today's commercial data network, telnet, is a porting to the commercial sector of ARPANET technology. With the Internet, and the concomitant World Wide Web, we have achieved a widely used network, *owned by no one*, that bypasses the "middle man" and provides inexpensive access to individuals anywhere. The book, "The Internet: A Guide for Chemists," edited by Steve Bachrach and published by the ACS early in 1996, lays out what was available and the extent of use by the chemistry community. Of course, that picture is changing rapidly so Bachrach's book must be considered a snapshot in time.

Finally, the last milestone I want to mention in this brief overview of computing, with direct impact on chemistry, happened in 1971 when a Japanese calculator company asked Intel to design an integrated circuit for a new product. An Intel designer realized that the arithmetic component easily could be made part of the chip, and the Intel 4004 was born. Handheld calculators burst on the scene. From that beginning came more sophisticated chips, personal computers, and, just a few years ago, the Intel Hypercube, a powerful scientific computer made up as an assemblage of 32 processors of the kind used in the IBM-AT personal computer. An interesting part of this story is that the concept and design of that highly concurrent multiprocessor computer was the brainchild of a physicist at Cal Tech—a computer user. But we would be remiss were we not to observe that the handheld calculator continues to evolve—indeed, a graphics display model is now required in the College Board's Advanced Placement in Calculus examination and, hence, in the course preparing the high school student for that examination.

Relevance of Computing to Chemistry

As computer-literate chemists it is important for us to display the various ways computers are impacting chemistry to the following three different audiences: (1) all other chemists, (2) those whose work overlaps with that of chemists, and (3) the interested layperson—especially young persons.

Of those disciplines dealing with the material world, chemistry is arguably the central science. As basic research in other disciplines dealing with the material world advances, increasingly molecular-based perspectives and chemical ways of modeling systems are used.

Chemistry is becoming increasingly interdisciplinary—and fragmented. For example, even though there are 34 technical divisions in the American Chemical Society (we created one of them, Computers and Chemistry, in 1974), large groups of researchers, including many chemists, have formed interdisciplinary societies separate from the ACS. I believe the advent of the computer not only gives us a major enhancement to the practice of chemistry but, in the process, offers *a path to a refocusing and renormalization of the structure of chemistry as a discipline.*

In my view, the core of chemistry is organic plus inorganic chemistry, and its practice involves determination of: (1) algorithms for synthesizing molecules, (2) mechanisms of chemical reactions, and (3) models of molecular structure.

Traditionally, chemistry is thought of as standing on four columns: analytical, inorganic, organic, and physical chemistries. However, physical and analytical chemistries are tools for doing chemistry, and although they have come to be essential tools in that they deal with modeling and with measurement, they are nevertheless not the same as core chemistry as defined here. Chemistry has its own language and models for representing the physical world, distinct from other disciplines. The laws of physics, the language of mathematics, and the instruments provided by engineers are adjuncts to—rather than inherently part of—the core of chemistry.

Additionally, natural language is included because the sharing of chemical knowledge is essential to the practice of chemistry. Machine-based methods for using natural language—in addition to the symbolism peculiar to chemistry—now go substantially beyond just word processing, spell checking, and word pattern recognition.

The use of computers offers better ways of doing chemistry, as follows:

- Design of the synthesis of complex molecules;
- Molecular structure elucidation;
- Literature and data information storage and retrieval;
- Chemical analyses, including all the spectroscopies;
- Real-time data logging and control of experiments;
- Simulation and modeling of complex chemical systems both apparently unchanging in time as well as changing in time.

Continuing application of computer-based methods to those lead toward:

- Greater opportunity for phenomenology;
- Standardization of well-accepted chemical problem-solving methods;
- Enhancements to technology transfer, including enhancements to global communication;
- Improved graphic, oral, and tactile input/output capability at the man–machine interface.

How did this all get started? Three complementary areas represented the major introduction of computers in chemistry: chemical information systems, theoretical chemistry, and crystallography.

Chemical Information Systems

Chemistry is a physical science based on the concept of a molecule. Just as we have alphabets of 25 to 30 letters from which hundreds of thousands of words have been created and, in turn, from which millions of books have been written, so do we have 100 elements from which millions of molecules have been created and, in turn, from which living and nonliving assemblages of molecules have been created. Just as we have dictionaries and card catalogs for words and books, so do we have a corresponding need to organize and to systematize information about molecules as well as the properties of bulk matter, which may be modeled as assemblages of molecules. Record keeping and report generation are fundamental to the functioning of society in general. Thus, the techniques and methods of file generation and handling created for accountants—and librarians—would be brought naturally to the service of chemistry as they evolved.

Punched cards early on became a useful technology for chemists interested in systematizing their literature reference files where stacks of notched cards could be sorted with a pointed rod. Mechanization of such systems, notably by IBM, accelerated the use of punched-card-based technology to the point where, by the mid 1940s, serious study began of designing punched card systems for correlation of chemical structure with physiological activity.

Within the ACS, machine-aided indexing, classifying, and coding techniques began at about the same time. So pervasive was the machinable punched card technology that serious consideration was given at that time to publishing *Chemical Abstracts* as boxes of punched cards!

The first major pioneer for establishing a search system was the Dow Chemical Company, which, in the early 50s, could search 1,000 compounds for desired structural features in a few seconds.

Of course, since that time the ACS Chemical Abstracts Service (CAS) has become an international resource, a behemoth with annual revenues of $200 million, serving scientists via networking worldwide and with a chemical registry system listing over 12 million compounds, with the number added every year steadily rising. One million new ones were added in 1995. CAS provides basic databases for other service organizations in addition to providing a range of services directly to the remote user at a computer terminal.

However, another computer-driven factor needs to be taken into account: namely, authors are creating their papers in a machinable form that routinely can be transmitted electronically over space and time. Disciplines other than chemistry routinely generate journals working from copy generated by authors. Indeed, at Los Alamos, a physics journal is in operation—in principle unattended by a human—to which authors contribute articles that are unrefereed. Why not chemistry?

Furthermore, high-density, extremely large capacity random access devices, CD-ROMs, now are available cost-effectively. As information sources, they can be read by programmable readers so that they can be searched by machine. *Dissertation Abstracts* was distributed to university libraries in CD-ROM form, which was improved by University Microfilms International. That product came to be widely regarded as the single most important key to recently completed scholarly research that has yet been made available nationally. When the test phase was completed, some of the participating libraries found it overpriced and discontinued the service. In the meantime libraries committed to bound books, reports, and serials are finding themselves in increasingly cramped physical space and increasingly cramped budgetary space. The U.S. NAS/NRC report, "Opportunities in Chemistry", displays a stack of books 14 ft high and notes that those are the 75-volume 1977–1981 10th Collective Index of Chemical Abstracts. Of course, they can be searched only manually.

Theoretical Chemistry

Theoretical chemistry generally spans the entire range of chemical modeling systems using the laws of physics and the language of mathematics. Generally, chemists view the physical world of bulk matter from a molecular perspective. Classical mechanics, quantum mechanics, and statistical mechanics applied to molecules—and to assemblages of molecules—lead to detailed descriptions potentially capable of high accuracy and fidelity. Generally, the complexity of the application of mathematics, including numerical methods, necessitates considerable simplification in order that the models can lead to equations involving physical observables that can be compared with experiments. One is restricted to fairly crude models of complicated systems or fairly good—but still approximate—models of extremely simple individual molecules.

The first major focus on computers in theoretical chemistry came within quantum chemistry. Here, representation of the wave function for a molecule generally was based on sets of atomic orbitals, each orbital a mathematical function with its coordinate's origin at one of the nuclei of the constituent atoms. Working through the Schrödinger equation for such systems, and including coulomb repulsion between pairs of electrons, leads to definite integrals, which may include in the integrand four different atomic orbitals with their mathematical origins at four different nuclei (so-called four center integrals)—hence, the integrals bottleneck to progress.

In 1951 an international conference was held at Shelter Island near Long Island in New York, N.Y. Most of the leading figures in quantum chemistry were present. Two persons there symbolized the phasing out of desktop mechanical calculators (Prof. Kotani from Japan) and the phasing in of electronic digital computers (Prof. Roothaan of the United States). That was the first major conference with a focus on the emerging computer in theoretical chemistry.

For me the transition was quite dramatic. In 1953, while a graduate student, I spent the summer building an analog computer to solve secular equations for up to 6×6 for the nonorthogonal problem. Two years later I learned how to program the newly arrived IBM 650, the first commercially available electronic digital computer. Four years after that, in 1959, I introduced programming in machine language for the UNIVAC 1105 as a regular part of a required undergraduate course (physical chemistry laboratory least-squares fit of vapor pressure data to the Clausius Clapeyron equation). That was the first introduction of electronic digital computing as a regular part of a required chemistry course. Four years after that, in 1963, I made programming accessible to high school students from the greater Chicago area on the IIT computer. Over the next 10 years, 15,000 high school students from some 200 high schools—and some 1,200 high school teachers—learned to program and were provided remote access to IIT's IBM 360 using ASR 33 teletypes from their schools.

Recently, there was a major symposium honoring John Pople's 40 years as a theoretical (quantum) chemist. His contributions exemplify theoretical chemistry at its best. He made major advances in adapting physics and mathematics to chemical problems. He then packaged his advances algorithmically in machinable and transportable form such that many other chemists could use those tools to deal with specific chemical problems. The integrals bottleneck referred to above was broken by Pople in two ways.

The first was through the invention (Pariser and Parr separately did the same) of the neglect of differential overlap approximation, NDO. That is, where a product of two atomic orbitals with origins at different atomic nuclei appears in an integral representing the charge distribution of one electron, that integral is set to zero, as the product of two functions over all space would be small if nowhere in space would both functions have significant magnitude. That enormous simplification, embedded in Roothaan's LCAO-MO-SCF method for finding optimal molecular orbital descriptions of molecules, became the basis for the widely and frequently used NDO+SCF methods such as CNDO and MINDO.

His second major contribution was to recognize that Boys and Handy's systematic use of gaussian functions to represent hydrogenic functions was well suited to breaking the integrals bottleneck for calculations requiring more accuracy than the various NDO methods could provide. Electron repulsion integrals over hydrogenic atomic orbitals are generally not amenable to integration. However, integrals over Gaussian atomic orbitals are integrable in closed form. Thus, even though a linear combination of a few Gaussians might be needed for an adequate representation of one hydrogenic atomic orbital, thereby greatly expanding the number of definite integrals to be computed, convenient access to computers rendered that approach eminently feasible and attractive.

Pople put together a corresponding computer program, Gaussian 80 (now up to Gaussian 94), which enabled workers all over the world to use the ab initio methods of quantum chemistry.

Regarding classical mechanics and statistical mechanics, the application of the method of molecular dynamics first developed by Alder in 1960, using liquid argon as the vehicle, was applied in 1971 by Rahman and Stillinger to the much more complicated system, liquid water. Although that work was done with the supercomputer of four computer generations ago, it was the right time in the evolution of computers to attempt a simulation whereby 216 model water molecules in a mathematical cube were allowed to move relative to each other under the influence of a simple potential. This potential was a combination of a Ne-Ne Lennard-Jones potential (using the united atom form of water plus an electrostatic potential due to the permanent electric dipole moment of the water molecule, represented here as four point charges embedded in each Ne).

Solving numerically, the classical equations of motion for each of the 216 molecules interacting with each of the others as a dynamic system, through some 5,000 cycles, led to a mechanical representation of bulk matter from a molecular perspective that gave considerable insight into the structure of liquid water. As simple as that mechanical model of water was, it even predicted a maximum in the density of liquid water as a function of temperature. That approach, only possible with the computer, builds a bridge between the limiting cases of a few molecules interacting with each other on the one hand, and the assumptions of statistical mechanics based on equilibrium in a system of Avogadro's number of particles on the other—from classical mechanics to thermodynamics.

A third major advance in theoretical chemistry made possible by the advent of the computer was made not by a theoretical chemist but rather by an organic chemist specializing in natural products. James Hendrickson, intrigued by the notion of using classical mechanics to model the structure of seven member hydrocarbon rings, found the sheer number of calculations with a slide rule to be overwhelming. Thus, he took a one-day course on FORTRAN and used the IBM 704 to do a conformational analysis of seven member hydrocarbon rings through transferring two-center interaction parameters from the well-characterized cyclopentanes and cyclohexanes. Hendrickson's paper, which appeared in 1961, startled the world of organic chemistry. It was cited almost 600 times from 1961 to 1980. Subsequently, other organic chemists went on to generalize that approach by introducing automatic calculation of the conformations of minimum energy. Two of them made their FORTRAN-based

TABLE I   Computer effort in dynamic simulations.

Year	Ref	System (Example)	Degrees of Freedom	No. of Steps	Approx. CPU hrs Supercomputer[a]
1960	6	Monoatomic liquids	759	500	0.05
1971	7	Molecular liquids (water, CCl_4)	2500	10^4	1
1974	8	Molten salts (KCl)	750	5.10^3	1
1975	9	Small polymers (decane)	1000	10^4	1
1978	10	Protein in water (BPTI)	1500	2.10^4	4
1980	11	Membranes (2×64 decane)	4000	10^4	4
1982	12	Protein in water (BPTI)	12000	2.10^4	30
Future		Surfaces, nucleic acids, large polymers, micelles, liquid crystals, organic systems	10^4	10^6 10^7	10^3 10^4
		Reactions	10^4–10^5	10^9–10^{12}	10^7–10^8
		Macromolecular complexes	10^5	10^{10}	10^8
		Protein folding	10^4	10^{12}	10^8

NATO Workshop on "High speed computation"; Julich, 1993; J. S. Kowalik, Ed., Springer Verlag; Molecular Dynamics on Cray, Cyber and DAP; K. J. C. Berendsen, W. F. van Gunsteren, and J. P. M. Postma; Laboratory of Physical Chemistry; University of Groningen; Nijenborgh 16: 9747 AG GRONINGEN, The Netherlands.

[a] CRAY or CYBER 205, assuming "brute force" dynamics

models available to other scientists—Boyd's MOLBD and the more widely used MM2 of Allinger—illustrating once again the importance of technology transfer via the vehicle of a FORTRAN program.

The NATO workshop, "High Speed Computation", in 1983 included a report, "Molecular Dynamics on CRAY, CYBER and DAP", by Berendsen, van Gunsteren, and Postma. Their "Table 1, Computer effort in dynamic simulations" constituted a brief history and a prediction of the extent of classical mechanics modeling of individual large molecules, as well as assemblages of molecules, as dynamical systems. That table (together with the reference context) is reproduced as Table 1.

Their prediction of 1 billion hours of CRAY 1 time to deal with protein folding has already been shown to be much too large with the subsequent refinement of the theory and the mathematical methods used.

Crystallography

Crystallography here is restricted to x-ray crystal structure analysis. While some regard this as a mature field with well-developed, fully automated structure determination systems commercially available, the fact of the two major white x-ray synchrotron sources under construction in the United States with high-energy polarized radiation opens up many new opportunities. The computer makes feasible realization of the potential of such sources, perhaps images of complicated molecules, in solution in real time.

In the early days of x-ray spectroscopy (before computers), single crystals were positioned in the path of an x-ray beam and the diffraction pattern observed on a photograph. The positions and the intensities contain information about the parameters defining the unit cell and the positions of the atoms within that unit cell. A major difficulty, however, derives from the fact that the phases of the reflections are not observable. A way to handle that problem is through some educated guesswork, several iterations, and, hence, a lot of computing.

In the 1950s the computer was being used by crystallographers to calculate diffraction angles for measurement. Computers were then used to generate paper tape to drive motors and encoders for the diffractometer positioning. By 1963 there were computer-controlled diffractometers and, by 1966, commercial diffractometers run by the PDP-8. For proteins, however, measuring 40,000 spots for a single protein crystal demanded staying with the photographic method whereby all reflections were observed simultaneously. By 1965 further progress was made when digitizing scanners were developed, and detectors were connected directly to a computer.

The computationally intense part of crystallographic analysis, however, is the Fourier synthesis, whereby the desideratum is the electron density at a "point". For a macromolecular structure there could be some 40,000 coefficients for the summation for each point! (There was an attempt to solve this problem through a dedicated analog computer, the X-RAC, at Penn State in 1950. The obsolescent technology and the emergence of general purpose digital computers made its utility short-lived.)

Once digital computers came to be used widely, the results came out as a grid of values printed on paper. One would draw contours by hand and then position sheets of paper over each other in order to see a three-dimensional structure. Of course, the computer entered at this point, and maps could be made showing peaks and valleys automatically. Thanks to the pioneering work of x-ray crystallographers in this regard, quantum chemists had a model for generating contour diagrams of charge densities working from approximate wave functions for molecules.

Indeed, A. Chris Wahl had a cover story in *Science* in February of 1966, "Molecular Orbital Densities: Pictorial Studies," in which a sort of periodic table of homonuclear diatomics was developed based on molecular orbitals instead of atomic orbitals and the results displayed as computer-generated electron density contours, much as crystallographers had been doing routinely.

Of course we would be remiss were we not to mention that Hauptman and Karle got the 1985 Nobel Prize for inventing direct methods of phase determination. That all began with the realization that the phases of the Fourier coefficients were not all independent because charge densities are inherently non-negative and negative the atoms should show strong charge density peaks. However, it was the computer program based on those concepts, MULTAN, written by Germain, Main, and Woolfson, that really revolutionized the use of direct methods, which now dominate the determination of small-molecule structures.

Graphics was mentioned earlier in the context of electron density contours. However, the major impact of crystallographers on graphics in chemistry was the making of stereo drawings of molecular structures at Oak Ridge National Laboratory in 1960. Carroll Johnson went on to generate a general-purpose computer program, the Oak Ridge Thermal Ellipsoid Plot (ORTEP) program, to show graphically the thermal motion of atoms in crystals. That was refined to show thermal excursions of oscillating rigid groups.

A major advance in graphics for crystallographers was FRODOR, by Alwyn Jones, for positioning a molecular chain on a three-dimensional Fourier map.

Finally, in listing the early pioneering work by crystallographers in introducing the enhancements of emerging computer technology to chemistry, we must mention crystal-structure databases. Not only did they pioneer in generating—and sharing—computerized databases (now over 100,000 unit cell parameters), but they also realized that having that data in machinable form meant they could sweep through the data and extract new information and new insight into chemical binding and molecular interactions (now more generally called data mining!).

A long time ago, a computer pioneer, Richard Hamming, said "The purpose of computing is insight not numbers". That has been amply demonstrated by the crystallographers who have been the pioneers in almost every aspect of bringing computers to the service of chemistry—structure elucidation, graphics, databases, correlations, data logging, and real-time control.

The International Conferences on Computers in Chemical Research and Education—The ICCCRE

Recent ICCCRE have focused on computer aids to organic syntheses, on computer-aided structure elucidation, on chemometrics, and to a significant, but lesser, extent on the quite closely related issue of chemical information storage and retrieval—subjects about which we have not had much to say so far.

The principal pioneers in the first three areas are: Shin-ichi Sasaki in structure elucidation of complex organic molecules (complementary to x-ray diffraction methods), with his first (and quite ambitious) paper in *Analytical Chemistry* in 1968; (E. J. Corey and) W. Todd Wipke, with the paper, "Computer-Assisted Design of Complex Organic Syntheses", which startled the world of science when it appeared in the October 1969 issue of *Science*; and Bruce Kowalski, with the first display of chemometrics (in all of its dimensions in an evolutionary context) perhaps best stated in his essay in *Analytical Chemistry* in 1978, "Analytical Chemistry: the Journal and the science, the 1970s and beyond".

All of these are important in terms of the economic impact of chemistry and are becoming important underlying tools for core chemistry (as defined earlier).

All of these illustrate the power of the computer as a vehicle for technology transfer from other disciplines to chemistry, primarily as follows:

1. most notably from control, sensor signal analysis, networking, and optimization techniques developed by engineers;
2. major porting to chemistry of applied statistics and applied mathematics—especially cluster analysis, nonlinear regression, multivariate analysis, fast Fourier transform, partial least-squares, and parametric and nonparametric modeling;
3. all the techniques of what is the most fundamental subject in so-called computer science, namely, sorting files—plus—expert system design first demonstrated in a chemical context with the appearance of DENDRAL (for DendRic ALgorithm), in 1969, and now being rapidly assimilated into chemistry.

Chemistry Curricula

As noted above, the computer was first integrated, as a required part, into a commonly required undergraduate course, Physical Chemistry Laboratory, in 1959.

By 1975 UNESCO had commissioned a study, "The Computer's Role in Undergraduate Chemistry Education". Some 300 copies of that 155-page report were distributed to developing countries. Shortly thereafter, another 500 copies were distributed by UNESCO in response to individual requests from the United States, Canada, and Europe. As it turned out I was accused of having written that report "tongue in cheek" as it became apparent that most of the chemistry departments in so-called developed countries would have been judged as "developing" by the standards in that report.

As that report (SC-76/WS/42) was considered a definitive statement of what the supporting technology made possible at that time, including descriptions of courses actually in place, the introductory statement titled Description is here reproduced:

Description

The computer is affecting all of chemistry in a comprehensive manner. The rapidly decreasing cost of minicomputer total systems and the emerging inexpensive microprocessors makes it possible to bring computer support to chemistry departments anywhere. In addition, telecommunications facilitate computer networking so that specialized resources may be shared, with the minicomputer providing the interface to computer networks. This report develops how the computer affects chemistry, and hence chemistry education, both as an aid to teaching and as an enhancement to the doing of chemistry. Three course outlines are given: the computer in experimental chemistry, chemical information, and numerical methods in chemistry. Numerous examples of computer programs in use in many college and university chemistry departments are described briefly, including a simple approach to computer-assisted instruction in freshman chemistry. The interface between chemistry with computer science and engineering is discussed. Computer related professional societies and their publications are listed as are sources of films and slide talks which teach various aspects of computing. A model chemistry department wide computer support system is sketched and information about minicomputer vendors is supplied. The question of faculty training is addressed and several schemes for resource sharing are outlined.

A much more comprehensive essay on this subject is the paper, "The 'E' in ICCCRE", which appeared in "Computer Applications in Chemical Research and Education", Ed. J. Brandt and I. Ugi, pp 1–6, Dr. Alfred Huethig Verlag GmbH Heidelberg, 1989.

An issue raised there continues to be relevant as is underscored by the following exchange between a dean of a major university and an assistant professor overheard at a conference: "What do you think about developing software for undergraduate education?" asked the assistant professor. Answered the dean, "Don't do it. Concentrate on research."

Nevertheless, chemistry faculty continue to produce chemistry education software, as the long-time ongoing series in the *Journal of Chemical Education*, the Computer Series, up to number 183 (March 1996) attests. To the theme of this book one should note *Journal of Chemical Education* (December 1994) Computer Series number 168, "Teaching Practical Computer Skills to Chemistry Majors".

Finally, one might note a significant event. Arguably, organic chemists are the least sophisticated of chemists vis a vis mathematics. On the other hand, computer-based tools created by physical chemists and others are increasingly available written in transportable programs and higher level languages and well interfaced to, and supported by, comprehensive molecular graphics software. All of that has been put together by a vendor in a package supported by a book, "Experiments in Computational Organic Chemistry". Thus, one has a commercially available package, supported by the vendor, adapted to run on a common table top computer widely used in academe, with the goal of supporting the teaching of undergraduate organic chemistry. The title and selections from the table of contents follow:

Experiments in Computational Organic Chemistry

By Hehre, Burke, Shusterman, and Pietro, Wavefunction, Inc., 1993 (a text to accompany the software package, SPARTAN, that started out on the Sun Workstation. A version was prepared for the Macintosh).

Selections from the Table of Contents

1. Structure and Energetics
   - Ring Strain in Cycloalkanes and Cycloalkenes
   - Glycine In the Gas Phase and in Water
   - Molecular Recognition. Hydrogen-Bonded Base Pairs
2. Conformation
   - Conformational Isomerism in 1,3-Butadiene
   - Conformational Equilibria in Substituted Cyclohexanes
   - Configurational Stability in Amines
3. Molecular Properties and Spectra
   - Infrared Spectrum of Acetic Acid
   - UV Spectra of Conjugated Carbonyl Compounds
   - Acidities and Basicities of Excited-State Molecules
4. Reactive Intermediates
   - Directing Effects in Electrophilic Aromatic Substitution
   - Free Radical Substitution Reactions of Alkenes
   - Stabilizing Reactive Intermediates
5. Reactivity and Selectivity
   - Rates of Electrophilic Additions to Alkenes
   - Activated Dienophiles in Diels-Alder Reactions
   - The Origin of the Barrier in the Diels-Alder Reaction

(Note that the name of the publisher is Wavefunction!)

Minicomputers Versus Supercomputers

In the beginning there were mainframe computers that did batch processing with punched card input at the computer. Shortly thereafter came rack-mounted dedicated laboratory computers. As brought out earlier, there has evolved a spectrum of computers from handheld programmable devices to $20 million supercomputers.

The minicomputer first entered chemistry in a visible manner when Professors Miller and Schaefer at the University of California at Berkeley received NSF support (I handled their proposal while on leave from IIT at the U.S. National Science

Foundation 1971–73) to buy a medium-size computer dedicated to support theoretical chemistry research for a group in a chemistry department. That started a chain of events whereby theoretical chemists routinely received support to buy VAX-like machines. The VAX became the default choice and, hence, enhanced transfer of software. Today VAXes come in all sizes, from Microvaxes, which even quite junior researchers can justify, to quite large and powerful machines substantially more powerful than the venerable VAX 780 of yesteryear.

In parallel with the evolution and distribution of minicomputers came the so-called supercomputers. (However, as Neil Lincoln said, "Supercomputer needs are always for computers two to three times more powerful than the most powerful computer available today!") Outside the nuclear weapons R&D arena, supercomputers were justified and heavily used by researchers working on fluid flow problems—especially in aerospace projects. Control Data dominated that industry and was heavily subsidized by the U.S. DOE labs, especially LLNL. Seymour Cray, designer of the CDC 6600 and the CDC 7600, left CDC to start his own company, and the successive models of the CRAY computer have become synonymous with the most powerful scientific computers available at any time. As was noted before, one of the early CRAY computers recently was decommissioned and is on exhibition in a museum. However, multiprocessor supercomputers such as IBM's 512 processor system are challenging the super CRAYs in terms of cost effectiveness.

The supercomputer has become a pawn in the economic competition between the United States and Japan. Trade barriers in the United States have prevented any significant influx of Japanese-made supercomputers, even though their performance as measured by common benchmarks should have made them quite competitive compared with U.S. machines. (Recently, the National Center for Atmospheric Research placed an order for an NEC supercomputer only to have the U.S. Department of Commerce contact the NCAR funding agency, NSF, with the opinion that the machine's production cost is substantially greater than NEC's bid and such a purchase would harm U.S. supercomputer makers.) In part because of a perceived global advantage in that regard, the United States is supporting national laboratories based on supercomputers and accessible by broadband communication networking.

The NSF was designated the lead agency to create, and for many years maintained, an NSFNET backbone to carry heavy data traffic and to interface some 19 networks. Earlier this year, that NSF "ownership" ended and the network transferred to the private sector.

Several years ago the United States funded five national supercomputer laboratories, in San Diego, Princeton, Urbana, Cornell, and Pittsburgh, attempting in the process to fund sites that were complementary in some important way. Having a variety of different vendors' products was a consideration except that Japanese machines were clearly not acceptable to the funding agencies. The current squeezing of federal budgets generally has led to a reevaluation of those supercomputer labs, with some decommissioning or redirection likely.

What Does this Mean for Chemistry?
Chemistry played a leadership role in recognizing the importance to chemical research of providing easy access for geographically distributed chemists to the most powerful computer available. In the United States that took the form of creating the National Resource for Computation in Chemistry (NRCC) in 1977. At about the same time the Daresbury Lab in England created a CRAY-focused national center

for computational chemistry, while in Japan the Institute for Molecular Sciences had as one component a supercomputer division to provide chemistry support nationwide, and in France there existed the pan-European CECAM.

The NRCC came after a long period during which a series of conferences, starting in 1970, were held to sharpen the focus on the concept of a shared supercomputer facility nationwide for chemists. Although the announcement of LBL as the winner of the contest to host the NRCC was made in July of 1977, the NRCC did not have a director in place and begin to function until February of 1978. Evaluation of the NRCC began in January of 1980. Six months later the NRCC was given the coup de grace by the cosponsoring agencies, NSF and U.S. DOE.

Clearly, the NRCC demise was more politically motivated than based on a clear-cut demonstration of a failure to meet a need.

Observers used the metaphor of early American settlers moving west in wagon trains. Wagon trains were susceptible to attack by American Indians and, when attacked, the wagons were pulled into a circle, with women, children, and horses within the circle and the men behind the wagons firing rifles at the attacking Indians. The similarity was noted that in the case of the NRCC, the chemists, unable to believe that new monies would be available to support a shared facility, perceived their individual research grants under attack and formed a circle—however, they pointed their rifles INWARD and fired!

Subsequent to the NRCC affair the aforementioned national supercomputing centers were formed, and some chemists—by now sensitized to the potential of supercomputers in chemical research—quickly became important users, thereby realizing much more computer power than would have been likely had the NRCC continued.

An important qualitative issue arises here: What is the usage profile of those supercomputers? How much of the system is used by an average mix of users? How many of the jobs done actually require that kind of machine? How many of the jobs processed actually saturate the machine and require a "bigger and faster" machine? How does the chemistry usage profile compare with that of all the other users?

Observation of the fluid flow user community suggests there is a qualitative difference: namely, the power of the supercomputers is such that models are now being created with scope and detail that transcend the competency of any one subdiscipline-focused researcher. Put another way, the level of detail and range now possible calls for an interdisciplinary team effort.

Chemists do not work that way. Chemistry is a cottage industry. At the National Center for Supercomputing Applications (NCSA) (University of Illinois at Champaign-Urbana), for-profit companies were given the opportunity to commit $1 million a year for three years in exchange for 1,000 cpu hours per year and the opportunity to have some of their R&D staff visit the NCSA in order to learn from students and others there more about how to take advantage of that resource. Several companies so committed. They included Amoco, Eli Lilly, and Eastman Kodak—all chemistry-based companies! Clearly, that commitment which is enhancing technology transfer from academe to industry played a major role in the recent decision to extend funding support. One could say the NRCC lives!

Another consideration is entering the picture: major computational support for those whose models need enormous computational support. Highly concurrent multiprocessor computers are fast becoming commercially available challengers to the traditional number cruncher. Starting with the hypercube mentioned earlier there came into existence such devices as the Connection Machine with as many as 64,000 processors—but the company just filed for Chapter 11 protection! At first the concept is quite appealing, as the cost of processors such as are used in personal computers is sufficiently small that having a large number of them working on parts of one's problem in parallel seems quite attractive. A problem with using parallel machines is that scientists and engineers doing number crunching are accustomed to using FORTRAN and developing correspondingly serial algorithms. Also, designing parallel machines to work on specific classes of problems seems more cost-effective than attempting a general-purpose scientific computer. It is enlightening to examine the subjects discussed at the long-running series, Annual Conference on Parallel Processing. Tutorials deal with Artificial Neural Networks, Parallel Supercomputing and AI Processing, and Pipelining: Parallelism in the Small. Technical sessions include subjects such as compiler data dependencies, synchronization and scheduling, coarse grain task allocation, and data flow. Many challenges remain for the chemist who wishes to find the most cost-effective way to use computer support.

A related consideration is that the computational power being put on a chip is also increasing rapidly with no diminution in that trend anticipated for several years. For example, Intel announced its first 64-bit microprocessor, the i860. At 80 million calculations per second it offers computing capabilities associated with supercomputing systems and three-dimensional graphics workstations.

Also of interest to chemists is that the news release about the i860 observed that applications would be fluid dynamics, molecular modeling, structural analysis, and econometric modeling. That tells us that chemists have been recognized as major users of computer power.

A Peek Into the Crystal Ball

Predictions about the future of computing tend to overestimate what will happen in one year and to underestimate what will happen in five years.

Of course rapid increases in power and greater ease of use—together with acculturation—are enhancing the infusion, and diffusion, of computer-based aids to chemical problem solving. We are on the rapidly increasing part of the normal growth curve. We are entering the golden age of computing in chemistry. For the individual chemist, the cost of the desktop personal computer (now workstation) has dropped to not much more than the cost of the old IBM PC but is of course much more powerful.

For the molecule-focused scientist, 3D graphics display is of paramount importance. Improved techniques of inducing the 3D effect in the viewer's mind continue to evolve—propelled by developments in DOD—and by manufacturers of arcade and home user games!

Improvement of the man–machine interface will expand to include tactile communication with molecular models. (Kent Wilson pioneered with his "touchy feely" touch-sensitive mechanical model of methane that he described, in an evolutionary context, in the Fall 1975 National ACS Meeting Symposium, Computer Networking and Chemistry.) Imagine wearing a pair of glasses, each "lens" a video display, and wearing a pair of gloves whereby you can feel the forces as you manipulate the molecular 3-D image you see.

Chemistry's storehouse of data will—itself—increasingly become a source of new information through the use of pattern recognition and other correlation methods, much as the crystallographers have for some time been doing.

A major old, but rediscovered, new development is, of course, the neural-net computer (suggested half a century ago, but only recently has the supporting technology—and significant support from the military—appeared to render it feasible). While having an adaptive device that "learns" from data is an exciting application of technology, as scientists we have a problem with such devices in that the successively improved "fit" of the model to the data does not easily reveal to us the features of the model in terms we are accustomed to using. On the other hand, such devices enable the chemist to work at the phenomenological level rather than be constrained to systems whereby models based on the laws of physics using the language of mathematics are used. A major player in the evolution of the neural-net computer is a chemist—John Hopfield at Cal Tech.

Chemists will not only contribute to extension and refinement of chemical problem-solving methods, but will play an increasingly active role in exporting those methods to other chemists, to other scientists, and to those who perceive they can apply those methods in their own work environment. So-called expert systems are already playing an important role in this regard and are expected to grow rapidly in the range and depth of application. Further improvement in the expert system–chemist interface should further accelerate such applications. Thus, *outreach* by chemists will be much enhanced—particularly via computer-enhanced chemistry-based instrumentation. A welcome aspect of such outreach is that it will more easily reach the nonchemist community!

Modeling and simulation is, of course, the most ubiquitous application of the computer to chemistry. Our models have gotten sufficiently sophisticated that they provide a single environment within which the wide variety of data one gathers about a particular system can be used to characterize the system in an internally consistent manner. The example of Peter Kollman and seven collaborators that appeared in 1984 in the *Journal of the American Chemical Society* is an excellent and early demonstration of that thrust. They developed a force field for simulation of nucleic acids and proteins building on the information harvested from simpler systems and moving on to the porcine insulin molecule as a test case. Porcine insulin has 500 atoms. A broad range of theoretical techniques and measurement types were accommodated within a single framework, further validating that approach but showing where additional refinement is needed. Chemistry is ever evolving!

Time-dependent phenomena involving bond breaking/making have not been discussed here so far. The rates and mechanisms of chemical reactions are generally difficult to determine experimentally—even more difficult to estimate theoretically. Whereas thermodynamic and spectroscopic data can be measured quite accurately, rates of reactions are inherently less well known. Theoretical predictions that agree with an experiment to within a factor of two or three are considered a success! A large amount of computer time is being consumed by chemical physicists (in the United States, most "chemical physicists" have their degrees in chemistry) in pursuing the dynamics of two- or three-atom molecules (using quasiclassical, time-dependent quantum mechanics, as those problems are impossible to handle quantitatively as many body problems from first principles).

The function *energy* is a major international concern; accordingly, burning fuel is an important issue. Indeed environmental issues may push into the background international tensions based on economic and military issues. Through a combination of theory and experiment, the mechanism of hydrocarbon combustion is fairly well worked out—there are hundreds of elementary reactions in the mechanism known so far. Those in charge of burning large amounts of fuel carefully tune their systems to a blue flame and avoid soot generation. Those manufacturing carbon black have a different objective. Both extremes are only fairly well understood in terms of the several hundred known steps. I have seen no evidence, however, that the information in that mechanism has been used to significantly improve either process. Indeed, a major modeling breakthrough was announced by General Motors Laboratory, whereby simultaneous solution of hundreds of chemical *equilibrium* equations was used to assess the combustion process in a gasoline engine!

Other very dilute gas-phase processes involving atoms and two- or three-atom molecules can be better understood with the aid of quantum molecular dynamics; however, a lot of chemistry is concerned with condensed matter made of larger molecules. Intuitively obvious classical mechanics models (hardly more than generalizations of the familiar wooden ball and stick models)—helped by CNDO, whereby charge distributions and geometries are needed—seem to be evolving as the most useful theoretical tools for the majority of chemists. However, even though chemical physicists are using an enormous amount of computer time, and hence must have a fairly well systematized set of algorithms and corresponding FORTRAN programs, technology transfer of the rather esoteric quantum molecular dynamics is not happening. Is it the transferees? The transferers? Or is it the remoteness from core chemistry as defined earlier?

Finally, we turn to the issue of the chemistry of computers.

The first chemist to pursue that issue in an unabashedly aggressive manner was Forrest L. Carter of the U.S. Naval Research Laboratory. He was stimulated by the following considerations:

1. If one extrapolates the density of circuitry on a chip as a function of time, one may expect that in 20 years or so the chip elements will be on the order of a few angstroms in size—that is, of molecular dimensions.
2. Work at the Naval Research Laboratory "on molecular wires" of linear polymeric conductors like doped polyacetylene and polysulfur nitride suggested that these made possible the communication between the switching components.

That work led to the Molecular Electronic Devices Workshop early in 1981. An updated summary of that workshop proceedings was given at the VIth ICCCRE. In 1988 there appeared the proceedings of the Third International Symposium on Molecular Electronic Devices. That thrust continues.

Photosynthesis research is providing insight to electron transfer and energy storage at the molecular level. Of course, molecular-level devices are subject to quantum effects, most notably tunneling. Recent work on electronic shift memory based on molecular electron-transfer reactions (by Hopfield, Onuchic, and Beratan) suggests that tunneling can be finessed, where it is a problem, by chemical molecular design. As practical working devices come to be fabricated, we can see significant technical advances coming; however, we are not likely to see major commercial application and use for many years (much like the high-temperature superconductor phenomenon).

Inspired, no doubt, by Carter's initiative, F. Eugene Yates of the Crump Institute for Medical Engineering of UCLA conducted a conference in October of 1983 on chemically based computer design. This conference was supported by the U.S. National Science Foundation in order to explore whether sufficient potential existed for the NSF to create a new funding program with that focus. There were four working groups focused on biology, computing, physics, and chemistry (Yates' ordering!). The underlying idea was to explore the potential of computer design using analog logic rather than digital. Thus "good enough" or "satisficing" engineering and computation, more biological in style, was being sought after. The language smacked of neural-net computing. And indeed this is where we are seeing interesting developments based on neural-net concepts but still working with relatively macroscopic assemblages when compared with single molecules, that is, at microscopic levels of mimicking living cells as computer elements.

The venerable *Wall Street Journal* picked up on that consideration and, in a mid-1983 issue noted, "But a small number of computer scientists are engrossed in trying to prove that someday it will be possible to create an organic computer, or 'biochip,' an incredibly small package of computer circuits assembled from molecules." Perhaps the writer had already anticipated that computer engineering is becoming chemistry and, hence, chemists are becoming computer engineers!

In parallel with those symposia, there has emerged an ongoing activity, the Foresight Institute. The leader, K. Eric Dexler, who has managed to conduct biannual conferences on nanotechnology, has attracted support for the institute and even the Feynman Prize of $250,000 for the first robot arm and computing component for an assembler to be used in the manufacture of materials and structures that measure up to 100 nanometers. The definition of nanotechnology includes synthetic chemistry! A recent piece in *Scientific American* (April 1996) by Gary Stix, "Waiting for Breakthroughs", the most recent contribution to the section Trends in Nanotechnology, gives an overview of the history and current status of Drexler's initiative.

A sign of the times is that almost immediately there appeared on the World Wide Web (http://www.foresight.org/SciAmResponse.html) a point-by-point critique of the tone and content of Stix's article.

Summary

Chemists' use of computers is perhaps one of the best examples to illustrate that revolutionary science occurs at the interface of disciplines. Not only computer science but also information science, applied statistics, applied mathematics, and engineering principles have been brought to bear on significant chemical problems.

Early efforts to solve chemical problems with the new computer technology depended on the computationally inclined chemist's ability to adapt business-oriented computers to the chemical domain or to build (from computer components) his or her own dedicated computer. Now, of course, many commercial programs, that run on a variety of computing platforms, are available to the bench chemist. The commercial sector sees a major market for chemistry-oriented hardware/software and professionally done systems in a highly competitive environment.

Chemical education also has benefited from computers and the efforts of computationally inclined teachers. The advances of computer technology in chemical education are evidenced by special publications of the American Chemical Society (and others), a long-running series in the *Journal of Chemical Education,* establishment of the Committee on Computers in Chemical Education in the Division of Chemical Education of the ACS, and of course the ICCCRE biennial series.

Chemistry is the molecular science. Computer-generated graphics display molecular structures. Algorithms modeling molecules and their interactions with each other and with radiation are conveniently shared—distant in time and in space—with students, with other chemists, and with other scientists. One of the most important features of the impact of the computer on chemistry is as a vehicle for sharing of computer-based problem-solving methods—as a vehicle for technology transfer.

Finally, the future of chemistry and computers seems to be merging with the emergence of ideas on chemically based computer design, analog logic computers (those that are more "biologic" in style), and nanotechnology.

3

Training the Practicing Chemist

Robert Megargle

The use of computers in chemical education falls into two distinct categories. First, computers can assist the educational process, helping to educate new chemists. They can be used for the preparation of educational materials such as visual aids, course syllabi, laboratory directions, announcements, and other materials. With suitable projection equipment, the color graphic capabilities of computers can be used to enhance lecture presentations. They can be used to provide tutorial drills, as well as prepare assignments, quizzes, and examinations. Computer modeling and simulation can provide insight into physical and chemical processes that is difficult to grasp otherwise. These same tools allow students to experience, at least in part, experiments that are too costly, time-consuming, or dangerous to actually perform in student laboratories. The computational capabilities of the computer allow an instructor to assign more difficult problems and exercises, ones that would otherwise result in students getting bogged down in the mathematics. Connection to the Internet provides students with access to information that might not be available at their school.

The second category deals with the computer knowledge students need to acquire to be effective practicing chemists. All professionals today rely heavily on computers to do their jobs, and chemists are no exception. To properly equip the next generation of chemists for the workplace they will face, it is essential to offer instruction in the use of this tool. Without an array of computer skills, employees cannot be as productive as other coworkers and will soon be left behind in the competitive arena of the modern workplace. This chapter presents an overview of some of the computer applications that students should know upon graduation. Some of these topics are covered in more detail in other chapters of this book. These are skills and tools that will serve the chemist in almost every type of chemical setting, including industry, academia, and government. They equip graduates for maximum effectiveness in all types of chemist jobs, including research, development, production, monitoring, information services, management, and consulting.

Word Processing

Virtually every professional in any field today should know how to use a word processor. The ability to edit documents and correct errors before the final copy is printed is by itself enough to justify the rather minor learning barrier needed to master a word processing program. Added to that advantage are spell-checking features, an on-line thesaurus, cut-and-paste features, and the ability to vary fonts and type sizes. Most of the newer word processors can now integrate drawings, graphs, and images with the text, but this requires a somewhat higher level of knowledge and experience. Word processors find application almost every place typewriters are used, including personal and business correspondence, technical reports, equipment operating instructions, maintenance logbooks, grant proposals, student laboratory directions, classroom assignments, examinations, homework reports, patent applications, and manuscripts for publication. Regular typewriters are better for filling out preprinted forms and possibly for addressing envelopes, although most word processors can perform the latter task as well.

Desktop Assistance

There are various personal assistance programs that chemists can use to improve their normal work patterns. On-line desktop calendars can take the place of a pocket scheduler.[1] They allow entry of appointments, meeting times and dates, and personal notes. The day's activities can be reviewed whenever needed, for example, to confirm the time and place of a meeting. Before leaving for home in the evening, one can review the commitments of the next day. Changes can be made easily, eliminating the scribbled notes and crossed-out entries that tend to clutter a paper reminder book.

Many personal-assistant programs can be used to keep records of important daily activities. They require some investment of time and effort, but can pay back with greater efficiency and productivity. In this scheme, summaries of important daily activities, such as topics discussed, are typed in, assignments given, agreements reached, and conclusions drawn. Later, with a few clicks of the mouse, one can review when meetings were held, who was present, and what was accomplished or decided. There are software tools for establishing milestone targets for projects, and then for recording actual accomplishments against those goals. Each new entry allows the user to enter the expected or actual start and stop times, a summary of the event, as well as a more detailed description. Regular daily, weekly, or monthly recurring events can be scheduled. Alarms can be set to remind the user of some important activity.

Most calendar programs allow different views of the contained data. One can tell the calendar program to show the details of a particular event or the entire day's schedule. They can be set to show a summary of the activities for the week, month, or quarter. Many have a to-do list function that lets the user record notes and check them off as they are accomplished. To-do lists can typically be organized by the description of the task, by project name, by priority, or by due date. This allows the determination of the most important tasks or those most likely to be overdue. Tasks that are scheduled for one day, but not completed, can be moved easily to future dates.

Some calendar programs can be set to remind the user of an appointment or a due date. The lead time is variable. Reminders may be implemented as a window that pops up or as a special symbol that appears on the screen whenever the computer is running.

Another useful program is the on-line address and telephone directory. This program holds the names, addresses, and telephone numbers of people frequently contacted. It serves as an on-line Rolodex. One can quickly scan through the names in alphabetical order to find a particular entry. Many telephone directory programs are capable of dialing the number (if the computer contains a telephone modem card and the desk phone is connected to that modem). Once the connection is completed, the user picks up the phone to listen to the rings (or busy signal), then talks when the called party answers. New entries are easy to add, and existing entries can be quickly corrected. An on-line system like this is not only faster, but it also tends to stay more up-to-date because it is easier to maintain than a paper system. One disadvantage is that the computer has to be running before it can be used, so in cases of power outage or system failure, the information is not available. However, the contents of the address and telephone file can be printed on request, so a paper backup copy can be kept in the drawer. The backup copy may become somewhat out of date over time, but it will usually suffice for the rare occasions when it is needed.

Literature Searching

Computers are used to assist in the preparation of most documents today, including important reference works like the *Chemical Abstracts,* the *Citation Index,* libraries of spectra, chemical handbooks, and databases of hazardous materials. The software for these projects is more sophisticated than typical word processors. They are capable, for example, of automatically generating the various indexes that make these reference materials so valuable. In these cases, it is relatively easy to build electronic information databases since the subject material is already available in electronic form. With the addition of suitable software tools, these databases can be used by scientists as a faster, easier way to search for information of interest to their work. It is not uncommon for an on-line search to reveal information in a few minutes that once took days or even weeks of library time using the printed versions of the reference documents.

To use on-line literature services, one must (1) understand the nature of the information contained in the database, (2) know the vocabulary and syntax of the searching rules, and (3) develop a search strategy. The techniques for using such on-line services are covered in Chapters 6 and 7 and are easy to learn. The usual strategy is to create sets of literature references that meet selected criteria, such as containing a particular keyword. These sets are combined with logical AND and OR operators to form new sets to isolate references of interest.

In some companies, professional librarians perform these on-line searches. Optimal searches, however, require the knowledge of the chemist. This individual is best able to select appropriate keywords, know about possible synonyms, and be able to judge how preliminary search sets should be combined to isolate the appropriate references. Therefore, automated literature searching is a valuable tool for students to learn, not only for their school projects, but also because it is an important skill to take into their jobs.

There are many on-line information services besides bibliographic databases. Spectral libraries, for example, have tools for locating spectra by chemical name, formula, and sometimes by chemical structure. They also have spectral matching programs that allow the user to input a complete or partial spectrum. The algorithm matches the unknown spectrum against the library and locates compounds with similar spectra. Often, a quality parameter is calculated showing the degree of certainty of the match between the unknown and library spectra.

Many chemical content databases have chemical structure features. They allow the user to define a complete or partial structure by entering atoms and connecting bonds. The software encodes this information into a chemical structure code system, which can be used with a structure index. Data on compounds that completely or partially match the user's structure can be located. This is valuable when one needs to find information about related compounds, or when the naming rules are complex or ambiguous.

Spreadsheets

Spreadsheet programs manage columns and rows of information. They were originally developed for accounting applications, but nearly all commercial versions today contain mathematical functions that facilitate their use in science. The mental approach to spreadsheets is somewhat different from programming languages and the other usual algorithmic designs normally used for problem solving. It therefore requires some instruction and practice to see how this tool can be applied to scientific problems (see Chapter 9, Spreadsheets for Doing and Teaching Chemistry, for a detailed discussion of this topic).

The appealing features of spreadsheets for doing chemistry include the ability to manipulate data using a variety of built-in statistical and mathematical functions, including the ability to invert and multiply matrices and vectors. The generation of a variety of different types of graphs is also relatively easy.

Several important applications of spreadsheets in chemistry are regression analysis, modeling chemical phenomena (the distribution of species of a system as it approaches equilibrium, chemical/enzyme kinetics), and finding roots of functions. The computational intensity and intricate graphing required make these daunting tasks if attempted by hand, but they become almost routine when spreadsheets are used.

Databases

Commercial database programs are useful to the chemist. At the personal level, they can be used to organize information for current activities. A good example is the collection of literature references for an ongoing project. Instead of saving these references on index cards, or worse, scraps of paper, they can be entered into a computer database. The database records would be set up to hold the author's name, bibliographic citation, title of the article, a few keywords, and a brief summary of the content of the article as it relates to the project. In return for the extra effort needed to type this information into the computer during the project, one is rewarded at the end with a set of references that are more complete and accurate. Ultimately, this approach is usually more efficient, since preparing the list of references for a final report or dissertation can be a lengthy and frustrating exercise. Some projects last several years, and locating paper records of literature work done at the beginning may not be easy. Often, the chemist might be forced to look up the references again, just to ensure that the citation is correct and the content of the article is appropriate for the purpose. During the project this bibliographic database can reduce duplicate work caused by looking up references twice, or by the needed information not being completely extracted the first time. In short, the database enforces good record keeping practices with regard to references during the project. Thorough and careful reference work is essential when patents may come from the project.

Databases have other applications, primarily when relatively structured information is being acquired. Many routine jobs and some research studies lend themselves to databases. They can be used to record observations or measurements. Spreadsheets are probably better when simple numeric data are being saved, but a database might be preferred when the information is textual or when there are several different ways one might want to use the information. Databases are also useful if the individual pieces of data have to connect to other data items in ways other than a simple one-to-one relationship.

The real power of a database comes from the indexes, which allow information to be cross-referenced in different ways. All of us have faced the problem of deciding how to save a certain piece of paper when it logically fits into two or more folders in our file drawer. In a database, the information is filed once, but references to it are included in different indexes so that the information can be retrieved by different criteria. These same indexes allow information in the database to be linked to other information in complex ways that are not possible with a paper record system. The main requirement is that the relationships be definable, and they happen often enough within the information set to make constructing the index worthwhile.

Like word processors and spreadsheets, databases require one to surmount a minimum learning barrier. Most of the popular versions come with tutorial exercises designed to help the new user through the process. In the end, however, it is the willingness of the chemist to invest a week or so learning to use the database that determines whether or not it will become a useful tool.

For simple applications, the steps are straightforward. First, one defines the structure of a record. Space must be identified to hold each individual piece of related information. A measured value, for example, needs a place for the result, the units used, the identity of the sample tested, the person who did the test, the date and time of measurement, and any error flags generated by the instrument. Coming back later to add a data field in an established database is usually possible, but often cumbersome. Next, indexes must be defined. First, one must determine how the information is to be used. Then, data fields can be selected so that the desired information can be located. In the preceding example, one might want to find records according to the sample tested, the person who did the test, or the date of measurement. Most databases allow compound keys consisting of more than one data field. This feature is useful to help sort retrievals in many situations. In our example, a compound index would allow records to be found according to who did the measurement on a certain date. It is less critical to get indexes correct at the beginning. In most database programs, indexes can be created after the data records have been entered. If a need for a new index arises later, it can usually be created easily with no loss of information.

Popular databases provide input and output tools. Although these are fairly generic, they can be used effectively in most simple applications. With relative ease, one can produce useful reports. While they may not always have optimal appearance, these reports are usually adequate. The output of a report can usually be redirected to a disk file instead of to the printer or terminal screen. This feature could be used in the bibliographic database described above, for example, when the project report is being prepared. The disk file so produced could be appended to the report file, then edited with a word processor to put the list of references in the appropriate format. The hard part, getting the names spelled right and the citations correct, is done automatically, assuming the original information was entered properly when the reference was first examined.

Database usage gets more complicated when the application requires more than a few simple record types, or when data in different records must cross-relate to each other in nonsimple ways. All popular databases come with a programming language that can solve these problems, but mastery of that language is considerably more difficult than using the standard input/output tools. It requires a level of effort similar to mastering a new computer language. Many of the popular relational databases have attempted to standardize on a database programming language called "Standard Query Language" (SQL). There are limitations and enhancements to this language for the database products of different vendors, so programs written for one database frequently do not transfer directly to a different one. Nevertheless, SQL enforces a similar logical approach, syntax, and vocabulary. This is better than the totally different approaches used prior to the standard, making SQL the preferred solution.

Programming

Commercial software exists today for many chemical problems. However, chemists cannot count on solving all their problems using only available programs. Sometimes, relatively simple questions arise that a short hastily written program can quickly resolve. At other times, issues are related to research problems for which no commercial programs have yet been written. In these cases, programming is a useful skill for the chemist.

Even when chemists choose a career track for which original software development is not required, an understanding of simple programming is valuable. Nearly all chemical measurements today are touched in some way by computer programs. Raw data are routinely manipulated inside instruments by computers. Instrument output is frequently imported into other computer programs for further processing and manipulation. Since chemists must be prepared to stand behind the results they report, it is essential that they have some understanding of how the raw measurements are turned into reportable data. This requires an understanding of the algorithm used by the processing software. The ability to understand an algorithm is facilitated by a knowledge of computer programming. What is important is not so much the details of the syntax and command sets of different programming languages. Rather, it is the mental approach one takes to crafting a program that produces correct results. Only the experience of writing simple programs gives one that appreciation of the strategies that are employed. Also important are knowledge of the methods of testing and verifying software, the meticulous care needed with minute details to successfully complete a program, and an appreciation of the limitations of any computer program.

Chemistry students should learn programming not so much because we expect they all will need that skill in their jobs. Rather, they should learn it because writing simple programs is the only way to understand both the power and restrictions of a computer. Because the goal is insight and understanding, it is not necessary for students to master a complex language like C++. Instead, they may be introduced to a language like BASIC, which is easy to use and debug, but still allows fairly complex computational problems to be solved. Elementary programming skills include input and output of data, arithmetic computational statements, string manipulations, loops, and conditional statements such as "if-then-else". Some instructors like to include screen graphics and simple disk read-and-write operations.

When beginning programming is taught, it is appropriate to also introduce the concepts of numerical analysis. Since the usual computer languages cannot directly process abstract concepts like the symbol manipulation of algebra, it is important that students learn how to approach the real problems encountered in chemistry. Such approaches can include iteration techniques for solving equations, matrix manipulation, estimating integrals, and other such strategies.

Math Packages

There are numerous commercial programs designed to solve algebraic equations and other kinds of math problems. They all have different capabilities, input strategies, and reporting mechanisms. Efficient solution of math problems is an important task in some chemistry careers, and the topic should be addressed in the education of chemists. This topic is discussed in detail in other chapters.

Instrument Interfacing

The ability to interface an instrument to a computer is a skill that only a few chemists will need in their careers. However, a large percentage of graduates will be using automated instruments. As with any endeavor, those who understand the principles of operation of the equipment they work with will be more effective users. One relies on such knowledge to make intelligent parameter selection, operate the equipment with less uncertainty, recognize problems when they occur and deduce likely causes, and generally assume responsibility for the workplace rather than be at the mercy of the instruments. It is always an unhealthy and inferior situation when chemists or technicians can only view their instruments as magic boxes that somehow spit out answers.

To avoid this, educators usually offer instruction in instrumental analysis, either as a separate course or bound into other courses. The students first need to understand the chemical or physical property being measured. They then need to know how the primary sensor of the instrument interacts with the sample. From this, the measured signal is developed or processed in some way to produce a result. Even if the course is not intended to impart instrument design and repair skills, it is important that students understand at a conceptual level the nature and strategy of the signal-processing system. It affects the results obtained from the instrument. The instrument may also include control functions to allow accurate determinations, such as the magnet control in mass spectrometers and NMR instruments. Students must learn about calibration and verification techniques for instrumental methods. It is also appropriate to discuss the applications and limitations of each instrumental method, and give some consideration of the advantages and disadvantages of the different techniques.

Computers are used within most complex instruments today, as well as within many of the smaller ones. In most cases, they are an essential part of the chain that links the primary measured value to the reported result. As such, they cannot be ignored in the presentation of chemical instruments. There are usually one or more sensor units that supply information to the computer. These include not only the sensor making the primary measurement, but also sensors giving the state of various other conditions or instrument parameters. Often, important control systems are attached to the computer.

In a polarography instrument, for example, the primary measurement might be cell current, but the computer may also be able to monitor cell temperature. The computer may have control over the applied voltage, a nitrogen purge valve, a stirring system, a mercury drop valve, and a capillary drop knocker. In this kind of system, almost all of the instrument functions are controlled by the computer. Different experiments can be performed by running different programs. At the instrument level, the only differences between normal scan, normal pulse, differential pulse, linear scan, cyclic voltammetry, and stripping analysis are the sequence in which the controls are used, their timing, and the measurements that are made.

In addition to interacting with the sample, the instrument computer also has a number of output mechanisms. These can include status lights, single-line numeric or textual displays, panel meters, full computer screens, chart recorders, printers, and communication ports to other computers. The instrument software connects the measurement process results to the output displays, often performing a variety of scaling and conversion functions. In some instruments, the data conversion process is quite extensive, such as with Fourier transform infrared (FTIR) spectrometers, where raw time domain beat patterns are converted to frequency domain spectra by Fourier transform calculations.

It is not obvious how the computer can be used for instrument control and measurement, even for people with experience writing computational programs. In a typical instrument system, raw data are usually created in analog form from the sensor. Only a few types of sensors, such as the scintillation counter in gamma ray detection or photon counting in low-light-level measurements, produce digital data directly. The others have to be converted to digital form, usually by a device called an analog-to-digital converter. Often, the instrument sensor produces continuous output, like a chromatogram or a spectra. In these cases, the analog-to-digital converter is used repeatedly to produce a set of numbers that represent the original data. More faithful representations will be obtained when more data points are measured at closer intervals. Often, the measurements are evenly spaced along the independent parameter. This would be time in a chromatogram, wavelength in an optical spectrometer, and mass/charge in a mass spectrometer. Even spacing is not required, however, and some instruments are designed to take readings more frequently when important features, such as spectral peaks, are present in the data, and less frequently when nothing important is happening, such as during a baseline state.

The result of the experiment is now a set of numbers in the computer memory. At this point, data processing routines are invoked to derive useful information. Baseline may be established to be used later as a correction. Peaks can be found from the result sets, often defined as groups of values that are significantly larger than the baseline. Peak height, peak center, and peak area can be determined by appropriate algorithms and sent for output display. Sometimes, the algorithms can be quite sophisticated. In chromatography, for example, peaks may be incompletely resolved, resulting in dips that do not go all the way to the baseline, or in worse cases, shoulders on the sides of other peaks. Considerable effort has gone into algorithm design to give useful results even in these difficult cases.

A general understanding of interfacing concepts is important for students to acquire. The scientific method demands that chemists accept responsibility for the integrity of their experiments and the quality of their results. This is impossible if the mechanisms used by computerized instruments remain a mystery to the scientist.

The instrument computer is in between the chemical measurement and the reported result, so what that computer system does to the data will affect the conclusions drawn by the chemist. Approaches to learning instrument interfacing are covered in more detail in Chapter 11.

Sample Preparation Robots

Computer automation of chemistry instruments and of processing of the resulting data is well established. The first of these systems appeared in the 1960s. Since the early 1980s, with the introduction of the Zymark robots, automation has also been routinely applied to the preparation of materials to get them ready for measurement or other experiments. Here, the sample is physically moved, treated, and processed in a variety of ways using a mechanical manipulator and a series of peripheral devices, all under the control of a computer. A robot is a mechanical device that handles objects in a way that duplicates the work of a human. A robot need not look like a human, nor does the procedure it uses have to exactly duplicate the steps a human worker would follow. The net result, however, is a sample that has been treated for measurement in an instrument or for use in some other experiment. The treatment is similar to that a technician would have performed.

Driving Forces

There are important driving forces behind the automation of sample preparation. The factors include the following:

1. Fewer people can perform more work. One person can monitor the operation of several automated sample treatment robots. This increases productivity and saves money.
2. People with a lower level of training can be used because the sample-preparation knowledge and manipulation skills are in the apparatus.
3. Twenty-four-hour-a-day operation is possible.
4. There is less variability caused by human limitations. This may yield higher precision and accuracy if sample preparation was previously a limiting step.
5. There are no procedural mistakes or human blunders. Robot mistakes are more obvious, like running out of supplies or failing to work at all.
6. People are removed from contact with dangerous materials. Included are poisonous, infectious, noxious, explosive, corrosive, and radioactive substances.
7. Contamination of samples is less likely when they are not touched as often by people.

Approach to Robotics

Robot applications are designed and implemented by chemists who must have a thorough knowledge of the application that is to be automated. When the application is fully implemented and tested, it may be used by technicians. To implement a robotic procedure, the chemist must thoroughly understand the nature and chemistry of the samples, the sample preparation steps that are needed, and the requirements of the instrument that will perform the measurements. Of course, the implementer must also know the capabilities of the robotic equipment and all the peripheral devices, learn the programming language used by the robotic system, and have system implementation skills to put it all together.

The approach is to carefully analyze a sample preparation method so that it can be broken down into a set of unit operations. Example unit operations include:

- weighing,
- adding,
- reagents,
- mixing,
- aliquoting,
- heating,
- filtering,
- centrifuging,
- decanting,
- solvent extraction.

Appropriate remote control devices must then be purchased or developed to carry out each of these unit operations. Each scheme for accomplishing a unit operation must be evaluated in terms of the current application. For example, a sample heating station designed for open-container solids may not be acceptable if the current samples contain significant volatile components. A different scheme may be necessary, one that keeps the samples in sealed containers. Today, there are commercially available solutions for most of the common unit operation tasks. It is rare that a custom device will be needed, but for those rare cases, the additional skills of equipment design and construction will be necessary.

Suitable automated devices are first placed around the robot to accomplish each unit operation. The sample preparation method is then implemented by putting the unit operations together in the correct sequence, moving the sample from one peripheral to the next. A program is written to sequence the sample through the necessary steps. In addition to the basic sample treatment, a variety of other techniques that humans usually take for granted must also be automated. These include such things as opening bottle screw caps, crimp-sealing vials, and reading sample labels. Peripheral equipment is also available to accomplish most of these tasks.

The last step of any robotic system implementation is careful testing. It is not sufficient to merely observe that samples are processed according to design. Samples with known results should be run through the system, sent to the instrument, and then measured, and the results compared to the known results. Sufficient numbers of samples should be tested in this way to establish the precision of the method. Samples containing possible interferences should be tested, as should samples at the upper and lower limits of analyte concentration. The robotic method should be operated long enough, with careful observation, to establish its reliability. Failure to properly complete the preparation of any sample is cause for concern. The cause of the failure should be identified and the apparatus modified if necessary to prevent its recurrence. As many failure modes as possible should be identified and tested (if feasible). In other words, testing should try to stress the system, not merely confirm that it works correctly under ideal conditions.

Types of Robots

Robotics are seen in two basic forms.[3] In one, the automated instrument itself is equipped with extensive sample-handling equipment to treat and manipulate materials.

The simplest example is the automated sample tray that uses a mechanical procedure to move samples one-at-a-time to an active station. Here, a tube or some other device dips into the container and extracts a sample that enters the instrument for testing. The mechanical movement of samples can be a simple rotation of a cylindrical tray with samples in one or more rows around the circumference. Alternately, it can be a more complicated conveyer belt, often folding back on itself several times, much like the line of people waiting to enter an attraction at a theme park.

Some automated instruments have sample processing schemes that are more complicated than mere delivery of material to the instrument. After a sample is picked up for testing, it may undergo additional preparation steps. It can be heated, diluted, filtered, centrifuged, have reagents added to it, or be processed in myriad other ways to get it ready for testing. These kinds of high-volume automated instruments are widely used in clinical and industrial laboratories where hundreds to thousands of samples may be measured every day. In these cases, the chemist has little opportunity to alter the process. The robotic operations are designed by the engineers and are an integral part of the instrument.

In the second major form of robotics, separate devices are purchased to automatically treat samples. These units are not designed for any particular instrument, and the sequence of operations is not entirely predetermined by the manufacturer. Instead, they are intended as general-purpose sample-handling systems. They are frequently attached to instruments, so that the final processed sample is delivered straight into the input section of the instrument. However, the instrument and the sample-preparation unit are usually purchased and implemented separately.

There are two kinds of general-purpose sample-preparation robots. In one type, the layout of the workspace is determined by the manufacturer. There may be limited options provided from the supplier, but the position of components that are included is rigidly defined. In the second category, the workspace layout is largely configured by the chemist. The first category is known as automated sample-preparation workstations; the second is typically described as laboratory robots. The naming is somewhat ambiguous, for in reality both types, as well as sample-handling units built inside instruments, use the same robot technology.

Most fixed-layout automated workstations consist of a work area that contains one or more racks to hold sample containers. Also within this work area are additional components, like waste liquid disposal, rinse functions, mixing stations, reagent delivery, and others. There is usually a mechanical working device that moves on an x–y plane over the workspace. The working device might be a sampling tube connected to a remote-control syringelike unit. It can pick up and dispense liquids. Alternately, the device might be a mechanical hand that can pick up objects, like test tubes, and move them around the workstation. Once in position over a given point on the work area, the working device can move up and down. With these degrees of freedom, most of the operations needed to prepare a sample can be accomplished by sequencing the necessary steps in the right order.

The more general laboratory robots have a mechanical manipulator in the center of a bench. This central robot can move a hand (or possibly other devices, like a syringe) to any location within its reach. The geometry differs from robot to robot. Some use cylindrical coordinates, others a more complicated articulated hand involving both elbow and wrist moves. In cylindrical coordinate systems, a central platform can rotate through some angle, usually covering all 360°, with possibly a little overlap. On that platform is a carriage that can move up and down on a set of rails. The carriage contains an arm that can move in and out. These moves are all independent, resulting in the ability to reach any location within a cylinder. A hand with fingers is typically attached to the end of the arm. The fingers can open and close to grasp objects or let go of them. Usually, the hand can rotate about its reach axis to allow objects to be tipped. If it were holding a test tube, for example, the rotation might be used to dump the contents.

The articulated-hand design has a spherical work volume, but at the outer limits, not all hand orientations are possible. There are certain tipping moves the articulated hand can perform that are impossible for the cylindrical coordinate system. In reality, however, these extra degrees of freedom are rarely useful for sample preparation. Straight-line movement in the articulated-arm robots is more complicated, usually requiring two or more motors moving simultaneously in a nonlinear way. The equations of such motion are complex. The natural movement of an articulated-arm robot is an arc.

Most robotic systems have no vision. They work only because everything in the work area is in a fixed location. The robot is "taught" the locations of these components. Teaching a robot involves moving the hand or working device to the exact place needed to perform a given operation with a set of manual controls, then "telling" the controlling computer to "remember" the motor coordinates of that position. This physical location in space is given a name and saved in memory (like a database). Later, a program is written to execute a sample-preparation procedure. In this program, whenever the sample preparation requires a return to that "taught" location, the memory name is specified. The computer looks up that set of motor coordinates and moves each motor to the correct setting, thus accomplishing the move to that location. Additional program commands are then used to open or close fingers, rotate the wrist, operate a peripheral device, or do whatever is needed next to process the sample.

Fixed-layout automated sample-preparation workstations are easier to use but less flexible. Since the type and position of all the features are predetermined, the workstation can be delivered with most of the control software prewritten and the locations pretaught. The chemist usually only decides what reagents will be placed in different containers, what the sequence of events will be, and how long various steps (like mixing) will be allowed to proceed. Only those operations that are supported by the delivered workstation hardware are permitted. Most of these systems use open sample containers, do not have uncapping equipment, and therefore are less suitable for volatile samples. Many do not support all the unit operations that might be needed. Many have limited programming options and may require, for example, that all samples be processed in the same way. This would eliminate applications in which different solutions sets are needed that follow some experimental scheme, such as a research problem. Nevertheless, when the application falls within the capabilities of a fixed-location system, it is the preferred solution. Such automated workstations are much easier to implement. They were designed on purpose to sacrifice flexibility in favor of programming simplicity.

The open robotic systems permit a wider range of applications, but are more difficult to set up and use. They are, however, fully within the capabilities of any chemist who is willing to spend a day or so to learn the procedures. With these systems, any suitable equipment can be moved within the reach of the robot, regardless of whether or not it was designed by the robot manufacturer. Some equipment, like sample racks, can be standard-issue labware. Other equipment may require simple interfacing to the robot. A vortex station, for example, needs to be turned on and off at the right time by the controlling computer. A standard oven can be used if a way to automatically open and close the door can be devised.

General sample-preparation robots require more user programming to define every step to be followed by the robot. This includes defining workspace locations, control of peripheral devices, making the appropriate moves, and determining all timing relationships. The task is not especially difficult, and can be mastered by undergraduate students for a well-defined laboratory experiment after two hours of lecture and one 3-hour lab period.[4] Assuming all supporting devices are available and the chemist is reasonably familiar with the programming techniques, even difficult applications can be set up with typically 1 to 3 days of work.

Automation is an increasingly important part of industrial chemistry. Chemists graduating from our colleges and universities will be better prepared to compete in the workplace and be valued contributing members of the team, if they are introduced to the approaches and techniques of such automation in school.

Simulation and Modeling

Sometimes, a mathematical model can more or less faithfully represent reality. In these cases, the behavior of that model can be studied under different conditions to gain insight into the relationships and stimuli that affect reality. A model is never a perfect substitute for actual observation since there is always a possibility that the model is incomplete or inaccurate in some details. Nevertheless, the model can often provide an important tool for understanding the reality, and it can suggest additional experiments that might be performed.

Simulation and modeling are valuable in education. They are a way for students to gain a conceptual understanding of a chemistry topic that surpasses textbook descriptions. They can substitute for laboratory experiments that are too costly, time-consuming, or dangerous for students to actually perform. Modeling gives a level of insight that is difficult to duplicate short of the actual experiments. Often, it is more efficient than experimentation, yielding a high return in knowledge and understanding for a relatively small investment of money, time, and effort.

Modeling is also important in the careers of students for the same reasons. Learning does not stop at graduation. Working chemists have the same need to understand new chemical processes on the job. If they have learned modeling and simulation approaches, they can apply these tools to their work problems. Anything that can help chemists learn faster with more complete understanding will make them better employees. The company will benefit from the greater efficiency of their workers, the higher quality decisions that are made, and the improved productivity of their operations. Modeling and simulation are treated more completely in other chapters.

Electronic Notebooks

In many companies, the laboratory notebook has the status of a legal document. In this book are recorded the daily activities, notes, and observations of the chemist. The laboratory notebook often contains the original data, the first written record of observations or conclusions that ultimately may affect important activities of the company. The laboratory notebook, for example, may be used to establish patent claims, justify a defense to a regulatory agency, or form the basis of advertising promises. Because of its importance, some companies have rigid requirements about laboratory notebooks. Some require that the notebooks stay on site and never be taken home. Some insist each page be countersigned each day by a supervisor. Some keep the notebooks locked up when not in use. Some companies make photocopies to be stored in a secure location.

In spite of their importance, most laboratory notebooks are handwritten, often in the laboratory itself. The pages become soiled with spills and dirt. The handwriting may be difficult to decipher. Sometimes the chemist adopts a private code, making interpretation by others difficult. The contents of the lab notebook often become unusable if the chemist is no longer available because of resignation, retirement, or death.

The Electronic Notebook

The electronic lab notebook is an attempt to apply computer technology to this document. The intention is to capture in a computer the information normally placed in the laboratory notebook. Once there, the information can be better protected in several ways. First, electronic lab notebook programs incorporate a carefully designed data protection scheme involving regular backup to a secure off-line media. Second, they have features that prohibit changing data without proper accountability. Typically, any change in original data does not erase the original, but only adds a corrected entry. This is accompanied by a record of who made the change, when it was made, and why the data were changed. The system allows the complete revision history to be reviewed should any item be challenged at a later time.

The electronic notebook enhances the usefulness of the information in several ways. First, the structure of the entry system forces the chemist to better identify the nature of each item put into the notebook. This makes interpretation by others easier since it eliminates the private codes and partially labeled entries. If the input system is well-designed, this advantage can be achieved without excessive extra work by the chemist. Second, the displayed or printed results are in typed or graphical form, avoiding the problems of handwriting. Printouts are on clean paper, not the soiled pages of a lab notebook. Third, it is now possible to electronically transport original observations to other data processing programs, or to other members of the project team. Thus, the chemists or their supervisors may be able to send results to spreadsheet programs, databases, statistical routines, or graphing programs. This transfer avoids the manual retyping of data, which is both inefficient and subject to transcription errors. Lastly, the electronic notebook provides a variety of indexing and viewing tools, allowing information to be cross-referenced or seen in different contexts. Significant insights can often be gained by viewing experimental results in different ways, organized by different criteria, or correlated with different supporting information.

Additional Features

Some electronic notebooks accept various types of information besides text.[5] Examples include graphs, spectra, images, and sound. One approach is to simply encapsulate this information in the electronic notebook, then call the software that generated or supplied this data when it is to be displayed, printed, or edited. The electronic notebook would not, for example, know how to process a photograph, but it would know how to start appropriate photoshop software and supply the image for display and manipulation. It is a significant advantage to the chemist to be able to use the electronic notebook as a launch pad to easily access different kinds of results related to each experiment.

The electronic notebook can be augmented by other features. It could be used, for example, to hold experimental protocols or standard operating procedures. These are textual descriptions, possibly with supplemental graphics, of how experiments or procedures are to be carried out. They are usually carefully planned, often with supervisory input and approval. They are used to help ensure that laboratory work will be productive, that fewer mistakes will be made, and that important considerations in the experiments and measurements will not be overlooked. When chemists are part of a team, it is especially important that everyone's duties be carefully defined to ensure all critical information is captured and to avoid duplication of effort. Standard operating procedures also help ensure good laboratory practices are followed, which in turn means the information generated can be legally defended.

Standard protocols help document the actual work accomplished since each chemist is required to sign off that they followed the protocol. Alternately, they must make appropriate notations in the laboratory notebook as to why variations in procedure were necessary. Part of the protocol might be an input data template for each experiment, where the chemist fills in various observations or results. The existence of a template makes data entry faster and easier, and also reminds the chemist of all the information items that are required to complete the experiment.

Standard protocols will, of course, be frequently modified over time as experience is gained and procedures, goals, and equipment change. The electronic notebook could maintain a history of all the standard protocol versions and the time periods during which they were used. This becomes an important part of the archive records. It documents the methods used to acquire the experimental results saved in the archives.

The electronic notebook might be used to better track the use of reagents and other supplies, keeping information such as vendor names, lot numbers, and expiration dates. They may allow entry of notes and reminders of future activities. The latter might automatically generate messages with appropriate lead time to help chemists organize their schedules.

Implementation

Laboratories that adopt electronic notebooks usually go through an initial period of dual use. The paper lab books are still maintained, even while the electronic version is used. This is necessary in the culture of many laboratories because of the high importance placed on the notebook. Only when everyone has full confidence in the integrity and reliability of the electronic notebook will chemists and supervisors be comfortable giving up the paper notebooks. The extra cost of dual record keeping is justified by the enhancements provided by the electronic version, and the recognition by management of the comfort and security provided by keeping the old system as well.

In some regulated laboratories, there is a fear that abolishment of the paper lab notebook is illegal. In some cases there may be real concerns, particularly because some inspectors may be less tolerant than others. In reality, however, the real issue for the government is that original laboratory observations be kept in a secure way, where data cannot be lost or changed indiscriminately or fraudulently. Once the integrity of electronic laboratory notebooks is properly established, these real or anticipated problems with regulatory agencies should disappear.

The Future of Computers

The computer revolution will continue its rapid sweep across the discipline of chemistry, as it is doing in nearly all other aspects of our society. The computer is already an integral part of the professional lives of most practicing chemists, encroaching on nearly all parts of the job. It enhances the collection, manipulation, view, and storage of chemical information. It presents information in different ways for review, and allows modeling and simulation to help people reach a better understanding of reality more quickly and easily. It assists calculations. It controls instruments and other parts of laboratory automation. It keeps normal office records and helps organize the workday. The more graduating chemists know about computer applications, and the more comfortable they are using them, the better they will fit into the chemical workplace of the 21st century. Educators have an obligation to prepare their students to be productive and competitive in this work environment.

References

1. User's Guide, Microsoft Works, Version 4.0 for Macintosh (Apple Computers, Inc.), Microsoft Corporation (1994).
2. Henry Freiser, *Concepts and Calculations in Analytical Chemistry. A Spreadsheet Approach;* 3rd Revised Printing; CRC Press, 1992.
3. Grandsard, Peter; Megargle, Robert; and Markelov, Michael; Robotics & Sample Preparation. Automation in Contemporary Labs; *Sci. Comp. & Auto.,* 1991, June, 15–30.
4. Megargle, Robert; Laboratory Robotics in the Undergraduate Laboratory; ACS Northeast Regional Meeting, Potsdam, NY, June 26–29, 1990.
5. User's Guide, ResearchStation (Helix Systems, Inc.), Megalon S.A. and Helix Systems, 1994.

4

Industrial Considerations—Information Technology and Its Effect on Corporate Culture

Nancy Woo

Introduction

The recent hype about the Internet and the Information Superhighway gives one the impression that computers are ubiquitous and that information technology is well integrated into the workplace. It is certainly true that since 1990, major corporations have rapidly evolved into "information-based" companies, viewing information as their competitive edge. The transition has involved the establishment of corporate-wide networks, both LANs and WANs, large-scale distribution of workstations, generally Intel-based PCs, and acquisition of huge volumes of software.

How has the adoption of information technology (IT) been implemented? Has it had any effect on corporate culture, and if so, what effect and to what extent? Who is affected and how? The discussion that follows will focus on these questions, particularly as they apply to technical staff. For those who like a forewarning before reading, no answers will be given here; rather, the objective is to pose questions that will cause readers to reflect and, quite possibly, raise even more questions. We will examine the effect of IT on:

- work flow and work processes;
- the type of expertise required of an individual;
- time differences, geographic distance and the working day;
- traditional organizational structures.

Students and faculty reading this may ask why they should be interested or concerned about these issues. The answer is very simple. Information technology has probably had its greatest impact in the laboratory, radically affecting both the nature of work done and the way in which it is done. Whole new fields of endeavor have been enabled by advances in IT. At the same time, laboratory personnel must now deal with questions such as: What constitutes raw data? How do I link my data (which exist as a set of computer files) and the conclusions I have drawn from them (which are given as a report written using a word processing package)? If I need to keep my data for at least 10 years, what medium and format should I use? Most scientists are not well prepared to address such questions because these issues are not considered integral to their fields of specialization. Nonetheless, the answers they choose may have serious consequences on, for example, their ability to protect intellectual property.

General Effects of IT on Corporate Professionals

We will discuss first the general effects of IT on corporate professionals and then some of the specific challenges that it raises for technical staff such as chemists. To provide a context for our discussion, let us begin by reflecting on a few of the milestones that characterize the last three decades. For example, it is striking to note that in 1972, the total number of computers in the world was 150,000. By the year 2000, Intel expects to manufacture 100 million processors annually. In 1970, most professionals wrote manuscript drafts by hand and either made final copies themselves on typewriters or gave the drafts to typists. When multiple copies of something were required, either carbon copies were made or stencils were cut for mimeograph or other duplicating machines....stencils that were quite unforgiving of errors! This work environment created a natural work flow—and also rigorously defined areas of expertise. Professionals were responsible for the content of a document; clerical staff were the experts in format or presentation. Because typewriters in those days were not computerized, formats were primarily limited to minor variations in layout and could not generally vary in font and type size. The resulting documents were distinguished not so much by physical appearance but by content. Since copies were so difficult to make, copies were made only for those who absolutely needed to have them. Others had to wait until the physical copy was routed to them. The result was a leisureliness in the review process that was necessitated by the requirement to route a physical object over real distances. (Note, too, that express delivery services had not yet come into existence.) Because of the difficulty in producing and distributing documents, individuals could generally manage the demands on their time.

By 1980, things had changed significantly with the advent of low-cost copiers, white-out, and self-correcting ribbons for typewriters. Now, multiple copies of documents could be made trivially. The copies often appeared "cleaner" than the original documents since correction fluids and ribbons disappeared in the photocopy process. The first stand-alone word processing units had also begun to appear by 1980. Professional staff, however, did not typically have access to these units, so domains of expertise remained well defined: professional staff dealt with document content whereas clerical staff took care of format. What had changed was the ability to distribute copies of documents to many recipients simultaneously and the ability of those recipients themselves to recopy and redistribute the document. Control of distribution began to erode.

The first word processing machines were very specialized units manufactured by companies such as Wang, Vydec, and Artec, and their use required substantial training with the reward lying in the ability to make modifications readily. The early machines were not typically connected and were not general-purpose computers, so inclusion of graphics or tables created by other computer programs into a document was generally difficult, if not impossible. Because of their cost and the training required, these machines were not widely accessible and remained the tools of specialists. As the 1980s progressed, word processors improved in usability and technical capability so that by the middle of the decade, proficiency on a word processor, generally Wang, was a requirement for many secretarial jobs. At the same time, the personal computer came into existence, and with it came a variety of software such as word processing packages, spreadsheets, and graphics programs. The PC gave professionals access to office automation technology.

By 1990, PC prices had dropped sufficiently for i386-processor-based workstations to be fairly common. Local area networks had also made significant inroads as people began to realize the benefits of device sharing and central storage. Microsoft's Windows interface was becoming popular and WYSIWYG (what-you-see-is-what-you-get) word processing software more common. This meant that the content and appearance of a document could be created together, allowing visual effects to be used to support the message. The old division of labor disappeared, and the creator of the document now was responsible for its presentation also. Not only had document preparation been radically changed, but with the advent of the fax machine and e-mail, so did the channels available for document distribution. While the copier allowed distribution of copies to everyone who was interested, those recipients who were geographically distant still were at a disadvantage because they had to rely on traditional distribution means (usually postal services) to receive their copies. Fax machines and e-mail eliminated distance as a barrier to collaboration.

In this very limited review of the last three decades, we can see immediately some of the changes that have occurred in the workplace:

Defined Domains of Expertise

In 1970, because of the difficulty involved in making copies, one domain of expertise was centered on the physical creation of the typed document and its reproduction. This domain was usually quite separate from the domain of expertise involved in creating the content. By 1980, copiers obviated the need for special media to be used for mass copying. Typewritten originals could be copied, and, therefore, content experts could venture into the domain of document reproduction and formatting. In 1990, with networked access to sophisticated laser printers and word processing software, content experts could now also deal with the look of the document and produce as many copies of it as needed. One result of this transition has been a reduction in the ratio of clerical staff to professional staff in most large organizations.

Since many professionals now use their computers to create documents of all sorts, they are also now faced with managing these documents—tracking different revisions, electronically filing (storing) each copy, and managing their distribution. None of these tasks appears to be overwhelming nor beyond the capability of most professionals. However, how many of us have thought beforehand about an electronic filing system, and how many of us simply have generic directories called "DOCS" or something of that sort into which everything is placed? Even if we have thought about it, how many of us know how to structure an effective electronic filing system? Because many of us have not developed efficient filing schemas for our electronic files, and because DOS limits file names to eight characters with a three-character extension, we spend considerable time searching for documents. Even the longer file names made possible by Windows 95 leaves file management problematic for professionals. The search can become quite laborious if storage on LAN servers is also available. What is even more significant is that the search task cannot easily be delegated to another person. This contrasts with pre-PC days when filing and searching for files were the domain of clerical personnel who generally had established an overall filing schema that allowed efficient retrieval of documents.

Even more important, paper file systems can readily be reorganized, archived, and cleaned—all tasks that are considerably more cumbersome with electronic files. Filing, retrieving, and reorganizing files in the electronic world are all typically done by the PC user. Delegation of these tasks requires relinquishing access to the PC and, therefore, is generally regarded as inconvenient. The end result is that professionals are now routinely engaged in time-consuming activities in which they have no special competence nor training.

Work Processes, Geographic Boundaries, and Time

IT has had a dramatic effect on work processes. It has enabled rapid dissemination of information and documents, requiring only that sender and recipient are connected in some fashion—perhaps by a telephone line. Geographic separation is no longer a factor in determining who should review a work, with whom to collaborate in authoring, or who should receive a copy of a work. Distribution tasks have changed from addressing and stuffing envelops, pickup, and delivery schedules, to determining e-mail addresses and fax numbers. The time rhythm of a workday is no longer specified by paper mail pickup and delivery schedules—e-mail and faxes arrive whenever one's machine is connected, regardless of whether or not the intended recipient is present. Expectations of senders are also changing. Because electronic distribution mechanisms are virtually instantaneous, responses to distributed items are expected in like time, ignoring other demands on the recipient's time, and ignorant of their possible absence. Many e-mail systems were designed as personal messaging systems and so have no mechanisms other than sharing of the ID and password for one to delegate the review of messages. An extended absence (because one decided to take a vacation, for example) can result in a large accumulation of e-mail. This will often be a mixture of general announcements and notices in the midst of which will be buried messages upon which one is expected to act. The task of dealing with the accumulated e-mail upon return to the office can often be daunting. Just reading the backlog can often take much of one's first day back—assuming that one can spend that day on the computer.

There is also an interesting ambivalence in user expectations about the reliability of electronic distribution mechanisms. Electronic mechanisms, specifically e-mail and fax, are expected to be totally reliable, so user tolerance for error is vanishingly small. At the same time, many have experienced e-mail messages that have bounced back (at least you know it didn't get through), messages that vanished into the ether, and faxes that simply never appeared at the other end. As a result, when it is important that a message or document be delivered, some have adopted the tactic of using redundant mechanisms and sending by e-mail, fax, and/or confirming hardcopy. Recipients end up inundated with multiple copies or they are sent messages that request immediate response acknowledging receipt of the message and its contents. Finally, most e-mail systems also offer a "file cabinet" function to help users sort and store their mail. While these are easy to use, they are usually only accessible through the e-mail software so that any search process must be conducted separately for e-mail files and non-e-mail files.

One of the significant changes in the work environment accompanying the introduction of electronic communication mechanisms is the elimination of time differences as a deterrent to dialogue. What has been required has been a modification of conversational style—responses may be delayed by several hours. Nonetheless, the recognition of the potential of electronic connections to lessen the complexities of multinational, multicultural collaborations has led to the aggressive adoption of these technologies in many corporations. Not only are time differences eliminated as major barriers, but problems created by language differences are also minimized. Telephone conversations involving parties whose native tongues are different always require someone to listen and respond in a language in which they may not be fluent. That often results in misunderstandings because of incomplete comprehension of what is said or the inability to phrase adequately what is meant. Electronic communications, by being written, allow more careful study of each message and allow more time to prepare a response. They also allow others to assist in the process and therefore result in more accurate communications.

Finally, electronic connectivity is beginning to allow the concept of a virtual office to be realized. Certainly, many e-mail systems permit dial-up connections so that whether at home or away at a conference or on a business trip, one can remain in touch with colleagues. Technology to permit secure dial-up LAN access is beginning to appear, allowing one to gain access to files without having to copy them to diskette. The net effect of this "anytime, anywhere" accessibility is simply that the defined work week is becoming less well defined. The challenge for corporate staff is to determine how to segregate work and personal demands on time to maintain balance. Since your office can literally now follow you around, not being at the office doesn't mean you can't work.

Organizational Structures

Electronic connectivity permits the creation of organizations that ignore geographic boundaries, and therefore, it presents novel management challenges. Corporations have addressed the problem with the creation of the "dotted line" reporting structure so that one may have several bosses. The difficulty this presents is understanding the domains of authority and who defines the boundaries of each domain. The practice is not new; it has simply become commonplace since IT facilitates the existence of virtual organizations.

In traditional companies, electronic connectivity may also mean that management is literally more accessible. Because e-mail is still relatively unsophisticated, management may be more accessible via e-mail than through more traditional routes such as paper mail or telephone because it is more difficult to have assistants filter e-mail than it is to have them filter telephone calls or paper mail.

Novel Problems

The examples just discussed demonstrate that the adoption of IT has resulted in fundamental changes in corporate culture that are only beginning to be noticed. In fact, one might contend that the changes themselves are just beginning because IT presents unique opportunities for ways to do business that have concomitant challenges and for which there are no parallels in a paper-based culture. An example of the latter can be found in the issue of how best to organize electronic files. One might contend that this problem does have a paper analog, since if one understands the basic principles of organizing paper files, these principles should be applicable to organizing any sort of file...or should they? The analogy is misleading, however, because with paper, the term "file" refers to one type of thing: documents. In IT the same term refers to document, spreadsheet, graphics, program, and system files—a veritable mixed bag! Even if one restricts the problem to user "created" files, that is, excluding program and system files, electronic files present problems that are nonexistent with paper. Specifically, so long as the physical medium is intact, paper documents can be read. With electronic files, the content is not directly accessible: one needs an agent, software, to examine it. The question is: Which software? With the rapid pace at which software changes, it is not unusual to find old files that can no longer be "opened", because one lacks the appropriate software. The problem is not only found with old files: Electronic collaborative authoring requires that all participants be able to manipulate the shared document, generally by requesting that everyone use the same software. Paper does not present such a problem.

The software industry is trying to address the interchange problem by creating products whose only function is to allow documents to be exchanged without requiring that all participants use the same software. These products, which include Adobe Acrobat, Common Ground, and WordPerfect's Envoy, among others, consist of two components. The first component creates a "portable document" by converting the output of an existing package—Microsoft Word, for example—to a proprietary format—Adobe Portable Document Format (PDF) (Acrobat), for example. The second component is a viewer that allows anyone who has it to view and print the document. The vendors have typically structured the products so that the viewer can be distributed at no cost to the sender or recipient. The output of these products is a "portable", shareable document. The products vary in the accuracy with which they reflect the printed document derived from the original package and in the editing functions permitted the recipient. While this begins to address the problem of exchange, it is a solution that still depends on the appropriate software existing at both ends of the interchange. One vendor's viewer doesn't work with the output of another vendor's product. Frustrated by the effort required to deal with documents electronically and the potential inaccuracies of conversion, many users elect to print and either fax or mail paper copies to each other. Note also that these products do not address the problem of gaining access to files for which one does not have the original creating software.

This is one small example of a novel problem presented by the adoption of IT. A related problem concerns archives—what should be archived and how to ensure continued access to archived material. In the paper world, if one wants to archive a document, the choices of what to archive are fairly limited: one can store the actual paper copy or one can store a microfilm (or microfiche) version. There are a number of schemes for indexing and cataloging archived material in either format to facilitate retrieval. Once printed material is retrieved, it is immediately accessible—no additional intermediaries are required to be able to read the material no matter how long ago it was archived.

In contrast, when the decision is made to archive an electronically created document, the choices of what to archive include paper (or microfilm), an electronic file that contains the image of the document in some standard format, such as TIFF, or an electronic file containing the codes and format of the software used to create it. In addition, if one elects to store an electronic copy, one must also choose whether to archive to magnetic tape or disk, or to optical tape or disk. Finally, in addition to selecting the physical format (tape or disk, magnetic or optical), one must choose which size: 3.5-in. floppy, microcassette, etc. The task has become more complex. Retrieval of a document can now require more than identification of its storage location—one must have appropriate devices on which to load the storage medium (disk, tape, etc.) and, if the document was stored in a binary format, software to gain access to its contents.

Most corporate planning for electronic archives addresses disaster recovery. Off-site locations are established that provide secured storage for critical information along with the requisite software and computers. These facilities are not meant to be true archives. General strategies for archiving are only beginning to be developed. Depending on their need to maintain records and documents for lengthy time periods, individual business units within corporations may or may not have established departmental strategies for archiving electronic files. Unfortunately, because the spread of computer use is such a recent phenomenon, it has been difficult to predict what will be stable or standard, and the rate at which things will change.

Thus, for example, those who were early adopters of office automation technology may have decided to store documents on floppy disks in the format used by their word processing packages. Those floppies were probably the 8-in. floppy disks, which were not uncommon as storage media in the late 1970s and early 1980s. The challenge for anyone who chose to archive documents on that medium and who wants to retrieve those documents today is finding the physical devices on which to load the diskettes and then finding the software with which to read the files and the hardware and system software required to run the application software! An alternative tactic that has been adopted by some organizations has been to convert all existing stored documents every time a major change in software and hardware is undertaken so that current software and hardware are always capable of reading stored files. It should be evident that this tactic, while it solves one problem, creates an enormous workload, as the volume of stored material increases.

One may conclude from this example that decisions about what to archive must now entail making educated guesses as to the lifetime of the storage medium (magnetic or optical) and the storage medium format (floppy, tape, size, etc.), as well as deciding what material to archive. As more factors need to be considered, it is clear that some knowledge of IT and the IT industry will be required of those who are making these decisions.

To complicate matters further, software now allows the creation of compound documents that have no counterpart in the paper world. These are entities that can comprise a variety of different types of objects: word processing documents, spreadsheets, graphics objects, databases, and pointers to other compound documents. Software technologies such as OLE (object linking and embedding) and hypertext links are the tools used to create these new entities. Now, the archiving problem has just been complicated further by the following questions:

1. What is to be archived: only the "envelop" document (the entity into which objects have been embedded or to which objects are linked) or the entire "compound" document?
2. If the latter, are only embedded objects included or are all linked objects also included? Note that linked objects may be remote and that nothing in today's technology ensures that a remotely linked object continues to exist just because

a document points to it. Those who "surf" the Internet are well aware of hypertext links that cannot be followed because the objects to which they refer no longer exist or the address (URL) for them no longer exists.

3. Embedded or linked objects may be of any sort, such as spreadsheets, graphics, word processing documents, databases, ASCII files, etc., and when inserted into a document, they are "activated" by invoking the program used to create them. Thus, for example, if one wanted to include a PDF object in this document, created using Microsoft Word for Windows 6, Word itself has no capacity to "present" the content of the embedded PDF file to the user but must launch Adobe Acrobat (or Acrobat Reader) to do so. The compound document is thus fully accessible only through the concerted use of all of the applications programs used to create the objects in it.

Note that file sharing of compound documents is also far more complicated than sharing of simple documents since all parties must have each software package needed to access the various objects in the compound document.

Finally, it must also be noted that the introduction of the 32-bit operating systems, Windows '95 and Windows NT, will also complicate file sharing and archiving during the period that 16-bit and 32-bit operating systems (OSs) coexist for the PC. Specifically, as more software is developed for the 32-bit environment, backwards compatibility will not be ensured. How long the 16-bit operating systems will persist is difficult to say, but with the size of the current installed base, it is hard to imagine that the transition will be accomplished in less than several years.

These examples are just a few that illustrate the unique challenges deriving from aggressive adoption of IT. The examples should not be taken as deterrents to using IT; rather, they illustrate the need to understand how pervasive the effects of IT are. The ability of an organization to use IT effectively will depend on its members having a good general understanding of information technology and being able to engage intelligently in discussions to determine what are the best IT-based work processes.

Specific Effects of IT on Corporate Technical Staff

Clearly, all of the changes discussed so far apply to all members of corporate staffs, including the technical or scientific staff. The latter have particularly benefited from electronic connectivity, which has enabled geographically separated colleagues to maintain much closer contact without regard for time or distance. Are there specific issues that are of more relevance to technical staff beyond those already described? The answer is yes simply because information technology has been so fundamentally integrated into laboratories through instrumentation, data analysis, and presentation. Some of the issues, such as copyrights and electronically published material, are very "hot" and are discussed widely, and will not be covered here. Rather, we will focus on one small set of examples dealing with daily life in a laboratory.

It is difficult to find instrumentation today that does not use computer chips as controllers, signal converters, and/or signal conditioners. In addition to the specialized on-board computers, most instruments also interface to a general-purpose computer for experiment control, data acquisition, processing, and reporting. The interface may be to a PC or other workstation or to a larger machine such as a VAX. Laboratory personnel will need some expertise in IT when:

- evaluating instrumentation for purchase;
- deciding what, when, and how to store and maintain results obtained from instrumentation;
- determining how to integrate information obtained from different sources.

The decisions that are made in each of these instances will depend on the type of laboratory involved (for example, academic research, industrial research, industrial development) and its requirements for being able to share information with other laboratories or departments. What kind of IT expertise is required?

Evaluating Instrumentation

Most instrumentation today uses computers in two ways: (1) specialized computers are built into the instrument to control its components and to process the raw signal, and (2) a general computer, usually a workstation (PC, Macintosh, or UNIX workstation), acts as the user interface. The latter function is accomplished through software provided by the vendor and usually includes modules for experiment setup and review, data review, and data processing and reporting. In addition, there will be control modules that communicate to the instrument's on-board computers to transfer experimental parameters and acquire the data. When assessing an instrument, one generally determines whether the instrument is technically adequate for the task in mind (Is the detector sufficiently sensitive? Can it operate under the physical conditions required?), whether the user software is sufficiently friendly, and whether the user software presents the data in the manner desired. Less often does one take into consideration the technical details of the hardware and software in the workstation. These latter, however, can cause an instrument to become obsolete earlier than desired and can limit the utility of an instrument.

For example, some years ago, when NeXT computers were first introduced, some instrument vendors thought that it would be a good machine to use as their system's external computer. It was relatively inexpensive, was designed to be "user friendly", and had an operating system that was a variant of UNIX. Unfortunately, the NeXT computer did not succeed in gaining popularity, and they are no longer manufactured. Both vendors and buyers of instrument systems incorporating NeXT computers have had to expend effort and money to accommodate the failure of the computer company. The lesson here is simply that one must take into consideration all parts of an instrument system when trying to project its useful lifetime, and that includes making some predictions on the commercial viability of the external computer.

This type of failure is fortunately rare. More common are software-related problems. In another example, one instrument vendor used an Apple Macintosh as the user interface machine. The user software was developed under Mac OS version 6 and was written in Pascal. Because the software is "generic", that is, designed to address the general audience of buyers of that type of instrument, it will necessarily fail to meet all of the specific needs of a laboratory, particularly if its needs change. In one instance, a laboratory that had purchased the instrument several years ago recently decided that it needed to modify the user programs so that it could more readily extract the needed data, which are not currently extractable in any simple fashion. The vendor informed them that (1) it could not do the custom programming for the laboratory, and (2) that the laboratory's software was designed such that new modules could be added on the proviso that they were written in Pascal. The laboratory does not have anyone skilled in programming on the Macintosh in Pascal and is thus faced with two options: continue to use an extremely labor-intensive process (copying values from the display) to extract processed data, or find someone to write custom programs using an old compiler on an outdated operating system (Pascal running under Mac OS version 6). The first option limits the utility of the instrument since it cannot be used for data collection when the workstation is running report software (required to get the data desired). The second option requires looking for and paying a programmer. If there is a reasonable chance that one will want to add new functionalities to the user programs that are not of general interest (and, thus, it is not likely that the vendor will supply them), then the evaluation of the instrument should include an assessment of the ease with which the software could be modified. This assessment would determine whether or not expertise in a programming language is required and, if so, how commonly found that expertise is.

Data Storage: What to Store and How

Adding external computers to instruments has also greatly expanded the possibilities of things to store. Prior to this, data acquired were printed out or recorded on a device such as a strip-chart recorder. In a laboratory that does chromatographic analysis, for example, the strip-chart recording was considered "raw data", and integration of peak area (area under curve) was accomplished by cutting out the peaks and weighing them. In this environment, data could not be reanalyzed and the experimental record consisted of the strip-chart recording and lab notebook pages, which give background material and the time and date of the work, describe the methods used, identify the samples and the order in which they are introduced, and tabulate the results of any calculations. If laboratory records need to be retained for long periods, then the notebooks could be archived.

With modern chromatographic systems, the same laboratory has the potential to store electronically raw data (the chromatographs), methods, sequence files identifying the samples injected, processed data, and reports. By doing so, one can reanalyze the data to see the effect, for example, of modifying the integration method by comparing the new processed data with the original. If one is going to store large numbers of these files, then it will be necessary to devise an effective scheme by which to identify related files and their locations. How complicated the scheme needs to be depends on what information is retained and how long it must be retained in electronic format. Raw data are typically stored in binary format and require the vendor's software for access and interpretation. Processed data files and reports are generally ASCII files whose contents can be examined and printed using general-utility programs. The decision as to what type of media to use for storage will obviously also depend on how much information is stored and for what time period. One should not forget that while much information is created and stored electronically, the physical laboratory notebook still exists in most corporations. Notebook pages need to be cross-referenced to related electronic files.

One of the complications that arises if raw data are stored for the purposes of being able to reproduce the calculated results lies in the fact that one must also be able to reproduce the computer environment in which the original results were obtained, including the same software, operating system, and hardware. Because instrument vendors are at the mercy of computer manufacturers, they cannot assure buyers of the longevity of their data systems. For this reason it is worth spending time thinking through the question of what needs to be archived (raw data, processed data?), in what format (electronic, paper?), for how long, and for what purpose. The latter questions will dictate the physical medium selected for storage. It should also be noted that laboratories that are required to retain data for long periods may feel inhibited from changing vendors because of incompatibilities between the original data system and files and a prospective new one. Thus, for these laboratories the initial selection of a particular instrument and its accompanying data system can have far-reaching implications.

Finally, the legal status of electronic records is still being defined and is complicated by the rate at which the computer industry introduces ever more complex functionality. There is a large and rapidly increasing body of literature that has developed on the subject of electronic records both in printed format and on the Internet. Suffice it to say that the discussions concern both the definition of records (what constitutes an electronic record) and record-keeping practices. These discussions are exceedingly relevant for laboratory scientists since they will determine how an electronic laboratory notebook is defined and what one must do to use it properly. Until such definitions exist and are commonly accepted, particularly by legal bodies such as the U.S. Patents and Trademarks Office, many individuals and organizations will continue the practice of printing out the contents of data files, pasting these in notebooks, and archiving the notebook.

Data Integration

At some time in most laboratories, one will want to export data from instrument data systems to some other computer. This may be done for further analysis, to free the data system so that the instrument can be used, or to combine results from different assays into one database. Whatever the reason, one should anticipate the need to extract information from the data system. Whether this task is easy or hard depends entirely on how the vendor has designed the software. In many corporate laboratories, central laboratory information management systems, to which each laboratory contributes, have been implemented. If one must routinely transfer data from one system to another, it is unreasonable to do so in any way other than automatically. Often, unfortunately, this requires creating a special program whose sole purpose is to extract data from one system and import it into another. The programs can be simple or complicated depending on how difficult it is to extract the information required, whether or not the two computer systems communicate easily, and how complex the data import process is.

Implications

These examples illustrate some aspects of how information technology has affected laboratories. In particular, they show that IT is an integral part of today's laboratories. Because of that, it is important that technical staff attain some level of IT literacy so that they will be capable of addressing the issues that will arise. Computer professionals can assist laboratory personnel to some extent, but they cannot make the ultimate decision of, for example, what data to retain, simply because they will not understand the implications of those decisions for laboratory operations. As computer technology advances, it will both enable and cause even more changes in corporate processes and, therefore, in corporate culture. It is increasingly important that all people who work in large organizations understand the implications of the technology introduced so that they can participate in determining what the new culture will be.

The future as it involves computers will not always be so problematic, however, for there will come a time when computers will be truly ubiquitous and fully integrated into one's work and home environment. One has only to recall that when the telephone was first introduced, it was regarded with great suspicion and not necessarily considered to be useful in a place of business. In fact, one should note that in the early years of telephony, there was much discussion as to what influence it would have on society and how it would develop as a technology. It is likely that few then anticipated the universally available system that exists today and that we all take for granted. It is quite probable that computers will also exert significant influence on societal organization and become equally integrated.

Suggested Readings

The references given here are meant as supplemental reading. Many cover topics not even hinted at in this discussion.

1. "School of Library and Information Science Project on the Functional Requirements for Evidence in Recordkeeping"; accessible via WWW at URL http://www.lis.pitt.edu/~nhprc/.
2. *The Telephone's First Century—and Beyond;* Thomas Y. Crowell Company: New York, 1977.
3. Katsh, M. E.; *The Electronic Media and the Transformation of Law;* Oxford University Press: New York, 1989.
4. Negroponte, N.; *Being Digital;* Alfred A. Knopf: New York, 1995.
5. Jones, P. B. C.; "Patent Rights Can Get Lost in Cyberspace"; *Genetic Engineering News* 1995, *15*(4), 4–5.
6. Reinhardt, A. "Managing the New Document"; *Byte,* 1994, *19*(8), 91–104.
7. Postman, N.; *Technopoly;* Alfred A. Knopf; New York, 1992.
8. Taubes, G.; "Science Journals Go Wired"; *Science,* 1996, *271,* 764–766.
9. Taubes, G.; "Electronic Preprints Point the Way to 'Author Empowerment'"; *Science,* 1996, *271,* 767–768.
10. Bearman, D.; "Managing Electronic Mail"; *Archives and Manuscripts,* 1994, *22*(1), 28–50.

5

Computing and Communications in Chemistry Education*

Martin S. Ewing

The Computing Marketplace

The technological position of the academic chemist/educator, and indeed the whole academic enterprise, is rapidly evolving within a larger economic matrix. Since the 1960s, digital computers have become central to chemistry, and have even spawned and supported whole new areas, such as computational chemistry and structural biology. However, while the role of computation in science has increased, the computing and communications industry has grown even more. With the introduction of the personal computer in the 1980s and the widespread adoption of the Internet and digital entertainment media in the 1990s, the scientific computing community is a relatively small influence among titanic market forces. Ironically, it is not hard to find high school students and homemakers working with computers that are equal to and sometimes more powerful than those that support academic groups in research universities. The trend toward powerful computing and information systems in consumer applications is accelerating, with multimedia home Internet connections, 500-channel television, and flexible digital video receivers on the horizon.

A quick newspaper survey will show that computing and communication are at the leading edge of a major social change. As Toffler[1] and others assert, wealth and well-being are increasingly dependent on production, manipulation, and dissemination of knowledge. In the pretechnological world, universities were the main centers of knowledge. With leadership and foresight, they may play an even more crucial role in the future. This will not happen, however, unless academics, chemists among them, fully engage the new information-rich interactive digital environment.

---

* An early version of this paper was presented at the national meeting of the American Chemical Society, August 22, 1994, Washington, D.C., at the Award Symposium honoring Galen W. Ewing with the ACS Division of Analytical Chemistry's Award for Excellence in Teaching.

This chapter will focus on the relationship between chemistry faculty and students and the general computing picture in the university. All chemists need to master the technology that they might use in the lab or classroom, but just as important and possibly more difficult, they must understand and exploit the information environment of their own institutions.

Computing Technology in a Nutshell

A few phrases and rules of thumb serve to point out useful concepts of computing science and technology. These brief vignettes may help chemists navigate the choppy computing waters at academic institutions.

"Computing power doubles every 18 months." This is a statement of Moore's Law.[2] Over 30 years or more, semiconductor technology has doubled processing power at a constant price every 12–18 months. As long as there is a large market for faster devices, and until still-distant physical barriers are reached, this progress will continue.

"Computing power is cheap." As a corollary to Moore's Law, the price of a unit of computer power drops rapidly. Thus, microprocessors have penetrated ever larger markets, to the point that they might be found in any laboratory instrument, coffee maker, or wristwatch.

"Bandwidth is cheap." A single fiber-optic cable can carry data limited primarily by the quality of the optical transmitter and receiver at either end. For years, the standard data rate between laboratory computers has been 10 Mbit/s (Ethernet), but with the explosion of processing power in each system, data rates are expanding dramatically. In new installations, 100 Mbit/s will soon be the norm. Advanced methods such as asynchronous transfer method (ATM) based on scalable bandwidth channels (155 Mbit/s or higher) offer a route to gigabits per seconds (10^9 bits/s) transfer rates.

Academic chemists are likely to want to work from home computers connected to campus systems. While standard applications like e-mail work well, the restricted bandwidth of typical home telephone connections limits the use of advanced applications that might use X Windows sessions or multimedia. Conventional analog phone lines are limited to 28.8 kbit/sec, for which inexpensive modems are now available. The same telephone wiring is also capable of 64 or 128 kbit/s digital transmission using ISDN (integrated services data network) service.[3] The most capable wiring system entering the home, however, is cable television, which could carry megabit digital streams. Although ISDN is now available in many areas, it is often expensive, and its acceptance has been limited. Many academic users will wait until the unfolding consumer market provides a low-cost wideband data service for the home.

"The network is the computer." This slogan has been used by Sun Microsystems, Inc., to make the point that a network of computers working together is much more powerful than an isolated system. Power is achieved by distributing certain functions, such as data storage, mail service, user authentication, etc., on "server" computers.

"The client/server paradigm." This is a particular style of computer system organization in which network "clients", smaller desktop user systems, access "servers" that offer specialized functions—file access, printing, mail, World Wide Web (WWW), and many others—to authorized clients. Client computers can have simpler configurations that increase reliability and save cost.

"Software is hard." The relentless increase in computing speed exceeds the improvement of software productivity. Operating systems and applications software have improved, along with programmer productivity, but academic faculty and staff who wish to create their own software rapidly run into a barrier: It takes time and very specialized programming knowledge to produce professional software products.

"Parallel software is very hard." Particularly important in computational chemistry, parallel systems, multiple processors that run simultaneously on one or more tasks, provide a cost-effective path toward increased computing performance. Many Unix systems now support symmetric multiprocessing (SMP), especially for server systems. This means that up to perhaps 32 processors may share a common memory space. The Silicon Graphics Challenger is an example of a popular SMP machine. Parallel computing can also be implemented through clusters of computers with physically independent memories, tied together with a fast communication system, for example the IBM SP2.

While parallelism offers the option to run multiple conventional programs simultaneously, dividing a single large application across multiple processors remains quite difficult, mainly because complicated interactions often occur between sections of a program. Additionally, when most of a program has been parallelized, performance will be limited by the remaining nonparallel code. This is one statement of Amdahl's law.[4]

"A balanced system has equal MIPS and megabytes." The Case/Amdahl principle[5] states that memory size and processor speed are proportional in a well-balanced system. For example, a system that processes 1 million instructions per second (1 MIPS) requires about 1 megabyte (1 MB) of storage. In the mid 1980s, a 1-MIPS, 1-MB computer was substantial. The rule still holds in 1995, when a 64-MIPS, 64-MB computer makes an excellent high-end personal computer.

"GUI." The graphical user interface (GUI; pronounced "gooey") is the common method of interaction between users and personal computers and technical workstations. It channels user communication through a graphic display using a "desktop metaphor" in which multiple windows are drawn on a high-resolution graphics screen as if they were sheets of paper on a desk. The GUI user "points and clicks" to manipulate windows with the aid of a mouse and push-button system. Since its first consumer introduction with the Apple Macintosh in 1984, GUIs have also been adopted by Microsoft Windows for DOS/Intel-based computers and by Unix vendors.

Extensions of GUI are in sight, including 3D displays (which are especially useful for molecular visualizations) and "virtual reality" (VR) displays that permit users to "move" themselves through a 3D space looking in various directions and perhaps manipulating objects with a "data glove". Sound is another extension to the user interface, including rendition of high-quality audio by the computer as well as input of user commands through speech recognition. Although systems now exist with one or more of these capabilities and they may be economically feasible, it is unclear how fast they will enter the mainstream computing market.

"Users will publish." Users with Internet connections are becoming publishers of their own data using the WWW system. In the academic environment, individuals, research groups, and departments establish "home pages" to describe themselves or their activities in order to provide their message to local or far-flung audiences. However, the ease of WWW publishing does not guarantee the quality of the product. Academic administrators are faced with a plethora of amateur self-publishers, not all of whom are able to present an appropriate view of their department or institution.

"The price of security is eternal vigilance." As the network is used for important personal and business transactions, the integrity of the network against assault from careless or malicious persons and systems becomes critical. Computer viruses are now the stuff of urban legend. Network-based attacks on academic systems abound. Users still connect to computer services with their user name and password, but this is increasingly inadequate to ensure a user's identity. Cryptography-based secure authentication and encryption of data will be required, especially when valuable data and cash transactions are passed on the network.

"There's no place like home—for computing." Personal computing systems are increasingly hard to distinguish from technical workstations. With improved network bandwidth to the home, many more scientific computing applications should be accessible from home computers. The hardware distinctions are also becoming blurred, although there remains a major difference in operating systems. High-end technical systems are commonly based on Unix, a complicated multiuser system that is unlikely to appear in home systems. Home personal computers can cooperate with university Unix systems through the client-server model, for example using X Windows emulation software. (X Windows is the GUI system most commonly found on technical workstations.)

The University Computing and Communications Environment

Brief History of Academic Computing

From the 1960s on into the 1980s, the typical university computing environment was a computing center where most users had to do their work. In the 1980s, many users migrated away from central systems to department-scale minicomputers (led by the Digital Equipment Corporation DEC VAX systems) that were less expensive and more responsive to local needs. Soon afterward the first personal computers were introduced to serve as word processors, spreadsheet calculators, and "dumb" terminal emulators.

In the 1990s, desktop workstation systems, usually based on the Unix operating system, have largely displaced department minicomputers, although there are often departmental server systems. For a majority of faculty, the desktop workstation provides all the computing power needed. Computing problems now revolve increasingly around support—software maintenance, networking, and operating system issues.

In the mid-1990s, the introduction of the Apple–IBM–Motorola Power PC and the Intel Pentium and Pentium Pro processors has nearly eliminated the performance distinction between high-end "personal computers" and low-end "workstations". Differences now are mainly in the operating system. Unix dominates the workstation market, while Microsoft Windows, Apple MacOS, and IBM OS/2 are the most common personal computer operating systems. Unix still provides much greater flexibility and power than the personal computer systems, at the price of more difficult operation and support. Microsoft Windows NT offers many Unix-like features while offering compatibility with older Windows applications. NT is rapidly gaining acceptance for technical and business computing.

The workstation–PC hardware convergence has led to an important development for universities: For the first time, students can and do own scientifically significant personal computers. With wideband networking (10 Mbit/s or more) available in student rooms on many campuses, students enjoy the same computing and communications power as many faculty. This can have an important impact on the curriculum, as we will see in a later section.

Current Institutional Needs

Academic institutions are facing serious operating challenges: They must provide their labor-intensive services despite rising costs and declining revenues. Computing and communications technologies can be a significant help, because they address the fundamental "information" character of the university and because their power is dramatically increasing while their cost is decreasing. Scientists in academia have an opportunity to exploit the new correlation between the scientific computing technology they directly require in their teaching and research, and the needs of their college or university.

Administrative Systems

Many academic institutions are seeking to "reengineer" their administrative processes to increase flexibility and efficiency while reducing overhead costs, along the lines recently promoted by Hammer and Champy.[6] The vision is of a "flattened", more distributed management structure taking full advantage of networking and distributed computing technologies. The Massachusetts Institute of Technology has an on-line description[7] of its reengineering program.

One expedient is to take advantage of the increasing computer literacy on campus. Many administrative functions are database transactions that end users, given the right software, can perform across a data network. Purchasing, grant accounting, address changes, and many other functions can be handled with minimal paperwork and staff support. The Mandarin Project,[8] a university consortium sponsored by Cornell University, is developing tools to allow institutions to design software clients that will handle their users' administrative needs securely and reliably on their own desktop systems.

Academic Systems

The "academic process" has evolved only slowly in colleges and universities. The present system of lectures, tutorials, written papers, and examinations would easily be recognized by faculty of 100 years ago. We can expect more rapid changes in the future, however, driven partly by financial pressures on the university, but also made irresistible by the growing capabilities of computing technology.

"On the Brink"[9] is a survey published by the American Association of State Colleges and Universities (AASCU), drawing on 230 member institutions. The most pressing issues identified by the AASCU study are presented in Table 1. There is a strong sense that the effects of technology will be pervasive and radical and will help academic institutions overcome fiscal and demographic challenges. Yet, in the short term, funding is very tight and acceptance of new systems is limited by inadequate support and training for faculty.

A view from Yale University supports these findings. At Yale, students are connecting with good bandwidth from their rooms and are purchasing personal computers that handle real-time audio and video; they are also accessing bulk data through CD-ROMs and other devices. Though not formally required, most students will seek systems with these capabilities. Some faculty already prepare multimedia lectures for electronic presentation in the classroom. Students have enough network information available to permit them to be very productive in their residences.

To a technology enthusiast it is just a minor step to bypass the lecture hall entirely in favor of network-delivered teaching. Indeed, ultimately the need for a physical

TABLE I Summary of issues recognized in the AASCU study "On the Brink."[a]

Issue	Conclusion
Student access and competencies	As students gain new computing and network abilities, education will shift toward a student-centered learning environment.
Distance education	Institutions must accommodate nontraditional students with "time and place independent" delivery of resources.
Institutional expectations for faculty	If faculty are to use new technology, they will require substantial support. New collaborative modes of teaching and research will be opened.
Institutional goals and plans	Rapid change makes technology difficult to build into long-range plans, but it is necessary to do so. Plans must include staff support and faculty development, but must also be sensitive to competing constituencies.
Connecting the campus	Network access and utilization should be encouraged across the whole academic spectrum, providing key common applications and training support.
Managing and financing information technology	Managing expenses of academic computing and networking is a challenge everywhere. The real goal is to find cost-effective means to support the institutional mission.

campus may come into question. However, the personal interaction between faculty and students is highly prized in the liberal arts setting, and technical developments that detract from the personal element will find strong resistance. The challenge for academic computing staffs and curriculum development committees is to enhance both the quantitative productivity in teaching and learning and the qualitative experience that transmits the spirit and values of the university.

Generational Issues

Exploiting new computing and communications methods is clearly limited by available support resources. Conventional teaching depends on instructors and teaching assistants. The quality of the delivered product depends on individual "chemistry" (the psychological kind) and rarely receives enough support from the institution. New electronic methods often require familiarity with software details that many faculty find difficult to manage. For such methods to be accepted, we need specialized support staffs, currently found in few academic budgets.

Another peculiarity of technology in teaching is that incoming students are often more adept with computer and networking tools than some faculty. Junior faculty often adopt new teaching methods faster than their tenured colleagues, even though this may not be a route to promotion. If a solution to the generational problem, other than the passage of time, is to be found, support resources will be the key.

Network Infrastructure

Contemporary academic computing is largely network-based and will be increasingly so in the future. A well-considered network investment strategy is crucial to the success of almost all computer-based services in the university, whether academic or administrative. While an individual laboratory computer is useful as an instrument controller, signal conditioner, or recording device, a computer's effectiveness is multiplied by attaching it to a communications network. This allows data to be easily moved to various locations, allows computer software to be updated, and allows remote users to connect to the lab system.

With good networking, academic computing clusters can be managed effectively at remote sites while preserving a common "look and feel" for student users. Dormitory clusters and even personal computers in student rooms can be supported in much the same way.

Network Technologies

A network connection can be of several varieties: direct-wired (such as Ethernet) or a telephone modem connection, using PPP (point to point protocol), ARA (AppleTalk Remote Access), or RAS (Microsoft Remote Access Server), that allows a user's computer to be a peer in a large packet-switched data network, such as the Internet. With a network connection, the user's system may act as a client of, or be a server to, many others on the network. The user may have multiple independent sessions (communications channels) active at one time. Many users are familiar with the older dial-up method using "terminal emulation" software such as Kermit,[10] which sends and receives simple streams of characters. Though not a full "network connection" in the contemporary sense, such a connection is still useful for simple terminal sessions and file transfers.

A quick survey of network software and communications methods will help chemists understand the options available in academic (or industrial) settings.

Physical Transport.   Networks are normally described in "layers" of functionality.[11] The lowest layer concerns the physical media and signaling scheme used to connect computers. Higher layers define global addressing, content, and error control, which are more specific to particular applications.

A number of physical transport schemes found in universities are presented in Table 2. A system based on previously installed telephone wiring is often most cost-effective. At Yale, most new connections are so-called 10baseT Ethernet, which uses existing telephone wiring and wiring closets. For most computers, a connection for 10baseT is either built-in or available as an inexpensive add-on.

Computer speeds are continuing to climb, following Moore's law, with inexpensive systems now delivering more than 60 MIPs. At such speeds, a single computer can saturate a 10 Mbit/s Ethernet segment, a situation that is common when transmitting large data files or displaying complex graphics. Transfer bottlenecks require faster data rates and methods to isolate computer connections with electronic switching. Several 100-Mbit/s network technologies based on twisted-pair wiring are becoming available at reasonable prices.

Protocols.   While a physical transport and signaling system forms the basis of a data networking system, it is the software protocols implemented "on top" of the hardware links that provide most of the flexibility and advanced features of a data network. Table 3 outlines some of the major network protocols found in university environments.

TABLE 2 Some campus network physical transport methods.

Common Designation	Bit Rate, Mbit/s	Media Type	Max. Length, m, Typical	Cost of PC Port, 1995, Typical
"Thick" Ethernet, 10base5	10	Coaxial	500	$100
"Thin" Ethernet, 10base2	10	Coaxial, RG-58	185	$100
"Twisted pair" Ethernet, 10baseT	10	Unshielded twisted pair	100	$100
FDDI (fiber distributed data interface)	100	Fiber-optic loop	2,000	$1,500
Fast Ethernet, 100baseT	100	Unshielded twisted pair	100	$200
100VG-Anylan	100	Unshielded twisted pair	100	$200
ATM (asynchronous transfer mode)	25, 155, and higher	Various	—	—

TABLE 3 Some common campus network protocols.

Protocol	Applicability
TCP/IP[a]	Internet, general campus networking, many vendors
AppleTalk	Apple computer (Macintosh), supported by various vendors
NetBEUI	Microsoft Windows 95, NT, LAN Manager
Novell IPX	Novell Networks, primarily for DOS/Windows-based PCs
DECnet	Proprietary to DEC (Digital Equipment Corporation), supported by various vendors

[a] TCP/IP = transmission control protocol/internet protocol

TCP/IP (transmission control protocol/internet protocol) is a common choice, since it is the basis of the Internet and is widely implemented as an "open", non-proprietary standard for Unix workstations and personal computers.[12] (An open protocol allows many vendors to compete on an equal footing. A proprietary protocol is controlled by a single vendor.) A TCP connection provides an error-free data "pipeline" between specific sender and receiver processes.

AppleTalk is Apple Computer's proprietary protocol and is often encountered in academic environments. It is the protocol for the built-in networking of the

TABLE 4 Dial-up connections.

Common Designation[a]	Primary Protocol Supported	Bit Rate, Mbit/s	Telephone Connection	Cost of PC Port, 1995
SLIP PPP	TCP/IP	Telephone modem rate (up to 28.8 kbit/s)	Standard analog	$75–300
ARAP	AppleTalk	Telephone modem rate (up to 28.8 kbit/s)	Standard analog	$75–300
RAS	various	Telephone modem rate (up to 28.8 kbit/s)	Standard analog	$75–300
ISDN	various	64, 128 kbit/s	ISDN digital	$300–500

[a] SLIP = serial line Internet protocol
PPP = point to point protocol
ARA = AppleTalk remote access protocol
RAS = remote access server
ISDN = Integrated Services Data Network

Macintosh. AppleTalk is frequently used over moderate-performance "LocalTalk" or telephone physical media (at 230 kbit/s) for printer interfacing and limited file sharing. While originally intended for office-scale LANs (local area networks), Local-Talk networks can be expanded to campus scale. The AppleTalk protocol is also useful over 10-Mbit/s Ethernet ("EtherTalk") or other media.

Many network protocols are available to DOS, Windows, and Windows NT personal computers. Two of the most common are NetBEUI and Novell IPX, which are used for file servers, printing, and other functions.

Digital Equipment Corporation's DECNET is another significant proprietary protocol. DECNET was the original networking protocol of the popular DEC VAX computers, and is common in labs with DEC equipment. DECNET normally runs locally over Ethernet or faster media, but there are implementations for slower serial lines for wide-area or point-to-point connections.

As university networks have grown, it is common to find many or all of these protocols on a single physical network. In some cases protocols are "encapsulated" into packets of another protocol to add flexibility.

Dial-up Connections.   Common methods for longer distance connections, to faculty or staff at home or on the road, are shown in Table 4. PPP (or its older cousin SLIP) provides TCP/IP connectivity across modems using standard analog telephone lines or a direct wire connection. ARA, proprietary to Apple Computer, allows Macintosh MacOS users to connect a remote computer as a member of a campus AppleTalk network; it can also carry TCP/IP data encapsulated in AppleTalk. Microsoft RAS for Windows 95 or Windows NT uses PPP to provide TCP/IP, NetBEUI, or IPX communications.

ISDN is a digital telephone technology that can carry various protocols. It is the first fully digital service practical for home use. The ISDN basic rate interface (BRI) provides two 64-kbit/s data channels that may be used for voice or data. The two channels can be operated in tandem as an effective 128-kbit/s data channel.

Competitive developments in communications are likely to bring faster home data connections using the same physical media as cable TV (coaxial cable or fiber optics) or perhaps even direct broadcast satellite TV. The available (one-way) bandwidth in these entertainment-oriented systems is very high, over 500 Mbit/s, and could clearly support very good computer-to-computer networking given sufficient market demand.

Campus Service Developments

The increasing use of computing by student, faculty, and administrators is resulting in broad synergisms in universities. The major technical enablers are fast data networks and powerful, universal, and affordable desktop computers. Chemists, as university citizens and as opportunistic technology users, stand to profit from these developments.

Current activities that are supported by network and computing infrastructure include:

- Academic services: Teaching options include classroom teaching support, computer laboratories, and computer classrooms. Academic software may be provided to students' personal systems.
- Administrative services: Students can call up grade reports and transact university payments from desktop personal computers. Faculty can download current financial status of research accounts. Researchers can place purchase orders or request other campus support services.
- Library services: Library catalogs, circulation processing, information services, etc., are provided over the data network.
- Campus information systems: Campus information is linked with worldwide data resources and offered to community members via World Wide Web, Gopher, etc.
- Conventional media services: The data network can be integrated with conventional cable TV service to offer closed-circuit or broadcast video and audio, including specialized academic programming such as SCOLA (satellite communications for language), recorded lectures, etc.
- Media on demand: Users may call up network-based audio, video, or digital multimedia for language teaching, film studies, etc. The "RealAudio" system[13] is notable as a method of providing audio services on demand using a modest bandwidth.
- Interactive media: The data network will soon support interactive video or audio teleconferencing, including special instructional material such as the WWW "Interactive Patient" system[14] from Marshall University.
- Computing support services: Computing support organizations offer client/server mail systems, file storage and backup, WWW publishing services, central user authentication, Unix or personal computer technical support, and on-line consulting.
- Other functions: Campus networks support building systems (heating and air conditioning, power, alarms, etc.), gate and doorway security (electronic card keys, emergency alarms), and dining hall operations.

TABLE 5 Instructional use of computing in the classroom.

Technology Level	Benefits	Concerns
Media preparation	Computer drawing tools—better quality	Initial equipment & software
	Easy editing	Color printer or film recorder
	Use existing data, images	May need photo lab or service bureau
Live projection	Bypass printing or film steps	Classroom equipment
	Easier data management	Reliability
	Integrate animation, video & audio	Preinstallation of data
Local interaction	Demonstrate live software/hardware	Preinstallation of data & program
Network interaction	Information sources	Network reliability
	Access instructor's lab system, higher power systems	Reconfiguration for portable systems

Teaching Technologies

Chemists who teach may interact with the academic computing environment in various ways. Methods that require the least change in the instructor's environment or habits will be adopted first. However, rapid change is being driven by institutional economics along with student demographics and expectations. University and college teachers may find themselves forced to deal with many new technologies sooner than they had expected.

Computer-based Lecture Presentations

Computing systems are now found in classrooms and lecture halls as extensions to traditional audio/visual systems. While many instructors still prefer the chalkboard or overhead projector for "real-time" expressiveness and flexibility, many are adopting computer methods. Table 5 summarizes some considerations for the use of computer technology in the classroom.

Preparation of Conventional Media.   At their simplest, computer graphics may be used to prepare conventional 35-mm slides or overhead transparencies. Commercial drawing programs such as CorelDraw, Microsoft PowerPoint, and others allow faculty to readily create and edit good-quality graphics. Data and images already in computer form are easy to incorporate.

Media preparation requires a computer system with software and an output device, such as a laser printer or color ink jet printer for transparencies or a film recorder for 35-mm slides. Commercial or campus service bureaus also provide high-quality print and film services.

Live Projection.   Live projection using a computer and classroom projection device provides more interactive possibilities and often produces a superior image. Software such as Microsoft PowerPoint or Aldus Persuasion is well geared toward live computer "slide shows". With simple commands, a lecturer can skip forward or back, place a pointer, etc. Video and audio clips may be inserted as needed.

Live computers in the classroom present a logistical problem. If no permanent, fixed installation of a computer and display device is available, hardware setup can take 10 minutes or more, which is unacceptable if the equipment is required frequently. A fixed computer, on the other hand, requires the lecturer to prepare and load media onto that specific system. A useful compromise is a fixed-projection system that can be cabled to the instructor's portable computer, but equipment must have very high reliability and availability for instructors to commit to this presentation medium.

Local Multimedia Interaction.   Using the same computing and projection systems, instructors can run scientific software and demonstrate programs, such as molecular editing and visualization tools, mathematical analysis programs, multimedia animations, etc. Such applications often require a more powerful computer than is provided for simple projection duty, and the application software has to be preloaded. Again, it is convenient for instructors to bring their own laptop portables.

Network Interaction.   A classroom or lecture hall computer installation can be made much more powerful if it is provided with connections to the campus data network. With such a connection, faculty may interact with research computers that provide more computational or graphics power than the classroom computer, perhaps using X Windows connections. Instructors may access such network services as the World Wide Web, file servers, etc.

Because of the extra complexity, network operation in class demands the highest level of reliability in support systems and software. Configuring portable computers for the network (assigning IP addresses, for example) often requires extra setup effort.

Computer projection can add a great deal to the presentation of technical material and is an aid to instruction. Besides being easily produced (once the software is mastered!), computer graphics often are better at holding an audience's interest than are traditional presentations. Projection is suitable for the largest lecture halls, especially with multiple TV monitors. During preparation, graphic elements are easily stored and reused, changes are easily incorporated, and external resources (images, animations, or audio) may be added. Interactive operation, including network access, adds dramatic value and richness to instruction and allows instructors to adapt to particular questions and problems that come up during class time.

Besides logistical setup issues and reliability, the main brakes on using computer technology in class are limited knowledge among faculty of the software and hardware systems and the limited support staff available to assist in media preparation.

The Computer Classroom

Rooms simply outfitted with a single computer and projection device are less expensive than "computer classrooms" outfitted with computers at student's seats. However, the computer classroom is invaluable to teach programming methods or to conduct computational laboratories, because it offers a much higher level of interaction with students.

FIGURE I   Classroom in Phelps Hall at Yale University. (Photo by Michael Marsland, Yale University Office of Public Affairs.)

The classroom shown in Figure 1 is a useful example. Twenty-four Power Macintosh 6100 computers with 15-in. color monitors and Ethernet connections are installed on desks, allowing one or two students to use each machine. An instructor's desk has a similar computer attached to a permanently installed projection system.

The Yale classroom is used in a variety of ways. Beginner calculus classes use the room in the first few weeks of a term to introduce Mathematica, from Wolfram Research, Inc. Afterwards, students are expected to prepare assignments on their own machines or in other public clusters. (The classroom doubles as a public cluster when not scheduled for classes.) Humanities classes, such as creative writing and modern languages, use the room more often than the sciences because key curriculum-specific software is available. Chemists at Yale use the computer classroom to introduce students to chemical modeling and analysis tools such as CambridgeSoft Corporation's ChemDraw and Chem3D. After an initial familiarization session, students are expected to work on their own using the classroom or other public cluster facilities.

The success of computer classrooms depends on good technical support—to teach the instructor to operate the facility, to keep hardware and software operational, and perhaps to explain system operation to students. We expect the computer classroom approach to be adopted by specialized computation-intensive courses and for certain sessions or laboratories of a wider group of courses. Because of its expense, however, the fully computerized classroom is unlikely to replace most conventional classrooms in the near future.

One challenge for the Yale computer classrooms and public clusters is to offer a convenient file space for students. Program distribution is provided by a comprehensive system of network file servers, but there is no convenient system for personal files. Floppy disks are adequate for files of up to about 1 MB, but data files incorporating graphics, video, or sound often exceed this size. Economical media are becoming available with 100 MB or greater capacity. These help, but a network file server with private space for each student will be more secure and convenient in many applications.

It is interesting to consider a hypothetical model of classroom interaction that would capitalize on students' increasing ownership of portable computers. With wireless networking (infrared or radio), student machines will be capable of downloading text and graphics from an instructor's machine in class, as well as copying notes or collaborating on assignments with other students. As these technologies become commonplace, teaching/learning models using dynamic assemblies of computers will become increasingly useful within the classroom, as well as across the campus and around the world.

The Automated Laboratory

The chemistry laboratory is increasingly dominated by computer-controlled instruments and digital data management, as described in other chapters. New opportunities for laboratories in education open up when laboratories are integrated with the campus computing environment. Primarily, this means giving labs full access to, and availability from, the campus data network. Lab students using personal computers or workstations can have access to information and computing resources to, for example, retrieve chemical data or literature, analyze molecular models, or perform reaction or process simulations on remote machines.

The Computing Cluster

Academic computing support organizations often provide general-access computing capabilities distributed around the campus. Most commonly these are personal computer installations that provide a wide spectrum of support for student computing at a controlled cost per user "seat". Until recently, the main capabilities of these facilities were word processing, electronic mail, time-sharing connections, and some simple scientific applications and programming tools. With increasing CPU power, however, we are able to provide software with much greater scientific capability. Chemical drawing, visualization, and modeling operate effectively, along with symbolic mathematics and simulation systems. Since 10-Mbit/s Ethernet networking is the rule, powerful network applications, such as X Windows sessions, are available in general clusters.

In order to take advantage of advanced computing in teaching, science faculty must appreciate what facilities are available. Often, such facilities are provided by an organization outside the department. Faculty may need to coordinate with the computing support staff to ensure that the right software is available at the right time. Care is needed to schedule classwork to avoid resource conflicts across departments and resource crunches during peak demand.

Class Logistics

As computing and communications technology become more readily available to faculty and students, it is natural to use the technology to automate class logistics and paperwork. At Yale we support a number of methods for class services.

Electronic Mail.   Since most students and faculty have e-mail capability, it is easy to require class members to be accessible by e-mail. The e-mail system can be used for one-to-one contact among students and faculty, or for one-to-many contact through an address list. Instructors can maintain a private list for their students or may create an automatic redistribution list (sometimes called a "listserv", "reflector", or "exploder"). Any authorized participant may then send mail to the whole group.

News Groups.   A computing administrator can create a local university news group for a class. The news group functions as a public bulletin board where students or faculty post messages that can be read by all the other participants. The bulletin board approach avoids filling user mailboxes with large volumes of mail. News groups commonly use an NNTP (network news transfer protocol) server that can be interrogated by user software on personal computers or time-sharing systems.

File Servers.   An instructor may set up an area on a network file server, such as an AppleShare or Novell system to hold class documents or software, much like a reserve shelf. Any student with a compatible system and network capability can access this "reserve shelf" freely by using a security password. Unfortunately, incompatibility between DOS/Windows computers, Macintoshes, Unix Systems, and other systems sometimes requires that data and software be provided in multiple formats on multiple servers to ensure that all students can access them.

   File servers may also be used to implement a "drop box", a disk area that students may write into but not read from. This provides a secure means of submitting homework or papers for grading. The instructor has a password to retrieve files.

World Wide Web.   The WWW, or "Web", offers another one-to-many communication channel. Faculty may create a set of Web pages on a Web server for class assignments or to provide needed information resources. Most Web browsing software (Netscape, Mosaic, etc.) has the advantage of handling images, animations, and other types of data, and can provide a convenient "shell" (command entry system) for access to other information services (FTP, gopher, telnet, news, etc.) At a more sophisticated level, the Web may be used for forms-based data entry and secure transactions, which might include collecting homework, registering for sections, or giving exams.

Creating Teaching Materials

   One long-standing problem with computer-based teaching has been the difficulty of developing curriculum-specific course materials. Although tools have improved dramatically, it is still rare to find faculty with the energy and resources to create such materials, especially those that might find a significant economic market. Rather, the emphasis in academic computing has been to provide a rich computing and network infrastructure and a set of widely useful tools that instructors call upon locally to support their teaching.

   While few faculty are potential authors of multimedia textbooks, many could use the technology to prepare better lectures and materials for their own classes. Small-computer software, such as Microsoft PowerPoint or Aldus Persuasion, is relatively easy to learn and inexpensive for still-projection images or to provide live media display. More elaborate tools, such as MacroMind Director or Adobe Premiere, are available for animation and video. Some faculty are writing Web pages and even using their personal computers as Web servers.

Support Services

Faculty prefer to concentrate on the subtleties of their subject material rather than on computer technology. The complexities of modern computing hardware and software limit the ability of many faculty to exploit new capabilities. Increased support services by staff or student assistants can be very helpful. However, departmental budgets rarely fund such services. The central university services that exist have typically been reallocated from staffs that formerly supported academic "computing centers" whose costly mainframe systems have been phased out.

Research Computing Technologies

While many of the laboratory systems described elsewhere in this book may operate well as isolated systems or as part of a small LAN, the effectiveness of expensive research resources and staff can be enhanced if they fully engage the campus environment.

A Data Management Model

One helpful view of the chemical laboratory is as a flow of data. A researcher may have, in the form of data inputs, a database of experimental parameters, sample descriptions, calibrations, schedules for staff and special equipment, etc. On the output side, experimental data from an instrument must be gathered in a raw or partially processed form and typically brought into the user's personal computer or workstation for display and analysis, and finally prepared for publication or other presentation.

The classical instrument, whether a pH meter or spectrometer, has no particular facility to receive or transmit data. Even when computerized, instruments often are implemented as isolated systems with limited ability to work in a general academic network, based on Ethernet and TCP/IP standards. It is still a problem to connect a particular instrument's data output channel with the input options of a user's personal computer or workstation. Using industry network standards is a big advance, but it opens up problems such as data security, resource allocation, and even physical safety. Even if networking is provided, data may be in arcane or proprietary formats that are not "open" to simple user-written software.

The Academic Desktop

A chemistry researcher/instructor faces the same computing issue as faculty in other disciplines: how to manage his or her electronic "life". Computing in the sciences still often means "number crunching"—analyzing data or performing theoretical calculations. However, a scientist's desktop system is increasingly communications-oriented: transmitting e-mail among colleagues and students, reading news groups, "surfing the Web". The desktop system must also support technical word processing and preparation of graphics. Faculty also have administrative roles: they communicate with administrators, submit grades, and manage budgets. Finally, faculty must be able to access student and teaching resources, including file servers, mailing lists, and Web pages.

Integrating such a variety of services on a single desktop system is a significant challenge for faculty and campus computing organizations. A single system is highly desirable for its physical compactness, simplicity of operation, and lower cost. Fortunately, high-end personal computers have a performance level that is adequate for the technical computing needs of many faculty. An efficient integrated desktop solution on a university scale depends on widely standardized and supported solutions for communications services such as e-mail, the Web, etc.

For faculty to efficiently carry out their administrative duties, the university accounting system should offer a modern "friendly" interface for the academic desktop. Such a system will show grant and contract balances and transactions, permit entry of purchase orders and campus service requests, and support normal faculty functions with a minimum of paper and staff intervention. Benefits to chemistry faculty will match benefits to the institution, fulfilling the technological promise of computers and communications in chemistry education.

References*

1. Toffler, Alvin; *The Third Wave;* Morrow: New York, NY, 1980.
2. *The New Hacker's Dictionary,* 2nd edition; compiled by Eric S. Raymond; MIT Press: Cambridge, MA, 1993.
3. ISDN URL = http://www.pacbell.com/isdn/isdn_home.html or http://www.sys.acc.com/acc/isdnprimer.html.
4. Patterson, David A.; Hennessy, John L.; *Computer Architecture: A Quantitative Approach;* Morgan Kaufmann: San Mateo, CA, 1990; p. 8.
5. Patterson, David A.; Hennessy, John L.; ibid; p. 11.
6. Hammer, Michael; Champy, James; *Reengineering the Corporation: A Manifesto for Business Revolution;* HarperCollins Publishers: New York, NY, 1993.
7. MIT Reengineering URL = http://web.mit.edu/afs/athena.mit.edu/org/r/reeng/www/.
8. Project Mandarin URL = http://mandarin.cit.cornell.edu/.
9. Anonymous; "On the Brink"; American Association of State Colleges and Universities: Washington, DC, 1995.
10. Kermit URL = http://www.cc.columbia.edu/kermit/.
11. Black, Uyless; *TCP/IP and Related Protocols;* McGraw-Hill: New York, 1992, p. 19ff.
12. Black, Uyless; ibid.
13. RealAudio, URL = http://www.realaudio.com.
14. Interactive Patient, URL = http://medicus.marshall.edu/medicus.htm.

* Note: A URL (uniform resource locator) is an address for Internet data that is accepted by World Wide Web browser software, such as Netscape and Mosaic. Unfortunately, information provided to the Internet is often transient. Therefore, although the cited URLs are valid at this writing, the data may have moved or disappeared since.

Part II
Accessing Chemical Information

6

Information Retrieval in Chemistry Across the Internet

F. S. Varveri

Introduction

For the past several years, the educational and research communities of all disciplines have increasingly entered into a phase of global collaboration, which employs the computer network facilities currently available.[1] At present, while the newly emerged information technologies continue to expand and improve, users of all specialties are offering their retrieved information to the services of their professions. Soon, if not already happening, the scientific community will become highly dependent on computer communications as a means to access data and information,[2] and no one can afford to be excluded from this exciting electronic world. Since the global information infrastructure[3] is already well underway, involvement in the new technologies has become essential for scientists. Noting this prerequisite, one will soon realize that, with the availability of increasing computational power and the growth of computer networks,[1,4] the emerging electronic communications society will strongly affect all aspects of life,[5] including our working environments.[6,7]

This article is addressed to chemists, those of limited Internet experience as well as adept users. The former should understand that even though users can acquire any information relatively fast with minimal training, the dynamic potential of the Internet is revealed only after they acquire a full knowledge of its capabilities. Nevertheless, it can be said that in these last few years, much has already been accomplished within the profession, but this is definitely only the beginning. A discussion of the broad topic of network information retrieval (NIR) is without doubt a serious responsibility to undertake, especially since NIR is still in its infancy compared to "traditional" information retrieval systems.[8] As for the search approaches to information retrieval using the NIR tools, one should not fail to take into account the continuously evolving Internet environment. However, with a clear understanding of the thus far offered Internet facilities, the chemical community will be more capable of a faster and fuller exploitation of the power of information technologies.

With the intention of highlighting the essentials of each topic, this chapter is presented in two key sections. The first section states some primary characteristics of the Internet and surveys its association to chemistry. The second section provides a brief account of the major Internet application tools and services, together with the searching methodology employed to retrieve information.

Internet

Internet, the Global Means of Exchanging Information

Networking can be defined as the ability to exchange and share resources and information, a process involving both humans and technology. All parts constituting a computer network[9] are designed so as to have a complete coordination and collaboration, in an extended distributed yet unified pattern, with other interconnected computer networks.[10] This will allow correct communications (interoperability[11]) for the purpose of human and computer interaction. With reference to the human interface, in order to overcome the fear of the complexity involved, the potential user should grasp some essential facts relating to: (1) the imperative need for adequate training,[12-15] (2) the documentation,[15-22] (3) the necessity of being familiar with the employed computing platform and operating system, and (4) the required "netiquette"[23] and ethics while practicing the art.

With respect to the Internet,[15,24,25] in a 25-year life span both its technology and physical growth have expanded beyond any expectation, and it now offers ready access to a plethora of information while becoming steadily an enormous and effective instructional, educational, and research tool. With innumerable users and countless physically separate yet intercommunicating Internet sites,[26] this evolving technology has succeeded in becoming a perfectly coordinated virtual communications unit. As for the stated Internets, these connect most major academic institutions and government and corporate sites for the purpose of user communication and user sharing of information across the Internet.

Upon joining the Internet community, one enters a world of literally having the globe at one's fingertips. The environment is friendly with a noble disposition and a moral responsibility to assist the newcomer and share its experience. Users cannot avoid being impressed with the quantity and quality of the available yet nonstationary resources, the speed of delivery, and the continuous improvement of facilities. The academic user should not be surprised to find works that are easily publishable in any reputable journal placed as files to be retrieved and distributed freely to the members of the academic community. These are only a few features of this accumulating, unself-centered world that has shaped networks into what they are today, a global community without borders, a community where everyone may exercise the right to learn, each member supporting the other. This practice has existed for years now. However, the question arises as to whether it will continue as such, since commercialization is now entering the Internet. The answer to this question remains to be seen.

As to the criticism that the Internet is somewhat disorganized, the situation is not exactly as claimed and definitely will not be so in the years to come. Although at present the Internet is impressive, extended services continue to develop in addition to efforts to resolve the most effective approach to managing this enormous volume of information. Intensive research on automated compilation of information has already lead to very impressive results, but the robots so far created must be improved further and aimed at better human-oriented responses to queries made.

Additionally, a point worth noting is that throughout the Internet a true democracy has been and continues to be practiced with courteous users, a concept hard to comprehend at the start.[27] For this, much credit must be given to those volunteering professionals who made the dream of a global community become a reality by offering their hard-earned data to be shared and exchanged.

Chemistry Over the Internet

The chemical community has always been a society that is readily adjustable to new processes and working practices. The use of computer networks is not an exception, and for the last couple of years, both paper and electronic publications have been devoted to the subject of using Internet tools in assisting chemistry.

The topic of chemistry-related information retrieval and the subsequent collection of resources across the Internet started relatively recently. In the early 1990s, T. C. O'Haver of the University of Maryland published his compilation of some excellent chemistry-related freeware and shareware software for both PCs and MACs.[28,29] The same author, in continuing his interest on the subject, proceeded into more diverse activities, such as electronic publications pertaining to the Internet,[30-34] as well as organizing the first on-line chemical education conference in 1993.[35-36] In those early days, P. Farrington,[37] R. Gillilan,[38] G. Wiggins of the University of Indiana,[39] and I[40-43] offered our first compilations of chemistry-related resources over the Internet.

In the following years, Internet technology offered more services[15,44-46] that, assisted by the increased human–computer interaction,[47] would effectively collect and transfer vast amounts of chemical information across the Internet, a small part of which is included in this article. Chemistry entered the Internet dynamically, most often through volunteering professionals who had the opportunity to see the potential earlier than others.

In recent times we have seen several additional compilations referring to Internet topic-specific resources, chemical databases, and indices.[39,42,48-53] A plethora of Gopher[39,42,53-59] and World Wide Web (WWW, W3, or the Web) sites,[39,42,48-53,55-64] as well as WWW chemical virtual libraries, were created.[50,64] For the past couple of years, both paper and electronic publications have been devoted to the subject of Internet services and their use by the chemical community. The use of Internet services[11,15,44-46] has been extended to offer numerous community mailing discussion lists,[39,40,65-68] electronic on-line conferences[36,51,69,70] and workshops, a large collection of special software[28,29,32,33,39,43,49,51,71] and hypermedia-associated documents.[64,72-77] With this rapid evolution, both the academic[78-81] and private[82-84] sectors recognized the challenges to be confronted since there is more information to satisfy the needs that one might have. It is important to note that, with the progress of the new technologies, colleagues have already seen new territories in which to expand, and we already have chemical applications on the Web,[64,72,75,77] and even the first proposed Internet Draft on Chemistry from H. Rzepa (Imperial College of London), P. Murray-Rust (Glaxo Research & Development), and B. Whitaker (Leeds University).[85]

In our history, our community has never previously been so globally united, offering and sharing so much information, at such a rate, with so many participants. The Internet is now part of our life and with time it will become more so. Besides being a means of communication, it also represents a global wealth of mainly charge-free information. To enter into a detailed account as to what the Internet can offer to every member of the community is an impossible assignment to undertake, considering both the limits of this chapter and the boundaries of chemistry. However, using the cited paper and electronic literature, we will attempt to overcome part of this difficulty by presenting available information that can be of service to a user's needs.

As for information retrieval in chemistry, there is plenty to be done in the future. A global search for a specific query will soon become more difficult as data continue to accumulate at an enormous rate. Most of today's search means[61,86-88] cannot meet tomorrow's needs, even if we have access to more sophisticated search engines. Along with the research conducted on the creation in "intelligent" robots, chemistry is also in need of its own effective specialized taxonomy, accepted by all parties involved (societies, authors, host administrators, etc.), to meet the requirements of a satisfactory query result. Such an ambitious project, with so many parameters involved,[89] will require close collaboration of chemists with the search engine creators, and it will take some time before the proposed robots will be able, if we are fortunate enough, to do those correlations that a professional can make.

As for the future, soon we will meet the 21st century and its challenges. For all of us, there is a duty to follow these evolving new technologies[90,91] and plan to meet future needs.[78-84,92] A prediction one may definitely make is that, in the years to come, much of the "classical" way of our professional behavior will alter to a great extent, if it has not already done so. These high technologies will change the manner to which each of us has been accustomed. The need for adaptation to the new environment is pressing, especially with regard to the effective communication with upcoming future colleagues who will mature with this new know-how "philosophy". As for the next generation of Internet applications, these will aim to further facilitate both collaborative and interactive access to information with the introduction of new environments for hypermedia, multimedia, and virtual reality.[8,93-95] The libraries of the 21st century will no longer be the way they used to be.[96,97] Electronic publishing will be transferred[98-101] from a print-based culture to a multimedia one: we will see more electronic journals, and multimedia information handling and transfer will be expanded.

Internet Information Retrieval: Services for Accessing Information—Information Searching Methodology

The early Internet was not the same as it is today. It started as a technology, and as networking and communication improved, accessible information expanded concurrent with the increasing need for developing friendlier services for the platforms people owned. Networking scientists met all these needs, and the Internet evolved into a global service offering a considerably efficient environment to the user.

In this section, we intend to familiarize the reader with the pursuit of information in association with the services and tools used today across the Internet. However, an Internet user must be aware that, although one is using the Net as an established technology, much of it is still under study and undergoing development. Apprehending this fact, along with the limits of this chapter, the text that follows will be brief and to the point. Those wishing more details are referred to the cited RFCs[8,44–46] and resources.[15,21,88,102–105]

Although technical concepts and specifications[9,11,15] will be avoided in this presentation, it is considered essential to underline the client–server dualism. According to this, communication is a request–reply pair, always initiated by the client. Both the client and the server are special software; the latter accepts requests, performs the service, and returns the result to the requester automatically over the network. In a like manner, via the client software, the local computer communicates with the server.

Tools for Collaboration and Communication

Primary information resources include electronic mail (e-mail) and group communication means, such as mailing discussion lists,[106] UseNet newsgroups,[45,46,107–109] and bulletin board systems (BBS).[110–111] As for e-mail, this is a fundamental Internet service, the most used one. To exercise the right to e-mail, one must be aware of its platform software and interface used, the offered on-line help, the existing netiquette,[23] and, most important, the recipient's address. Worth noting is that with e-mail, the user cannot only send mail to any random network,[112] but can have information delivery from any Internet site in lieu of an interactive session.[113]

The electronic mailing (discussion) list is an extension to e-mail and is an activity that offers information exchange between people with common interests who are usually served by the proper software (listserv, majordomo, etc.), with an automatic distribution of messages sent to every member of the list. In order to join and use mailing lists, one must know only a few basic commands for each of the existing different mail-server programs.[114,115] It is imperative also to recognize that a keyword search of the list's archives,[116] using Boolean combinations, can be executed.

The UseNet newsgroups are remotely related to mailing discussion lists. They resemble a big bulletin board where anyone can post a message or give information for everyone to read. These newsgroups are organized in a treelike hierarchical structure according to the specific areas of interest. Along the major categories the "sci" mainstream is included, which covers the general subject area of research and application of the established sciences. Chemistry is under this category. The UseNet servers are mostly UNIX-based, and to read the news the user must have a connection to such a server. Also, one can reach the newsgroups via e-mail, Gopher, and the WWW, and if the user is not connected to the UseNet, it is possible to post messages via e-mail.[113]

Recently, electronic collaboration and communication has extended to electronic conferences, workshops, and teletraining. We have participated in a number of these and have been impressed by both the quality offered and the zest of those involved, who offered their time and effort to have us attend a conference or trained us in our offices where we were not physically away from our work. Furthermore, if we are to consider electronic publications as a communication means, a tremendous burst of activity is also occurring in this area. Well-known professional journals and publishing houses are undergoing transformation to meet these needs. It will not be long before we see much more in this respect.

To conclude this topic we emphasize that, as can be seen from the above narrative, both mailing discussion lists and newsgroups serve as primary sources of information to any Internet user, the former being more indispensable than the latter.

Services Used for File Transfer

Transferring files over the Internet can be accomplished with most major Internet services and application tools. A most important element prior to file transferring and its subsequent use is the knowledge of file format (text, nontext) in order to handle it.[117] The "traditional" service employed for a file transfer between connected machines is the use of the file transfer protocol (ftp) and e-mail. File transfer can also be concluded via Gopher and the WWW. Lately, to overcome problems encountered with multimedia mail (binary data, image, sound, video, audio), Multiple Internet Mail Extensions (MIME) mail[118] was developed, which is a service of great potential.

Discussing ftp, one of the most powerful TCP/IP protocols, one should be aware of the difference between and use of both the regular and the anonymous ftp.[119] Closely associated with the anonymous ftp facility is the Archie file and directory locator service,[44,46,120] which is an indexing database service maintained in several locations. Archie, being the oldest of the Net search tools, has been used by practically everybody on the Internet, and it can be accessed by a local client, an interactive telnet session, e-mail, as well as by Gopher or the WWW.

Remote Machine Connection (Telnet)

The telnet protocol permits a remote connection and work on another system inside the Internet.[45,46] As in regular ftp, the telnet service requires permission to access the remote computer, which is often offered freely to the public. Through the telnet protocol one can have access to databases, library catalogs, Campus Information System (CWIS), BBSs, and to Gopher and the WWW servers. As with everything else, information for these publicly Internet-accessible facilities most often comes through mailing lists and frequently updated resources.[88,102–105] Closely associated with the telnet is the Hytelnet protocol,[45,46] which is an indexed service associated with public accessible databases and libraries. Hytelnet started as a PC resident program and has now evolved into a hypertext browser that gives full instructions for logging onto telnet-accessible sites across the Internet.

Interactive Information Delivery Applications

Gopher and the World Wide Web.   Gopher is a hierarchical menu-driven search tool that is used to find information on the Internet and as a means to download search results once they are found.[8,45,46,121] The Internet Gopher protocol is also based on the client–server model. In the absence of a client, one can use this service through e-mail, via telnet, or by the WWW. Gopher, its capabilities and weaknesses, are discussed at length in the references cited. Since it was developed (1991) users have found it useful and easy to handle. In Gopher, the unit information is the entire text, and it has the capability to link to other Gophers on the Internet in a chain-progression-like fashion. Associated with Gopher is the indexing search engine called Veronica,[122] which operates keyword and Boolean searching in the Gopherspace, that is, the global Gopher web. Related to Veronica is the Jughead[123] facility, which also utilizes Boolean algebra. However, while the former searches the whole of Gopherspace, the latter limits its searches to the menus of a single Gopher or a small Gopherspace.

In conclusion, we should mention the Gopher Jewels project conducted by D. Riggins that was terminated in late 1994. This project succeeded in cataloging by subject the best of Gopher features, and until its termination it was a primary source of information.

The World Wide Web developed lately is an easy and powerful global information system where, in a space (the Web), information accessed by earlier protocols (Gopher, WAIS) waits to be retrieved with the same ease as later files created with newer methods.[8,44–46,124–126] The WWW relies on interlinking pointers to other non-linear "texts" (hypertexted documents), which are created with a special language (HTML, hypertext markup language)[124,125] that allows admittance to a location through a special syntax (URL, uniform resource locator[125]). Like many of the Internet services, the WWW is based on the client–server concept where the respective protocol (HTTP, hypertext transfer protocol[127]) permits a browser (client) program to communicate with other Web servers. The most frequently used browsers include Internet Explorer, Mosaic, and Netscape, which can capture any document, ranging from a plain text file to a display of graphics, sound, and video (hypermedia).[126] Finally, as with most tools, the Web can be accessed by telnet or via e-mail.[113]

Even though the Web is complicated in its construction, from the user's point of view it is easy to handle. Its multimedia capabilities, that is, the ability to combine text and nontext files, have opened a whole range of new possibilities, with its potential and implications.[129] The Web is ideal for retrieving any information since it also offers the prospect of using robots and engines to scan the cyberspace and perform the search for the user. Pointers to Web tools (engines and indexing) can be easily found across the Internet.[87,88,130]

Text-Based Indexing Services (WAIS) and Other Internet Application Tools.
WAIS allows the user to search through archives of files for which indices exist.[8,45,46,131] After the results are presented, one can then select the items one wishes to see. The major drawback of WAIS is that it can only search those documents compatible to the WAIS software.

In references cited[8,15,44–46,102–105] one will find many more services and application tools users need to surf and explore the Net. In this text, of all the listed utilities and protocols, those selected are of more use to the retrieval of information—without, of course, demoting the rest.

Information Retrieval and Information Searching Methodology
Information retrieval can be defined as the identification of an information unit at a location within a collection, which can be reached and thereupon be accessed for usage. Unfortunately, as any Net user would know, the process of searching over the Internet is far removed from a library-like model, where the organizational environment was long ago set and had plenty of time to improve. Here, the resources are not only decentralized to a great extent, but they are also created and maintained by humans who do not necessarily follow established disciplines and exact organizational rules, even though the source creators have every intention of accommodating the end user. This applies especially to the quality and "aging" of a document.

Regardless of the somewhat disorganized Internet environment, the case is not as grave as stated if the end user is well familiar with the owned system capabilities and has a clear understanding of the existing retrieval tools to use. Furthermore, the user must be aware of the Internet's broad space, with its nonstatic nature, and comprehend the search logic needed to access information per each of the available utilities. Special difficulties arise for those who are trying to discover something they do not know while interacting with what may be an unfamiliar computer system.[132] Apart from the said parameters, one must always be aware that Net knowledge does not only come through reading manuals and guides alone, but with hands-on practice. As any experienced user would know, if from the beginning one spends time to discover and become familiar with the environment, the process of searching, accessing, and using retrieved information will be facilitated to a great extent.

A major problem regarding the access to electronic information is that computers cannot "read" (that is, comprehend) documents (records) as humans do; therefore, they cannot form all those index terms (search keys) needed to satisfy a human-type search (logic query). To satisfy part of that need, the Net frequently offers contents to a specific collection, which is definitely beneficial to the end user. Those indices, together with the available Help files, give adequate descriptions to enable users to formulate an opinion or a decision regarding the value of the source found during a search. Indices, even though helpful, are not sophisticated enough, and efforts are now centered on a more efficient approach to searching, namely, to move from the concept of based-indexing to that of content-based algorithms.[133]

Regarding ways to search, one method closely associated with locating relevant information is browsing. In browsing, users most often start with criteria partly defined or not known in advance and with the absence of any specific search method. Browsing for a newcomer can be of both help and confusion, most often resulting in disorientation and information overload. However, for someone with experience, intentional or at-random browsing most often can be considered a planned search strategy since it provides the means for a user to scan an area of interest, to search for an explicit goal, and to determine or evaluate the extent of the interior.[134] Schwartz et al.[135] discuss the conceptual relationship between the various developed searching tools and present a lengthy taxonomy of resource discovery systems. In their taxonomy they analyze the interrelated issues of organizing, browsing, and searching for information, focusing their proposals on the problem of uniformity in information access and search. As it seems, browsing in itself is not sufficient for information access, and searching is considered more advantageous if the application supports it as an automated process.[135] Lately, of course, with the development of hypertext technology, browsing is considered a valuable asset and a central feature when users can have access to machine-supported links.

From another point of view, Fung et al.[136] point out the problems involved in information retrieval, and along with their proposal they briefly discuss the classes of retrieving models in use. Boolean search logic, which helps to narrow down the number of hits using combinations of the search terms "and", "or", and "not", is mostly used. In the vector space method, documents and queries are presented in a vector space, and the relevance of a document to a query is computed as a distance measure. The probabilistic method estimates the probability that a document representation matches or satisfies a query. The authors continue their discussion, stressing how important it is for the user to comprehend the topic of interest and the qualitative and quantitative relationships among relating topics, and to know how to state the information needed in a format the system can understand and act upon.

In conclusion, the major Internet services have been discussed and comments made for each one stated. To facilitate one's effort to retrieve information, we advise the reader to (1) master each of the major facilities and know its capabilities and limitations, (2) appreciate the power of mailing lists as primary means of information, (3) acquire knowledge for each indexing service available, and (4) be attentive to present and future resource compilations for assistance in browsing through an already investigated area. A special note is made for the Web since its capabilities are beyond conception. The user must not only know the existing robots, engines, and indexing services in the WWW, but be alert of those yet to come.

The topic of information retrieval is broad and complex. Taking into account all factors involved,[132–138] one must arrive at the conclusion that even though we have all that computer power close to the human brain, the fact still remains that much more progress in retrieval technology is yet to come. Better automatic mechanisms are badly needed, as well as resource transporters, since it becomes increasingly harder for humans to manage resources by hand.[139] Considering the nonstatic nature of computer networks and in particular of the Internet, there is much to be seen in the future, especially now that multimedia and virtual reality are part of the Internet.[93,140,141]

All URLs cited here were checked and active as of December 1995. A current and comprehensive set of URLs for chemistry on the Internet can be found at http://macedonia.nrcps.ariadne-t.gr/ (http://143.233.2.971).

Acknowledgment

The author wishes to thank both the anonymous scientist whose hard-earned data was placed there to be retrieved, and Panagiotis Telonis, the ARIADNE Network Manager, for his helpful comments while reading this manuscript.

References

1. "The computer in the 21st century"; Rennie, J., Ed.; *Scientific American* (Special Issue) 1995, 272 (3).
2. Jennings, D. M.; Lawrence H.; Landweber, L. H.; Fuchs, I. H.; Farber, D. J.; Adrion, R.; "Computer networking for scientists"; *Science* 1986, *231,* 943–950.
3. Borgman, C. L.; "The global information infrastructure as a digital library"; Agre, P., Ed.; *The Network Observer* 1995, *2* (no. 8, August); http://communication.ucsd.edu/pagre/tno/august-1995.html#borgman.
4. Paxson, V.; "Growth trends in wide-area TCP connections"; IEEE Network 1994, *8* (July/August), 8–17.
5. Palmquist, R. A.; "The impact of Information technology on the individual"; Williams, M. E., Ed.; *Annual Review of Information Science and Technology;* Learned Information, Inc.: Medford, NJ, 1992, *27,* 3–42.
6. "Information: How the revolution is transforming Chemistry"; Heylin, M., Ed.; *Chem. & Eng. News* (Special Issue) 1995, *73*(13), 22–82.
7. *The Internet, a Guide for Chemists;* Bachrach, S. M., Ed.; American Chemical Society: Washington, DC, 1996.
8. "A status report on Networked Information Retrieval: tools and groups"; IETF/RARE/CNI Networked Information Retrieval—Working group (NIR-WG); Foster J., Ed.; RFC1689 (RTR: 13, FYI: 25) 1994, August (pp 204) (with references cited therein).
9. Tanebaum, A.; *Computer Networks,* 2nd edition; Prentice-Hall International, Inc.: Englewood Cliffs, NJ, 1989.
10. Corovesis, Y.; Telonis, P. E.; "Network Architecture"; ARIADNE Network Technical Report; 1993, March.
11. Comer, D. E.; "Internetworking with TCP/IP, Vol. I—Principles, Protocols and Architecture", 2nd edition; Prentice-Hall International Editions: USA, 1991.

12. IETF/TERENA Training Materials Catalogue; 1996; http://coolabah.itd.adelaide.edu.au/TrainMat/catalogue.html.

13. Enns, N.; "Internet resources"; 1995, June; http://www.brandonu.ca/~ennsnr/Resources/Welcome.html.

14. Lunde, A.; "Online guides to the Internet"; http://nuinfo.acns.nwu.edu/world/online-guides.html.

15. Comer, D. E.; *Everything You Need to Know About Computer Networking and How the Internet Works;* Prentice-Hall International, Inc., A Simon & Schuster Co.: Englewood Cliffs, NJ, 1995.

16. Malkin, G.; Marine, A.; Reynolds, J.; "FYI on questions and answers: Answers to commonly asked 'experienced Internet user' questions"; RFC1207 (FYI: 7) 1991, February (pp 15).

17. Martin, J.; "There's gold in them thar networks! or searching for treasure in all the wrong places"; RFC1402 (FYI: 10, Obsoletes: 1290) 1993, January (pp 39).

18. Sellers, J.; "FYI on questions and answers: Answers to commonly asked 'primary and secondary school Internet user' questions"; RFC1578 (FYI: 22) 1994, February (pp 53).

19. Marine, A; Reynolds, J.; Malkin, G.; "FYI on questions and answers: Answers to commonly asked 'new Internet user' questions"; RFC1594 (FYI: 4, Obsoletes: 1325) 1994, March (pp 44).

20. RFCs: e-mailto rfc-info@isi.edu, body_txt: help: ways_to_get_rfcs; ftp://nis.nsf.net/internet/documents/rfc/rfcnnnn.yyy (nnnn = number, yyy = txt or ps) and/or http//www.cis.ohio-state.edu/hypertext/information/rfc.html.

21. "Internet documentation (RFC's, FYI's, etc.) and IETF Information": http://ds.internic.net/dspg0intdoc.html.

22. "Internet book reviews"; Milles, J., Ed.; 1995, April; e-mailto listserv@ubvm.cc.buffalo.edu, body_txt: get nettrain reviews.

23. Hambridge, S.; "Netiquette guideline"; RFC1855 (FYI: 28) 1995, October (pp 21).

24. Krol, E.; Hoffman, E.; "FYI on 'What is the Internet?'"; RFC1462 (FYI: 20) 1993, May (pp 11).

25. Dusold, L. R.; "History of the Internet"; ref. 7, pp 3–24 (with references cited therein).

26. Internet (capitalized) refers to the worldwide connected Internets, whereas Internet (lower case) pertains to the TCP/IP Internets (ref. 11).

27. Doctor, R. D.; "Social equity and information technologies: moving toward information democracy"; Williams, M. E., Ed.; *Annual Review of Information Science and Technology;* Learned Information, Inc.: Medford, NJ, 1992, 27, 43–98.

28. O'Haver, T. C.; "PD_science.txt"; 1991; gopher://gopher.inform.umd.edu:70/11/Education Resources and Support/Faculty Resources and Support/ChemConference (CHEMCONF)/BackgroundReading/file: PD_science.txt.

29. O'Haver, T. C.;. "PDChemistry_91–92.txt"; Summer 1992; ibid., file: PDChemistry_91–92.txt.

30. O'Haver, T. C.; "Computer Networks and Communications: A Primer for Chemical Educators"; 1992, July; ibid., file: NetChemEd.txt.

31. O'Haver, T. C.; "Computer Graphics Primer"; 1992; ibid., file: Graphics_tutorial.txt.

32. O'Haver, T. C.; "Catalogs of Chemistry-Related Software on the Internet"; 1992; ibid., file SoftwareSources.txt.

33. O'Haver, T. C.; "Internet resources for mathematics and science education"; 1993, October (v. 1.2); gopher://life.anu.edu.au/bioinformation, journals, standards, FAQs/faq/file: math_sci_education.faq.

34. O'Haver, T. C.; "Chemistry and science on the Internet"; 1994, January; ref. 28, file: Internet_Chemistry.txt.

35. O'Haver, T. C.; "Online Conferencing: Sitting on a virtual table"; 1992; ref. 28, file: OnlineConferencing.txt.

36 O'Haver, T. C.; "Electronic conferencing"; ref. 7, pp 185–199 (with references cited therein).

37. Farrington, P.; "Internet Chemistry resources"; CHMINF-L chemical discussion list at listsev@iubvm.indiana.edu; Wiggins, G., Ed.; 1993, February.
38. Gillilan, R.; "Chemist's guide to network resources"; ibid. 1993, October.
39. Wiggins, G.; "Some Chemistry resources on the Internet"; ibid. 1995, February (v. 10); http://www.rpi.edu/dept/chem/cheminfo/chemres.html.
40. Varveri, F. S.; "Information retrieval in Chemistry: Chemistry related mailing lists and newsgroups"; 1994, November (v. 8); ftp://leon.nrcps.ariadne-t.gr/pub/chemistry/, file:/m_lists.chem/mailing.09-94. [mirror: ftp://ftp-chem.ucdavis.edu/go_chem/ Index/ChemSites_ac/leon.nrcps.ariadne-t.gr and/or gopher://gopher-chem.ucdavis. edu/Index/ChemSites_ac/leon.nrcps.ariadne-t.gr and/or http://www-chem.ucdavis.edu/Locally Maintained Resources/Chem Sites/leon.nrcps.ariadne-t.gr].
41. Varveri, F. S.; "Information retrieval in Chemistry: Collected information on fields related to Chemistry"; ibid., file:/related_fields/relat.09-94.
42. Varveri, F. S.; "Information retrieval in Chemistry: Chemistry related Internet servers and databases"; ibid., file:/sites/servers-dtbases.09-94.
43. Varveri, F. S.; "Information retrieval in Chemistry: Anonymous ftp sites"; (v.6); ibid., file: /sites.chem/ftp-sites.07-93.
44. Adie, C.; "Network access to multimedia information"; RFC1614 (RTR: 8) 1994, May (pp 79) (with references cited therein).
45. "Guide to network resource tools"; EARN Staff, EARN Association; RFC1580 (FYI: 23) 1994, March (pp 107) (with references cited therein).
46. Kessler, G.; Shepard, S.; "A Primer on Internet and TCP/IP tools"; RFC1739 1994, December (pp 32) (with references cited therein).
47. Grudin, J.; "Interface, an evolving concept"; *Commun. of the ACM* 1993, *36*, 110–119.
48. Balbes, L.; Richon, A.; "Netsci: Science related resources on the Internet"; *Netsci Newsletter* 1996, March; http://edisto.awod.com/netsci/resources.html.
49. Laaksonen, L.; "Yet another resource list (YARL) on resources of information in the Internet" 1995, July; http://laaksonen.csc.fi/docs/stuff.html.
50. Bachrach, S. M.; Pierce T.; Rzepa, H.; "Chemistry on the Internet: The best of the Web 1995"; 210th ACS Meeting; Chicago 1995 (August); http://www.ch.ic.ac.uk/ infobahn/Paper38 (with references cited therein).
51. Martindale, J.; "Martindale's 'The reference desk'"; 1995, April; http://www-sci.lib.uci.edu/~martindale/Ref.html.
52. O'Haver, T. C.; "Internet resources: Chemistry: A selection of resource links"; http://www.inform.umd.edu/UMS+State/UMD-Projects/MCTP/Technology/Chemistry.html.
53. Varveri, F. S.; Telonis, P. E.; Anagnostopoulos, A.; Christodoulou, J.; Boyatzis, S.; "Information retrieval in Chemistry"; http://macedonia.nrcps.ariadne-t.gr/.
54. Bachrach, S. M.; "Chemistry and Gopher"; ref. 7, pp 243–247.
55. Heller, S. R.; "Analytical Chemistry resources on the Internet"; *Trends Anal. Chem.* 1994, *13*, 7–12.
56. Edvarden, O; "Using the World-Wide computer network, Internet, in chemical sciences"; *Acta Chem. Scandinavica* 1995, *49*, 344–350.
57. Willis, L.; "Internet snapshots"; *Chemtech* 1995, *25*, 8–11.
58. Noble D.; "Analysing on the Internet"; *Anal. Chem. 1995*, *67*, 255A-259A.
59. Chowdhury, J.; "Taking off into the world of Internet"; *Chem. Eng.* 1995, *102*, 30–35.
60. Tissue, B.M; "Distributing and retrieving chemical information using the World Wide Web"; Heller, S. R., Ed.; TrAC/Internet Column © Elsevier Science B. V.: Amsterdam; 1995, July; http://www.elsevier.nl/freeinfo/trac/intntod.htm.
61. Stembridge, R.; "Chemical information resources on the World Wide Web"; COMP Division, 210th ACS Meeting; Chicago 1995 (August); http://www.ch.ic.ac.uk/ infobahn/.
62. Krieger, J. H.; "Chemistry sites proliferate on the Internet's World Wide Web"; *Chem. & Eng. News* 1995, *73*(46), 35–46.

63. Mounts, R. D.; "Chemistry on the Web"; *J. Chem. Edu.* 1996, *73*, 68–71.

64. Rzepa, H. S.; "Chemistry and the World Wide Web"; ref. 7, pp 249–275 (with references cited therein).

65. Labanowski, J. K.; Bender, C. F.; Pisanty, A.; "Electronic lists"; ref. 7, pp 65–74.

66. Varveri, F. S.; "Information retrieval in Chemistry"; *J. Chem. Edu.* 1993, *70*, 204–208.

67. Labanowski, J. K.; Bender, C. F.; Pisanty, A.; "Electronic lists for Chemists"; ref. 7, pp 203–233.

68. Boulez, K.: "Overview of Chemistry mailing lists"; 1995; http://bionmr1.rug.ac.be/chemistry/overview.html.

69. Rzepa, H. S.; "Electronic Chemistry conferences"; Heller, S. R., Ed.; TrAC/Internet Column © Elsevier Science B. V.: Amsterdam 1995, June; http://www.elsevier.nl/free-info/trac/intntod.htm.

70. Krieger, J. H.; "Organic Chemistry conference on Internet sets new pace"; *Chem. & Eng. News* 1995, *73*(34), 35–38.

71. Varveri, F. S.; "Information retrieval in Chemistry: Chemistry related anonymous ftp sites"; *J. Chem. Edu.* 1994, *71*, 872–873.

72. Rzepa, H. S.; Whitaker, B. J.; Winter, M. J.; "Chemical applications of the World-Wide-Web system"; *J. Chem. Soc. Chem. Comm.* 1994, pp 1907–1910.

73. Rzepa, H. S.; "Hyperactive Chemistry: A global scientific enabler"; Liverpool University; 1995, March; http://www.ch.ic.ac.uk/talks/.

74. Murray-Rust, P.; Rzepa, H. S.; Leach, C.; "CML—Chemical Markup Language"; abstract 40; 210th ACS Meeting; Chicago 1995 (August); http://www.ch.ic.ac.uk/infobahn/cml/.

75. Casher, O.; Chandramohan, K.; Hargreaves, J.; Leach, Ch.; Murray-Rust, P.; Rzepa, H. S.; Sayle, R.; Whitaker, B. J.; "Hyperactive molecules and the World-Wide-Web information system"; *J. Chem. Soc. Perkin Trans.* 1995, *2*, 7–11.

76. Rzepa, H. S.; "The World Wide Web"; ref. 7, pp 105–161 (with references cited therein).

77. Tissue, B. M.; "Applying hypermedia to chemical education"; *J. Chem. Edu.* 1996, *73*, 65–68.

78. Haderlie, S.; "Chemistry teaching with new technologies and strategies"; *J. Chem. Edu.* 1994, *71*, pp 1058–1062.

79. Wolman, Y.; "Chemical education on the Internet"; TrAC/Internet Column © Elsevier Science B. V.: Amsterdam 1996; http://www.elsevier.nl/freeinfo/trac/intntod.htm.

80. Zielinski, T. J.; Shibata, M.; "The education Internet connection: What shall it be?"; TrAC/Internet Column © Elsevier Science B. V.: Amsterdam 1996; in press.

81. Kieger, J. H.; "Chemical research faces opportunities, challenges from information tools"; *Chem. & Eng. News* (Special Issue) 1995, *73*(13), 26–41.

82. Kirschner, E. M.; "Chemical companies discover a weapon for globalization and reengineering"; *Chem. & Eng. News* (Special Issue) 1995, *73*(13), 62–71.

83. Krieger, J.; "Process simulation seen as pivotal in corporate information flow"; *Chem. & Eng. News* (Special Issue) 1995, *73*(13), 50–61.

84. Pierce, T. H.; Cozzolino, T. J.; "Chemical industry and the Internet"; ref. 7, pp 277–308.

85. Rzepa, H.; Murray-Rust, P.; Whitaker, B.; "A Chemical primary content yype for Multipurpose Internet Mail Extensions"; Networking Group Internet Draft 1995; ref. 76, cited reference no. 28.

86. Indermaur, K.; "State of the art: Baby steps"; *Byte* 1995, *20*, 97–104.

87. Winship, I. R.; World Wide Web searching tools—an evaluation"; *Vine* 1995, *99*, 49–54, and/or http://www.hamline.edu/library/links/comparisons.html.

88. "Scout Toolkit Directory"; SCOUT discussion list at scout@internic.net (http://rs.internic/scout); http://rs.internic.net/scout/toolkit/.

89. Bowman, C. M.; Danzig, P. B.; Manber, U.; Schwartz, M. F.; "Scalable Internet resource discovery: Research, problems and approaches"; *Commun. of the ACM* 1994, *37*, 98–114.

90. Kriger, J.; "Revolution comes to the Chemical community"; *Chem. & Eng. News* (Special Issue) 1995, *73*(13), 22–24.

91. Rzepa, H. S.; Leach, C.; "The future of Chemistry and the World-Wide Web"; abstract No. 20; COMP Division, 210th ACS Meeting; Chicago 1995 (August); http://www.ch.ic.ac.uk/infobahn/futures.html.

92. Lepkowski, W; "Policy permeate efforts to create information infrastructure"; *Chem. & Eng. News* (Special Issue) 1995, *73*(13), 72–82.

93. Newby, G.B; "Virtual Reality"; Williams, M. E., Ed.; *Annual Review of Information Science and Technology;* Learned Information, Inc.: Medford, NJ, 1993, *28*, 187–229.

94. Furht, B.; "Multimedia systems: An overview"; *IEEE Multimedia* 1994, *1*, 47–59.

95. Chan, V. W. S.; "All-optical networks"; *Scientific American* 1995, *273*, 56–59.

96. Milstead, J. L.; "Invisible Thesauri: The year 2000"; *Online & CDROM Review* 1995, *19*, 93–94.

97. Webb, T. D.; "The frozen library: A model for twenty-first century libraries"; *The Electronic Library* 1995, *13*, 21–26.

98. Lesk, M. E.; "Electronic chemical journals"; *Anal. Chem.* 1994, *66*, 747A–755A.

99. Borman, S.; "Electronic publishing increasingly offered as alternative to print medium"; *Chem. & Eng. News* (Special Issue) 1995, *73*(13), 42–50.

100. Heller, S. R.; "Publishing on the Internet—A proposal for the future"; Heller, S. R., Ed.; TrAC/Internet Column © Elsevier Science B.V.: Amsterdam 1995; http://www.elsevier.nl/freeinfo/trac/intntod.htm.

101. Rzepa, H. S.; "The future of electronic journals in Chemistry"; Heller, S. R., Ed.; TrAC/Internet Column © Elsevier Science B. V.: Amsterdam 1995; http://www.elsevier.nl/freeinfo/trac/intntod.htm.

102. Yanoff, Sc.; "Special Internet connections"; 1996, March; e-mailto inetlist@aug3.augsburg.edu, empty subject_line and body_txt and/or ftp://ftp.csd.uwm.edu/pub/inet.services.txt and/or http://www.uwm.edu/Mirror/inet.services.html (with references cited therein).

103. Maas, R. E.; "The MaasInfo Indexes", 1994–1995; ftp://NCTUCCCA.edu.tw/documents/Internet/MaasInfo/(with references cited therein).

104. December, J.; "Tool tips"; 1995, April; ftp://ftp.msstate.edu/docs/words-l/Net-Stuff/internet-tools.txt (with references cited therein).

105. December, J.; "Information sources: the Internet and Computer-Mediated Communication"; 1995, September (v. 3.999); http://www.december.com/cmc/intd.toc3.html (with references cited therein).

106. Kovacs, D. K., Ed.; "Directory of scholarly and professional electronic conferences"; 1996, March (10th Revision); e-mailto listserv@kentvm.kent.edu, body_txt: get acadlist readme and/or gopher://gopher.usask.ca/1/Computing/Internet Information/ Directory of Scholarly Electronic Conferences and/or http://www.mid.net/KOVACS.

107. "What is Usenet?"; 1992, June; ftp://ftp.msstate.edu/docs/words-l/Net-Stuff/whatis.usenet.

108. "Collection of FAQs and files from Usenet newsgroups"; ftp://rtfm.mit.edu/pub/usenet/.

109. "Available FAQs"; e-mailto listserv@cc1.kuleuven.ac.be, body_txt: get netfaqs filelist.

110. De Presno, O.; "On line world" (v. 2.4); 1995, December; http://login.eunet.no/~presno/index.html (with references cited therein).

111. Emailto inetlist@aug3.augsburg.edu, empty subject_line & body_txt_line.

112. Yanoff, Sc.; "Inter-Network mail guide"; 1995, December; ftp://ftp.csd.uwm.edu/pub/internetwork-mail-guide and/or ftp://rtfm.mit.edu/pub/usenet/comp.mail.misc/file: Updated_Inter-Nework_Mail_Guide.

113. Rankin, B.; "Accessing the Internet by e-mail"; 1996, March (v. 5.34); e-mailto mailserver@rtfm.mit.edu, body_txt: send usenet/news.answers/internet-services/access-via-email and/or ftp://rtfm.mit.edu/pub/usenet/news.answers/internet-services/access-via-email.

114. Milles, J.; "Discussion lists: mail server commands"; 1996, January (v. 1.27); http://lawlib.slu.edu/training/mailser.htm.

115. Viehland, D.; "Listserv guide"; 1993, July; e-mailto listserv@arizvm1.ccit.arizona.edu, body_txt: get listserv guide and/or ftp://ftp.msstate.edu/docs/words-l/Net-Stuff/list-serv.guide.

116. Thomas, E.; "Listserv system reference library"; 1988 (v. 1.5n); e-mailto listserv@umdd.umd.edu, body_txt: get listdb memo.

117. Lemson, D.; "File compression, archiving, and text—binary format"; 1995, March; ftp://ftp.cso.uiuc.edu/doc/pcnet/file: compression.

118. "MIME (Multipurpose Internet Mail Extensions)" (RFC1521 & RFC1522); http://www.oac.uci.edu/indiv/ehood/MIME/MIME.html.

119. Rovers, P.; "Anonymous FTP: frequently asked questions (FAQ) list", 1995, November (v. 3.1.2); ftp://rtfm.mit.edu/pub/usenet/news.answers/ftp-list/faq.

120. Maas, R. E.; "How to use 'Archie': How to use the 'Archie' service to find out where files are available on the Internet"; ref. 103; 1995, February; file: MaasInfo.Archie.

121. Lindner, P.; "Gopher frequently asked questions (FAQ)"; 1994, December; ftp://rtfm.mit.edu/pub/usenet/news.answers/gopher-faq and/or gopher://mud-honey.micro.umn.edu:70/00/Gopher.FAQ.

122. Foster, S.; "Common questions and answers about veronica, a title search and retrieval system for use with the Internet Gopher"; 1994; June; ftp://veronica.scs.unr.edu/00/veronica-docs/how-to-query-veronica and/or gopher://veronica.scs.unr.edu/00/veronica/veronica-faq.

123. Jones, R.; 1994; ftp://boombox.micro.umn.edu/pub/gopher/Unix/Gopher-Tools/jughead/file: jughead.ReadMe.

124. Bernes-Lee, T.; Cailliau R.; Luotonen A.; Nielsen, H. F.; Secret A.; "The World Wide Web"; *Commun. of the ACM,* 1994, *37*(August), 76–82.

125. Fielding, R. T.; Frystyk, H.; Berners-Lee; "Hypertext Transfer Protocol—HTTP/1.1"; (HTTP Working Group) 1996, January; ftp://nic.merit.edu/documents/internet-drafts/file: draft-ietf-http-v11-spec-01.txt.

126. Boutell, T. "'World Wide Web' Frequently Asked Questions (with answers, of course!)"; 1996; http://www.boutell.com/faq/www_faq.html.

127. "Uniform Resource Locators (URL)"; Berners-Lee, T., Masinter, L., McCahill, M., Eds.; RFC1738 1994, December (p 25).

128. Schatz, B. R.; Hardin, J. B.; "NCSA Mosaic and the World Wide Web: Global hyper-media protocols for the Internet"; *Science* 1994, *265*, 895–901.

129. Pesce, M. D.; Kennard, P.; Parisi, A. S.; "Cyperspace"; 1995; http://vrml.wired.com/consepts/pesce-www.html.

130. Erez, S.; "An update of network resoures: Web indexes again"; NIC-NEWS 1995, 14, 07 (June 8); http://www.washington.edu/nic-news/.

131. Kellem, J. "Wais: A sketch of an overview"; 1991, September; ftp://sunsite.unc.edu/pub/docs/about-the-net/libsoft/wais.txt.

132. Shaw, D.; "The human-computer interface for information retrieval"; Williams, M. E., Ed.; *Annual Review of Information Science and Technology;* Elsevier Science: Amsterdam, 1991, *26*, 155–195.

133. Kantor, P. B.; "Information retrieval techniques"; ibid. 1994, *29*, 53–90.

134. Chang, Sh.-J.; Rice, R. E.; "Browsing: A multidimensional framework"; Williams, M. E., Ed.; *Annual Review of Information Science and Technology;* Learned information, Inc.: Medford, NJ, 1993, *28*, 231–276.

135. Schwartz, M. F.; Emtage, A.; Kahle, B.; Neuman, B. C.; "A Comparison of Internet resources discovery approaches"; *Computing Systems* 1992, *5*, 462–493.

136. Fung, R.; Del Favero, B.; "Applying bayesian networks to information retrieval"; *Commun. of the ACM* 1995, *38*, 42–48 & 57.

137. Doszkocs, T. E.; Reggia, J.; Lin X.; "Connectionist models and information retrieval"; Williams, M. E., Ed.; *Annual Review of Information Science and Technology;* Elsevier Science: Amsterdam, 1990, *25,* 209–262.

138. Lynch, Cl.; Preson, C. M.; "Internet access to information resources"; ibid. 1990, *25,* 263–312.

139. Weider, C.; "Resource transporders"; RFC1728 1994, December (pp 6).

140. "Next-Generation. Database Systems: Achievements and opportunities"; Silberschatz, A.; Stonebraker, M., Ullman, J., Eds.; *Commun. of the ACM,* 1991, *34,* 110–120.

141. Rooholamini, R.; Cherkassky, V.; "ATM-based multimedia servers"; *IEEE Multimedia* 1995, *2,* 39–52.

7

Coping with the Transformation of Chemical Information

Carol Carr
Arleen N. Somerville

Introduction

In this chapter we will explore how computers have changed the way chemists obtain information and the implications of this electronic revolution for educators and for chemists trying to function in this new environment. We will concentrate on the traditional information sources such as abstracts, indexes, and journals. We also discuss some sources, such as structure and reaction databases, that use the power of the electronic medium to provide new searching capabilities for chemists. The wide array of Internet resources is discussed in Chapter 6, Information Retrieval in Chemistry Across the Internet (Varveri).

Although we will discuss some of the major chemical information resources, this chapter does not describe them in detail. Detailed information on these sources can be found in guides to the chemical literature.[1-4] We also give examples of databases available from commercial online services. We primarily cite examples from STN International and Knight Ridder Information's DIALOG Service, since these are the most frequently used services in the educational setting. However, several other services such as Chemical Information Systems and Questel/Orbit offer chemistry databases. (Contact information for the major organizations and products mentioned in this chapter is provided in Table 1.) Access information for the electronic resources described in this chapter is provided in Table 2. Resources are listed in alphabetical order.

TABLE I   Major suppliers of chemical information mentioned in this chapter.

Beilstein Information Systems Inc.
15 Inverness Way East
Englewood, CO 80112
800-275-6094
http://www.beilstein.com
PRODUCTS: CrossFire, Current Facts, MOLTERM

CARL Corp.
3801 E. Florida, Ste 200
Denver, CO 80210
800-787-7979
http://www.carl.org/carl.html
PRODUCTS: UnCover, UnCover Reveal

Chapman & Hall USA
115 Fifth Ave., 4th Flr
New York, NY 10003
212-260-1354
http://www.chaphall.com/chaphall.html
PRODUCTS: Dictionary of Natural Products, Dictionary of Organic Compounds

Chemical Abstracts Service
2540 Olentangy River Rd.
Columbus, OH 43202
800-848-6533
http://www.cas.org
PRODUCTS: Chemical Abstracts, CA on Disc, Caplus, CASurveyor

Chemical Information Systems (PSI International, Inc.)
810 Gleneagles Ct. Ste 300
Towson, MD 21286
800-247-8737

CRC Press
2000 Corporate Blvd NW
Boca Raton, FL 33431
800-272-7737
http://www.crcpress.com
PRODUCTS: Properties of Organic Compounds

Derwent North America
1420 Spring Hill Rd.
McLean, VA 22102
800-451-3451
http://www.derwent.co.uk
PRODUCTS: World Patents Index Files, Chemical Innovations

DIALOG (see Knight-Ridder)

InfoChem GmbH
Landsberger Str. 408
D-81241 Munich Germany
+49-89-58 30 02
PRODUCTS: ChemReact

TABLE I (continued)   Major suppliers of chemical information mentioned in this chapter.

Institute for Scientific Information (ISI)
3501 Market St.
Philadelphia, PA 19104
1-800-336-4474
http://www.isinet.com
PRODUCTS: Current Chemical Reactions, Current Contents, Index
Chemicus, Reaction Citation Index, Science Citation Index

Knight-Ridder Information Inc.
2440 El Camino Real
Mountain View, CA 94040
800-334-2564
http://www.dialog.com
PRODUCTS: KR ScienceBase, DIALOG database system, DialogLink

Micropatent
250 Dodge Ave
East Haven, CT 06512
800-648-6787
http://www.micropat.com
PRODUCTS: Retrochem

MDL Information Systems Inc.
14600 Catalina St
San Leandro, CA
800-635-0064
http://www.mdli.com
PRODUCTS: MACCS, REACCS, ISIS, Index Chemicus

Springer-Verlag New York, Inc
175 Fifth Ave.
New York, NY 10010
1-800-SPRINGER
http://www.springer-ny.com
PRODUCTS: European Journal of Biochemistry, Chemical Educator,
ChemReact

STN International
2540 Olentangy River Rd
Box 02228
Columbus, OH 43202
800-848-6533
http://info.cas.org/stn.html
PRODUCTS: SciFinder, STN Express, STN database system

Questel/Orbit, Inc.
8000 Westpark Drive
McLean, VA 22102
800-456-7248
http://www.questel.orbit.com
PRODUCTS: Pharmsearch

Pharmsearch information:
O'Hara Consulting Inc.
215 12th St SE
Washington DC 2003-1427
800-949-5120

TABLE 2   URLs for WEB sources mentioned.

ACS Division of Chemical Information
   http://www.lib.uchicago.edu/~atbrooks/CINF/cinfhome.html
ACS/SLA Clearinghouse for Chemical Information Instruction Materials
   http://www.indiana.edu/~cheminfo/cciimnro.html
American Chemical Society http://www.acs.org
American Institute of Physics http://www.aip.org
American Physical Society http://www.aps.org
Annual Review Web site http://www.annurev.org
Beilstein Information Systems, Inc. http://www.beilstein.com
Biological Abstracts http://www.bios.org
Chapman & Hall USA http://www.chaphall.com/chaphall.html
Chemical Abstracts Service http://www.cas.org
Chemical Educator http://journals.springer-ny.com/chedr/
Chemical Information Sources http://www.indiana.edu:80/~cheminfo/
CICOURSE http://indiana.edu/~cheminfo/400home.html
CIStudio course http://www.rpi.edu/dept/chem/cheminfo/cistudio/index.html
CRC Press http://www.crcpress.com
Derwent North America http://www.derwent.co.uk
Electronic Computational Chemistry conference
   http://hackberry.chem.niu.edu/ECCC2/homepage.html
Elsevier Science http://www.elsevier.nl/
Engineering Information Village http://www.ei.org/
Entrez molecular biology subset of MEDLINE http://www3.ncbi.nlm.nih.gov/Entrez/
Harvard Biological Laboratories http://golgi.harvard.edu/journals.html
Institute of Physics Publishing http://www.iop.org
Institute for Scientific Information http://www.isinet.com
Journal of Chemical Education (JCE Online) http://jchemed.chem.wisc.edu/
Journal of Chemical Education: software (JCE Software) http://jchemed.chem.wisc.edu/
Journal of Chemical Physics Express http://jcp.uchicago.edu
Knight-Ridders's KR ScienceBase http://krscience.dialog.com
Materials Research Society http://www.mrs.org/
McGill physics-related journals http://www.physics.mcgill.ca:8081/physics-services/
MDL Information Systems Inc. http://www.mdli.com
Optical Society of America http://www.osa.org/
Questel/Orbit, Inc. http://www.questel.orbit.com
Springer-Verlag http://www.springer-ny.com
STM electronic journals survey http://journals.ecs.soton.ac.uk/survey/survey.html
STN International http://info.cas.org/stn.html
UnCover http://www.carl.org/carl.html
Wiggins' Chemistry Resources on the Internet
   http://www.rpi.edu/dept/chem/cheminfo/chemres.html
WWW Virtual Library: Chemistry http://www.chem.ucla.edu/chempointers.html
Yahoo: Chemistry http://www.yahoo.com/Science/chemistry

History of Electronic Information

Databases became available in electronic form in the mid-1960s, when publishers converted to computerized photocomposition. This process resulted in magnetic tapes, which, along with being used to print information, could also be searched to manipulate and extract that information. Chemical Abstracts Service was one of the first publishers to offer an electronic version of its database. In 1965 *Chemical*

Abstracts issued *Chemical and Biological Activities* simultaneously in printed form and on magnetic tape.[5] Chemists were quick to take advantage of the power and efficiency that computer-based searching provides. Today, *Chemical Abstracts* is one of the most heavily used online databases.[6]

The early search systems were designed for and primarily used by information specialists who mastered the command syntax and searched on behalf of researchers. Today, with faster, more powerful computers on everyone's desk, information seekers do not want to wait for someone else to locate information for them. They want to do it themselves. Today's search systems are designed to make this easier. The Mac and Windows point-and-click mode lessens the need to know a complex command language. Several products have recently appeared that are designed to facilitate searching for scientists: SciFinder, Beilstein CrossFire, and KR ScienceBase. These are described below (see Abstracts/Indexes and Data/Information Compilations sections).

Current Developments

Search systems are beginning to move from commands to natural language. Today's systems also offer more sophisticated ways of manipulating information. For example, a searcher can look for patents on monoclonal antibodies and rank the answers by the assignee to get a list of companies that have the most patents in this area. Faster, more powerful computers are enabling the development of more complex systems, such as computer neural networks, designed to emulate the pattern-recognition and parallel processing capabilities of the human brain. This new generation of tools may finally fulfill the promise of artificial intelligence techniques, which have long been touted as a way to facilitate the search for information. We may indeed soon have intelligent "know-bots" that can analyze our needs, surf the various networks, and retrieve that elusive nugget of useful information.[7]

Along with more sophisticated methods of searching, faster, more powerful computers can store and retrieve more information. Many early databases contained citations only. Today the full text of journal articles (plus images) is available. Modern systems can search and analyze large text databases as well as collections of chemical structures, crystallographic data, and sequence data.

Electronic databases are becoming an integral part of how chemical information is shared and distributed. Researchers enter data into electronic laboratory notebooks, creating in-house databases. Scientists now submit their crystallographic and sequence data to electronic databases along with their journal article. Many spectrophotometers now include a library of spectral data that chemists can use to compare experimental results against a database of known spectra. Industrial in-house information systems integrate a company's research with published results so that chemists can seamlessly search both using one system. It is now possible to have the entire record of a research project available in electronic form, from its beginning in a laboratory to the archival record in a journal article or patent.

What advantages does electronic information offer? Speed is one. Users can search through over 35 years of *Chemical Abstracts* and download the citations in minutes. The downloaded information can then be transferred to a database management application such as EndNote to create a searchable personal database. Electronic databases also offer greatly enhanced access to information. All of the information can be searched, not just an index. Searchers can get very specific information directly by constructing Boolean statements (e.g., articles on C60 microtubules originating from Cal Tech or MIT). Searchers can also extract information from an electronic source in ways almost impossible to do using print (e.g., search the *Merck Index* online for all pesticides with molecular weights between 200 and 300 that do not contain phosphorus).

According to Moore's law, computing speeds and densities double every 18 months.[8] Computer chips will become cubes; neural computers are predicted to match human brain capacity early in the 21st century. Fuzzy logic, pattern matching, and artificial intelligence will enhance access to electronic sources far beyond the confines of systems based on character-string matching and Boolean logic. Perhaps these advances will change the way information is communicated and collected, but many current electronic resources are still based on the traditional printed methods of exchanging information. Today's chemist needs to know the characteristics and utility of these sources in order to use them effectively either in print or in electronic form.

The electronic revolution has made it possible for anyone with a powerful enough computer to become a publisher. Researchers can indeed offer their papers from a personal home page. Publicly available Internet searching tools (such as Lycos, Alta Vista, and the Web Crawler) are available. In the mid-1990s, however, for chemists, the freely available information on the Internet cannot replace the traditional sources, although it is increasingly becoming an important supplement to those sources.

Electronic Status of Chemical Information

Journals

Journals and patents are the primary means of communicating chemical information. These publications are the first opportunity for those not immediately involved in the research to see a description of the work. Other publications also provide a "first look" (preprints, published conference proceedings, reports), but we are limiting our discussion here to journals and patents since these are the most common vehicles in chemistry. See the guides listed[1-4] for discussions of these other formats. The journal has a long tradition in the sciences. The Royal Society began publishing its *Philosophical Transactions* in 1665. The first chemical journal (*Chemisches Journal*) began in 1778. Today, *Chemical Abstracts* covers approximately 15,000 chemistry-related journal titles.

Electronic Status/Trends.   In 1982 the American Chemical Society (ACS) became the first publisher to offer journal articles electronically on a commercial database system. Only the text was available; images followed in 1992. In the mid-1990s journals are the most volatile segment of the information spectrum. Many speak today of a crisis in scientific publishing. The proliferation of journal titles plus the high cost of subscribing make it impossible for an individual to subscribe to many titles, and make it difficult for an academic or company library to own all of the journals a chemist will need. Also, the staggering number of articles being published (Chemical Abstracts Service indexed nearly 750,000 items in 1995—a good percentage of which were journal articles) makes it impossible for a chemist to browse journals to keep abreast of a field. The storing of print journals is also creating a space crunch in major research libraries as well as in chemists' offices.

All of these factors have created a demand for electronic versions of journals. Publishers, information companies, and technology firms are testing a range of electronic journal products. Products range from complete electronic issues to selected electronic articles, tables of contents (some with abstracts), and preprints. A few chemistry-related initiatives are described below to indicate the range of activity in this area. For more information consult the articles by Borman[9] and Taube.[10]

Electronic Journals.   Some journals provide electronic versions of their print publications, while others are only published in electronic form. Electronic publishing offers users earlier access to an article plus a high potential for enhanced access to information, for example, links to supplementary material such as extensive experimental data, links to related articles, and links to the full text of the citations.

As this chapter is written, several chemical titles are available on the Internet, some as a limited-time free trial, others via a paid subscription. Titles include all of the Royal Society of Chemistry and American Chemical Society journals and others, e.g., *Journal of Biological Chemistry, Journal of Molecular Modeling, Journal of Computer-Aided Molecular Design,* and *The Chemical Educator.* Preprints for the *Journal of Chemical Physics* are also available. However, the Internet is a constantly changing landscape with titles appearing, disappearing, and changing addresses constantly. For a current list consult one of the Internet indexes such as Yahoo-Chemistry or Internet Chemistry Resources, compiled by Gary Wiggins (see Table 2). These indexes are discussed in Chapter 6, Information Retrieval in Chemistry (Vaveri). A 1995 survey of STM (science, technology, medicine) electronic journals by Steve Hitchcock, Leslie Carr, and Wendy Hall was published as a Web document (see Table 2).

Several chemistry journals are available on CD-ROM. The ACS offers three journals as of 1996: *Journal of the American Chemical Society, Journal of Organic Chemistry,* and *Biochemistry.* The ACS plans to add others. Springer-Verlag offers the *European Journal of Biochemistry.* CD-ROMs offer an easy search interface plus a solution to journal storage problems. However, a user can only search one year of one title with current CD-ROM technology.

Electronic Journal Collections.   In addition to individual journals, several collections of electronic journals are available. STN International offers the full text of journals from the ACS, the Royal Society, plus several commercial publishers via its online service. Full-page images from ACS journals (1992 on) are also available. Searchers must use a separate graphical front end to view these images. The text is searchable, allowing users to search a number of journals simultaneously over a range of years (most titles are available from the mid-1980s on). STN's SciFinder (described below) offers scanned images of ACS journal pages. Future enhancements will include searchable text plus journals from additional publishers.

The collections described above are commercially available products. Several projects tested the feasibility of electronic journals. These projects are described in a recent *Chemical & Engineering News* article.[9]

In 1996 electronic journal collections began to appear on the Web. Publisher collections include Academic Press (IDEAL program), Springer-Verlag, Royal Society of Chemistry, and American Chemical Society. OCLC plans to offer a common interface for electronic journals from several publishers in 1997.

Alerting/Document Delivery Services. These offer searchable or browsable tables of contents with electronic ordering. Delivery of the full article is by fax, e-mail, or mail. These services have become more important to scientists since the rising cost of journals is causing many university and company libraries to cancel subscriptions. Current Contents, from the Institute for Scientific Information (ISI), is one of the oldest of these services. It is available in print, on disk, on tape, and online from several information companies. The information in the electronic versions is searchable. Document delivery for most titles in Current Contents is available via their Genuine Article service. Current Contents is available in seven editions, three of which include chemistry journals—Physical, Chemical and Earth Sciences; Life Sciences; and Agricultural, Biological and Environmental Sciences.

The UnCover service from CARL Corporation offers searchable tables of contents over the Internet. Approximately half of these are science titles. Articles can be ordered while online. STN's SciFinder offers browsable tables of contents for 1,300 journals. Full text is available for the ACS journals; more publishers will be added. Several publishers provide tables of contents for their journals on the Internet. Springer-Verlag and Elsevier are examples (see Table 2).

In addition to alerting services, other document delivery options are available. Many databases and database systems have document delivery services. Full texts of items found in their databases can be requested by phone, fax, or online. *Chemical Abstracts* has such a document delivery service and recently implemented a Web request form (see Table 2). Knight-Ridder recently acquired the CARL Corporation and its subsidiary, the UnCover Company, which will enhance its document delivery service.

Implications for Chemists. What does this electronic access to journal articles mean for chemists? Although chemists will probably continue to use abstracting and indexing services (discussed below) to locate relevant journal information for the foreseeable future, electronic access to the full text of journal articles offers several advantages. Full information can be available immediately at the user's desktop. Searchable journal articles allow chemists to retrieve all of the information in an article, not just what indexers have decided to include.

The myriad products offer chemists both a choice and a challenge. Until technology and the marketplace normalize how the information in journals is best distributed, becoming aware of what is available and how best to use the alternatives can be a time-consuming task. Many issues remain to be resolved: In the fluid electronic environment, how can an authorized, archival version of an article be ensured? If researchers self-publish, will peer review disappear? How will the review and reward system in academe change? Is the journal a functional entity in the electronic medium? What role will publishers play? Will copyright laws be modified?

Patents

Patents are another primary source of information, especially for information on industrial research and development. Patents have the added complexity of being legal documents. A patent is an exchange. The inventor describes his or her discovery in exchange for protection against others using that discovery for a specified period. Naturally, an inventor wants to obtain the greatest protection legally possible. This goal often produces very broad claims, and it is sometimes difficult to discern the specific example within the patent that is most useful. However, patents are important sources of information. Patents are often the only documents that describe the research and development activities of a company. According to Derwent Inc. (an information company specializing in patent information), approximately 70% of the information in patents is never published anywhere else.[11] Patent searching for patent protection or infringement is still the purview of the information specialist, since millions of dollars can ride on the information obtained. For chemists seeking the information found in patents, however, a range of tools exists both in print and electronically. Information on patent databases can be found in the guides listed.[1-4] For a recent journal article on electronic patent sources, see Thompson.[12]

Electronic Status/Trends.   The electronic products that exist for patents are similar to those for journals. Full-text patent collections are available on online systems via CD-ROM and on Internet. U.S. patents are available in full text back to the 1970s on several online services. Patent images (drawings, reaction schemes) became accessible as of the mid-1990s. A separate graphical front-end application such as STN Express or DialogLink is needed to view the images currently. However, as the emergence of graphical online search systems such as SciFinder and KR ScienceBase (discussed below) indicates, integration of images with text will become the norm in short order, not only for patents but for all types of electronic information.

CD patent products range from total collections (e.g., all U.S. or European patents) to subject-related subsets, and from full text plus images to abstracts or claims. Several companies will create custom CDs based on a client's interest. Most products have searchable indexes or abstracts plus scanned full text and, increasingly, scanned images. CD-ROM collections of chemical patents include the Chemical Innovations series of U.S. patents from Derwent Inc., and RetroChem (U.S. chemical patents from 1976 to the present) from MicroPatent. In addition to chemistry CDs, a range of products covering drugs and pharmaceuticals is also available.

Patent information is also found in several online chemical databases. *Chemical Abstracts* includes patents, as do several more specialized databases such as APIPAT (petroleum) and Paperchem. Other databases are entirely devoted to patents of all types. One of these, Derwent's *World Patents Index,* has a well-developed online chemical information search system available to subscribers for their *Chemical Patents Index.* In addition, other electronic patent databases cover specific subjects, for example, Pharmsearch (available from Questel), which covers pharmaceutical patents from the U.S., France, and the European Patent Office. Patent information is also available on the Internet. Several free sources exist as of 1996. The Internet Multicasting Service offers searchable full text of U.S. patents back to 1994; Source Translation & Optimization (STO) offers the Internet Patent Search System, patent titles back to 1975 searchable by patent classification.[13] The U.S. Patent and Trademark Office offers searchable abstracts of patents for the last 20 years. Chemical Abstracts Service launched Chemical Patents Plus! in 1996. All U.S. patents from 1974 are searchable and Chemical Abstracts indexing (including registry numbers) has been included for chemical patents. A fee is charged for full display or delivery of the full text (see Table 2 for addresses for these sources). In 1996 the IBM patent database on Internet offered U.S. patents' claims with images (http://www.ibm.com/patents/).

Implications for Chemists.   Most chemists do not realize the importance of patents until they work in industry. Many faculty and students overlook them as a source of information. With the full text available electronically, chemists can get the entire description of an invention. Patents can sometimes give step-by-step directions for doing a procedure.

Electronic access to the full text of journals and patents is a valuable tool for chemists. However, not every journal is available electronically, and patent databases for the most part only go back to the 1970s. No chemist can possibly keep track of all the patents or journals that cover his or her specialty. The size and complexity of this primary literature gave rise to a second type of chemical information source—abstracts and indexes.

Abstracts/Indexes

These services review primary sources such as journals and patents, and facilitate access to these sources via indexes (lists of key concepts and other access points, such as authors) and abstracts (brief summaries). Chemists saw the need for these resources early. Less than 75 years after the first chemical journal appeared, the first chemical abstracting service, Pharmaceutisches Central-Blatt (later called Chemisches Zentral-blatt) began in 1830. Approximately 60 chemistry-related abstracting and indexing sources exist today. Most have both print and electronic versions. These sources range from literature-based, which describe the contents of a publication, to data and structure sources, which extract information from a publication and format it for easy retrieval. This section covers literature-based sources.

The most comprehensive chemical source of this type is *Chemical Abstracts*, which contains information on all aspects of chemistry plus relevant information from related areas. Other sources are more specialized. Examples include *Analytical Abstracts, Chemical Safety Newsbase,* and *Metadex* (metals). These sources provide more detailed indexing in their subject areas than does a general service such as *Chemical Abstracts.* For example, *Analytical Abstracts* has three types of index entries: analyte, matrix, and concept (for reagents, techniques, and parameters).

In addition to the chemical abstracting and indexing sources, sources in related areas such as *Medline* (medicine), *Inspec* (physics, materials science), and *EICompendex* (engineering) can contain useful information. Many of these are described in guides to chemical information.[1-4] Database catalogs from online services (especially Knight-Ridder's DIALOG Information Service and STN International) are also a good source of information for the electronic versions.

Electronic Status/Trends.   Most of the chemistry-related abstracting and indexing services are available online; several are also available on CD-ROM. An advantage of large online systems is that many databases can be searched using the same search commands. Also, several sources can be searched simultaneously. Up to the mid 1990s most services use Boolean logic to retrieve information from text. Search commands and conventions are hard to remember if not used frequently. However, online systems have begun to offer alternatives such as relevance searching and graphical interfaces. For example, DIALOG's TARGET searching and SciFinder allow searchers to input terms without Boolean operators. In addition, user-friendly systems are being developed for scientists.

Facilitated Science Searching.   Several products specifically for the practicing scientist have recently been introduced. These use menus and graphical interfaces instead of commands, making it easier for the occasional user to retrieve information. Built-in expertise is also provided for some common tasks such as searching for publications by an author or searching for information on a compound. Some of these products also offer fixed-cost access to the information via annual subscriptions (rather than the pay-as-you-go pricing mode used for traditional online services). This pricing format encourages frequent use, since the cost has been paid up-front.

SciFinder from STN allows chemists to search Chemical Abstracts Service databases by keyword or structure, browse the table of contents of over 1,300 journals, and view the full text of journals without lengthy training and practice. As of 1995, full text of ACS journals are available, and more will be added in the near future.[14]

Knight-Ridder's KR ScienceBase provides search assistance and access to 20 science databases from Knight-Ridder's DIALOG service. Searchers use a point-and-click interface to get information on a range of predefined topics such as information on specific chemical substances, new drug-approval announcements, or patents from a specific company. The product is available via the Internet at a Web site.[15] In 1996 Knight-Ridder and STN introduced Web-based searching of their databases.

CD-ROM.   A range of chemistry-related databases is available. In 1996 *Chemical Abstracts* began offering a CD version of the entire publication. The CASurveyor CDs contain *Chemical Abstracts* references for selected subject subsets; 15 are available in 1996.

Other databases available on CD-ROM include *SciSearch* (*Science Citation Index*), *Inspec, Compendex, Chemical Business Newsbase,* and *Petroleum Abstracts* to name only a few.[16-17]

CDs offer user-friendly searching at a fixed cost and also offer libraries a solution for their space crunch. However, large CD systems require an investment in hardware, which must be purchased, maintained, and upgraded. (In 1996 *Chemical Abstracts* on CD will generate 4–5 disks/year). Each CD source must be searched separately, often using a different search system. Users lose the advantage of searching many files or many years simultaneously, which is possible on the major online systems.

Implications for Chemists.   Chemists need to keep abreast of the electronic options available. Often, users rely on one source for all of their information needs. The smart searcher will review sources to identify what special indexing or search features might facilitate his or her task. For example, it may be easier to locate detailed analytical information using *Analytical Abstracts* than it is using *Chemical Abstracts.*

Electronic databases can make searching easier and faster; however, users still need to know the structure of the database to use it effectively. Users should know the topics and sources covered, the indexing guidelines, and the style guidelines. Knowing the style guidelines (such as abbreviations, British versus American spelling) is especially important when using electronic sources. Although algorithms exist to automatically search for singular/plural, British/American spelling, they are not in general use as of the mid-1990s. The structural features mentioned above are the same in both the print and electronic versions of a database. Therefore, the effort spent learning how to use a print source will reap rewards when the electronic version is used.

Data/Information Compilations

As was mentioned above, another method of organizing information is to extract specific information from many sources and compile it to facilitate its use. Some compilers also evaluate the information, selecting the best values or optimal method. Again, chemists have had a long history in this area. Beilstein began extracting preparation, reaction, and property data on organic compounds in the 1800s. Everyday chemists' tools such as the *CRC Handbook of Chemistry and Physics* and the *Merck Index* are examples of this type of information source. The new electronic data banks are more current and complete than their print predecessors. Scientists now submit their crystallographic and sequence data to electronic data banks as they publish a journal article. These sources allow chemists to quickly find a needle in a haystack, e.g., a straight-chain aliphatic alcohol with a dynamic viscosity of 0.14 g cm^{-1} at 25 °C. This type of source can also be used to do more complex analyses, such as the pattern matching or similarity searching possible in the sequence databases.

Just as with the literature-based services, chemists must know the inclusion policies and the formats of these sources to use them effectively. These sources are at once the easiest and the most difficult to search. Since the information needed is clearly defined (e.g., the bp of a compound), it is easy to decide what you need and to recognize it once you've found it. However, finding it in some of these databases requires knowledge of an often complicated database architecture (Beilstein Online has hundreds of data fields). Searchers must also know the units used to express the data, plus the search commands for the system. For the specialist these skills become second nature, but for the occasional user this can present a difficult barrier. However, just as with abstracting services, user-friendly products are being developed to make this complexity transparent to the user.

Electronic Status/Trends.   Many compilations are available electronically from commercial online services, on CD-ROM, and, increasingly, over the Internet. Traditional print sources such as the *Merck Index* and the Beilstein and Gmelin handbooks are available online. In addition, wholly electronic sources exist, such as the Cambridge Structural Database and a range of sequence databases. Some basic tools available on CD-ROM include the Chapman & Hall Dictionaries and CRC's *Properties of Organic Compounds*. Beilstein offers Beilstein Current Facts on CD (this has no print counterpart).

Facilitated Science Products.   Beilstein Information System's CrossFire databases use a graphical client to provide users with easy access to property, reaction, and citation data from the Beilstein and Gmelin handbooks. Chemists can search by property, by substance (using structures), and by reaction. Information on CrossFire is available from Beilstein's home page (see Table 2).

Implications for Chemists.   It is very important that students be exposed to these sources along with the traditional literature databases. Some property data can be found by searching *Chemical Abstracts,* but it may be more efficient to use compilations. Students also need to be aware of the types of analyses possible with these tools. A few exercises during organic laboratory class using CRC's *Properties of Organic Compounds* CD can be a valuable experience for undergraduates. For example, students can search for all compounds having a certain set of properties, e.g., a melting point greater than 100, IR peak at 1,700, and a density greater than 2. As with the other types of sources discussed, chemists must know that these tools exist, what conventions or units are used, and what their limitations are.

Structure and Reaction Databases

One of the biggest payoffs of the increased power of computers for chemists has been the creation of structure-based databases. These allow chemists to use the natural language of chemistry—structures—to ask questions. These tools contain structure and reaction information that has been extracted from primary sources and formulated to facilitate retrieval.

For the most part these sources are electronic. Chemists have long tried to capture information about structures and reactions. The *Beilstein Handbook of Organic Chemistry* is organized by chemical structure, and sources such as Theilheimer's *Synthetic Methods of Organic Chemistry* and ISI's *Current Chemical Reactions* collect and organize reaction information. Electronic databases have increased the utility of this information by allowing chemists to search not only for exact structures (has this compound ever been reported) but to find information on substructures. The same power extends to reaction information. Instead of being limited to searching for a specific reaction, chemists can look for classes of reactions or reaction conditions.

Electronic Status/Trends—Structure Databases.   Several online systems offer structure searching. On STN, structures from *Chemical Abstracts, Beilstein,* and *Gmelin* are searchable. On DIALOG, structures in the *Beilstein Online* file are searchable. Ring data and chemical name segments are searchable in CASearch, DIALOG's *Chemical Abstracts* file. Other online systems that offer structure searching include Questel/Orbit and Chemical Information Systems. Graphical front-end programs have been developed to assist with online structure searching.[18] These translate the structure drawn into the commands necessary for searching it. Examples include MOLTERM (available from Beilstein Information Systems), STN Express (available from STN International), and ISIS/Draw (from MDL Information Systems). As was mentioned above, the advent of new products, such as SciFinder and CrossFire, that include graphics will lessen the need for separate structure drawing packages. Several CD-ROM products offer structure searching, including Chapman & Hall's *Dictionary of Natural Products* and Beilstein's *Current Facts.*

Electronic Status/Trends—Reaction Databases.   These sources add reaction information to the structures. Online databases include STN's CASREACTS (reactions selected from *Chemical Abstracts*) CHEMINFORMRX (reactions selected from *Chemischer Informationdienst*) and CHEMREACT (reaction types selected from the VINITI/ZIC structure database) as well as BEILSTEIN. SciFinder will include reaction searching in the near future. Reaction databases are also available for leasing. Examples include CrossFireplusReactions (see Data/Information Compilations above), InfoChem GmbH's ChemReact (see CHEMREACT above), and ISI's *Reaction Citation Index.* The ISI product includes reaction information (structures and bibliographic data) plus information on who has cited the article describing the reaction.[19] Both of these databases run on MDL Information Systems software (see below).

MDL Information Systems Products.   MDL offers structure and reaction searching products that have become widely accepted in chemical and pharmaceutical companies. These products include structure searching and database management software, which allow users to store in-house structure and reaction information. MDL software platforms such as ISIS allow users to search both in-house and open-literature databases. MACCS and REACCS are MDL's mainframe structure and reaction database management systems. Information on these products is available from the MDL home page (see Table 2). Databases available in MDL-compatible formats include Current Synthetic Methods, *Journal of Synthetic Methods,* Theilheimer, as well as the ChemReact and *Reaction Citation Index* files mentioned above.

Implications for Chemists.   Thinking of structural and reaction parameters is natural for chemists. However, users of structure or reaction databases must still know the inclusion and editorial criteria to search them effectively. It is vital that students be exposed to these specifically chemical tools during their education. If fledgling chemists know how to use these resources, they will have a running start on the job market.

Students may get some experience using electronic literature-based sources from the databases provided through the library. However, they can only become familiar with these valuable chemistry resources through instruction within their chemistry curriculum, within a course, in a guided research project, or via a seminar or workshop.

The Transformation of Information—Implications for Educators

The electronic revolution, therefore, has important implications for chemistry educators. Chemists today need information skills along with their technical skills if they are to function in the highly competitive chemical marketplace. What skills are needed and how are tomorrow's chemists to get them? The next sections will discuss these questions.

Information Skills Needed By All Chemists

The electronic revolution that is transforming chemical information by creating new and everchanging information sources affects all chemists—in industry, government, and academe. At the same time, chemists face a changing environment: decreased funding; increased productivity requirements, along with reductions in staff; decreased product development cycle time; lack of job security; and increased emphasis on applied research.[20] An important way to increase productivity is to acquire relevant information in a timely, cost-effective way.

In industry, reductions in personnel have eliminated chemical information specialists in many companies. Increasingly, chemists in all companies and at all levels of responsibility must become proficient information seekers.[21] Managers are no longer willing to train new hires to do such tasks; they expect new employees will have these skills. Chemists today can expect to change assignments, jobs, and even careers several times. To succeed under these conditions, chemists must be flexible and effective users of information.

In academe, chemists face increased teaching responsibilities and significant reductions in grant and contract funding to support research and teaching. Information searching skills are needed to function effectively as teachers and researchers. Faculty must develop and maintain these skills personally to pass on information skills to their students.

What Skills Should Be Taught?

If future chemists are to acquire these skills, it is important that chemistry students at all levels get a basic grounding in using information sources, especially because industry expects all chemists (entry level as well as PhD) to know how to use information effectively on arrival. To aid in this task, the authors developed the following model for a chemical information curriculum.[22]

Undergraduate Students. Every undergraduate chemistry major should know that an extensive chemical literature exists, e.g., that there are specific scientific and chemical dictionaries, encyclopedias, indexes, and data compilations. In addition, students should learn the structure of chemical information, basic searching skills, some techniques for using electronic sources, and evaluation skills to select the optimum sources.

A. Structure of chemical information:
  1. The primary mode of published information transfer in chemistry is the journal article or patent.
  2. A range of sources exists. Students should know when and how to use each source.
     a. journal articles (the types of articles—"letters", full article, review; sections of a typical full article)
     b. patents
     c. books (entire books exist on topics simply mentioned in their texts)
     d. handbooks (i.e., *CRC Handbook of Physics & Chemistry, Merck Index, Dictionary of Organic Compounds*)
     e. abstracts/indexes: bibliographic (e.g., *Chemical Abstracts, Physics Abstracts, Biological Abstracts, Beilstein, Gmelin, Current Contents, Science Citation Index*); numeric and other types specific to chemistry (e.g., properties, reaction, structure, sequences)
     f. citation indexes
     g. review publications (journals, book series)
  3. Electronic versions of print sources often exist, and some sources are available only electronically (online, CD-ROM, Internet, or via online catalogs on library servers in academe).

For the most important sources, students should learn subject coverage, indexing policies, sources covered, information provided, time period covered, style policies (abbreviations, etc.), as well as questions that can and cannot be answered by a specific source. This knowledge is essential even when easy-to-use electronic search options are available. Students should learn to search these important sources in an efficient way and learn how to evaluate sources to select the best options to answer specific questions—for example, whether to use *Beilstein* or *Chemical Abstracts* to obtain physical properties information for an organic compound.

B. Basic search skills that all undergraduates should learn:
  1. Locate background material, such as review articles and encyclopedia articles.
  2. Compile a list of publications by an author.
  3. Know the value of, and be able to use, a citation index.
  4. Find information on:
     a. subjects (in *Chemical Abstracts* and other indexes, e.g., *General Science Index, Biological Abstracts, Physics Abstracts, Medline*)

        b. properties (spectra; chemical, physical, and toxicological/safety, etc.)
        c. compound preparation
    5. Know that reaction databases exist.
    6. Appreciate the power of structure searches.
    7. Understand the importance of patents, their organization, and how to locate chemical patents in the literature.
  C. Techniques for using electronic sources:
    1. Know electronic searching techniques, such as using Boolean search logic (AND, OR, NOT); choosing relevant search terms (including synonyms, abbreviations, codes); mapping search terms between databases.
    2. Know special searching techniques unique to retrieving chemical information, e.g., name segments, molecular formula, and structure and reaction queries.

Graduate Students and Advanced Undergraduates.   Students at these levels should learn to review the literature before starting any project and should develop good information habits, such as monitoring the current literature and employing cost-effective search techniques. Students should take advantage of sources on the Internet: preprints, sequence databases, listservs, etc.

In addition to the basic skills acquired during undergraduate years, these students should acquire additional skills so they can:

1. complete a comprehensive subject search using a variety of sources—e.g., physical chemists should be able to use *INSPEC* (*Physics Abstracts, Computer and Control Abstracts, Electrical and Electronic Abstracts*) as well as *Chemical Abstracts.*
2. complete a comprehensive search for information about a compound.
3. conduct reaction searches.
4. conduct structure searches.
5. monitor current literature to stay up-to-date on a topic.
6. evaluate sources to select those most suitable to answer a specific question.

Chemical information resources are increasingly available in electronic form—online, CD-ROM, Internet, and, in academe, often available along with the library's online catalog. Newer formats, such as hypertext, chemistry-oriented graphics, and natural language searching (examples mentioned earlier in article), offer great opportunities to educators. While these technologies may require substantial funding, searching is easier, more fun, and more likely to be used by students. Educators can then spend less time teaching which buttons to push and spend more time teaching content of sources and how to evaluate sources best suited to answer a question. The importance of understanding content is worth stressing. Early users of SciFinder easily learned to search *Chemical Abstracts,* but soon had questions about its content, organization, and indexing policies.[14] This is basic information that should be learned during student days. The World Wide Web offers mesmerizing searching, which can consume large amounts of time. However, once the initial excitement is over, searchers tire of rummaging through screens of chaff looking for the elusive kernels of wheat. Here, also, techniques for evaluating the quality of information and identifying useful sources are crucial.

Teaching Options.   Chemical information skills can be taught in a variety of ways: in a separate information course, by integrating information assignments into courses, in a workshop or short course, by informal communication from faculty/colleagues/librarians, or a combination of the above. The ACS Committee on Professional Training recognizes that "Students preparing for professional work in chemistry must learn how to retrieve specific information from the enormous and rapidly expanding chemical literature. The complexity of this task is such that one can no longer easily acquire the necessary skills without some formal instruction."[23] A 1993 survey of all chemistry departments found that information skills were taught in the following ways: 162 schools taught a separate course, 166 integrated information into two or more courses, and 40 offered workshops or short courses. Instruction in information skills was most often integrated into organic courses (for both undergraduates and graduates), followed by physical, inorganic, analytic, biochemistry, and polymer courses. In addition, information instruction was often an integral part of seminars.[24]

A separate course has the advantage of being able to introduce the full range of skills listed above. On the other hand, integration into courses has the pedagogical advantage of helping students recognize that efficient information gathering can contribute to success in their course (with the potential extention to their success on the job). Integration also enables information instruction to be incorporated into the curriculum as early as the freshman year and sequentially throughout their four years. However, coordination of effort among instructors is essential to avoid gaps in students' knowledge. Workshops, short courses, and information communication can achieve a good deal, but it can be difficult to cover all of the desired topics. Optimum reinforcement is achieved when a combination of methods is used throughout students' education.

For faculty members and other instructors the process of staying abreast of changing sources described in this chapter can be daunting. However, aids for instructors are available and are listed later in the section How to Cope with the Information Explosion. Of special interest is the Clearinghouse for Chemical Information Instruction Materials, which includes syllabi, lecture materials, practice questions, tests, etc.

Once chemists leave their student days, they continue to use information sources. The next section reviews ways in which electronic information can help chemists answer typical questions.

How Electronic Information Can Help Chemists

Developing Courses and Assignments.   Electronic sources offer an increasing array of locations for new ideas that supplement traditional methods of staying current with scientific—especially chemical—education trends, updating courses, developing new ones, and planning new curricula. Entire courses and textbooks are offered via the Internet. Lectures and lab experiments, including audio and motion graphics, are appearing increasingly on the Internet, as well as newly available chemical education software (see Varveri's Chapter 6 for information on Internet resources). The chemical education listserv is a lively, active discussion of all issues across the entire spectrum of chemistry education (e.g., lab experiments, software, textbook evaluation, organic chemistry course, freshman course, foreign languages) (Send to LISTPROC@ATLANTIS.CC.UWF.edu and state SUBSCRIBE CHEMED-L). The *Journal of Chemical Education's* Web site, JCE Online, offers a variety of information that supports instruction: table-of-contents of the journal, abstracts of its articles, selected full-text articles, forum for curricular changes, news about what's happening in chemical education, and evaluated chemistry web sites. JCE software listings and descriptions span the entire spectrum needed by chemistry instructors and can be found at the same Web address as the *Journal of Chemical Education*. The *Chemical Educator*, begun with the March 1996 issue, is published only via the Web. It aims to provide timely information needed by chemistry instructors: classroom activities and course work, laboratories and demonstrations, use of computers in chemistry, news, a forum for educators, and Internet resources. An upcoming feature will send information to subscribers' e-mail addresses based on their specified interests. Educational software is also found at Web sites of computer companies, which can be located via major Web sites.[25] The ERIC database, available via Internet, online (e.g., Knight-Ridder Information's DIALOG Service and STN International), First Search, and CD-ROM, is an excellent source for articles on trends in scientific education and for specific chemical curricula.[26]

Beginning a Project

Initial steps in beginning a project include locating review articles, identifying key researchers and reading relevant articles, and using citation searching to identify how past research has been updated.

Review articles published in journals and annual review publications can be identified readily in *Chemical Abstracts* (by searching for the word "Review" as a document type), in *Science Citation Index* (by limiting to "Review" as a type of publication), and by checking the Annual Review Web site. This last source offers keyword searching of titles and keywords for articles published since 1984 by the publisher, Annual Reviews, Inc. At this time, the Annual Review index is somewhat limited, but is expected to expand its coverage. More Internet resources of this type will appear in the future. Chemistry-related citation searching can be accomplished via *Science Citation Index* and *Reaction Citation Index*, both published by the Institute of Scientific Information and mentioned earlier in article, as well as by patents citation indexes (Claims files from IFI/Plenum for U.S. patents; World Patent files from Derwent).

Keeping Up-To-Date on a Topic

Chemists need to stay abreast of new trends in chemistry overall, in the broad area of chemistry of interest to them, and, in depth, in topics of immediate interest.[2] The array of current awareness sources continues to increase with new electronic options appearing continually. A good way to identify useful titles is to review the sources listed below and determine which best meet your subject needs and which technical resources are available to you. The optimum sources that answer a question will change over time as new ones appear or current ones become available in yet another electronic form or with other cost configurations.

Some sources cover many more subjects than chemistry but are also useful to chemists. UnCover from CARL Corp. (Tables 1 and 2) provides tables of contents to over 20,000 journals in all subjects via the Internet and can be searched by author, title keywords, and journal titles as soon as journal issues are published. UnCover also offers the option of receiving tables of contents automatically through its REVEAL service.[27] *Current Contents*, mentioned earlier in the article, is updated weekly and provides access to author names, journal title words, added keywords, and journal titles. While the overall database covers the most frequently used science and technical journals, a chemistry subset of approximately 850 journals can be acquired separately. A companion publication, *Science Citation Index,* which covers approximately 3,300 of the most frequently used science and technical journals, is also updated weekly online and via Web.

CAplus, an online database produced by Chemical Abstracts Service and available on STN International, provides access to authors, title keywords, and journal titles for 1,350 core journals within one week of receipt, in addition to all information found in *Chemical Abstracts.* CAplus also offers a table-of-contents service for these core titles. The 1,350 journals are also available via SciFinder's current awareness feature (described earlier in Abstracts/Indexes section). Users can display tables-of-contents, which can be tagged to identify which titles are available in one's organization library. Knight-Ridder Information Service (which offers the Dialog search services) developed KR ScienceBase, described earlier, which also includes a current awareness feature via the Web.

Automatic searches (called Selective Dissemination of Information, or SDI) of online databases as they are updated have long been an option in many databases on commercial (i.e., Knight-Ridder's Dialog Information Services, STN International, Institute of Scientific Information) and in-house computer systems. A Chemical Abstracts Service's version, titled *CA Selects* (CAS), currently provides weekly updates of the bibliographic information and abstracts for approximately 20 subjects. New electronic delivery options will be available soon. Watch for new ways to stay current from CAS via its Web site. The Institute of Scientific Information's "Research Alert", changed to "Research Alert Direct" in 1996, sends the results of SDI searches to one's e-mail address. This new electronic version will offer information by journal title (i.e., to see tables-of-contents), by topic, and as customized by the requestor.

Index Chemicus (available weekly in print, CD-ROM, and via MDL, Inc.) and *Current Chemical Reactions* (published monthly in print and available via MDL, Inc.) offer excellent ways to review developments in organic chemistry. These sources, both published by the Institute for Scientific Information, have the added advantage of providing structures and reaction diagrams in each entry.

Tables of contents (TOCs) of journal articles are becoming increasingly available via the Internet. In addition to more comprehensive products like SciFinder and KR ScienceBase, many TOCs are available via professional societies' or commercial publishers' Web sites. For example, the American Chemical Society, Springer-Verlag, Elsevier, Institute of Physics, American Institute of Physics, and American Physical Society all publish TOCs on the Internet. Some Web sites, such as EI Village for engineering, have gathered these TOC addresses together in one general category, so users can find them more easily. These sites are likely to change over time, so are best identified by checking the best overall Web home pages[25] or by monitoring the chemical information listserv. Preprints of papers accepted by a few journals can be found on the Internet now (such as the *Journal of Chemical Physics Express*) and more are likely to appear in the future.

Some resources that cover disciplines which overlap with chemistry also offer current awareness services via their Web sites. Some are provided by publishers (such as *Biological Abstracts*) or professional societies (such as the American Institute of Physics, American Physical Society, Optical Society of America, Materials Research Society, and Institute of Physics). Other resources are gathered together by subjects (such as physics journals from McGill University, bioscience journals and newsletters from Harvard Biological Laboratories, and the molecular biology subset of *Medline* called *Entrez*).

Sometimes the first public indication of research is presented at conferences. Several successful conferences, such as the Electronic Computational Chemistry, have been conducted via the Internet and more will be scheduled. Announcements of upcoming electronic conferences appear on many subject-related listservs, including the chemical information listserv, and on the major chemistry Web home pages.[25]

Finding Specific Information

The first step is to evaluate your question and decide what sources best answer the question. Then, determine the methods of access available to you: paper, online, CD-ROM, Internet, or via your library's online catalog. This requires staying current with options available within your organization and from the library, because the options change much more often than in the past. Your librarian or information specialist can offer advice about local options and, in general, provide guidance in locating needed information.

The best sources can be determined in several ways: review guides or other publications listed in this article, such as the series of three articles on information sources for organic chemistry;[28-30] search the specially created indexes to databases available from Knight-Ridder's Dialog Information Service (DIALINDEX) and STN International (STNindex); or check catalogs from Dialog and STN International that list databases by category (i.e., chemical substances, patents). In addition, several organizations/resources (such as Cambridge Life Sciences, Inc., and *Engineering Index*) have begun to organize access to chemistry-related sites on the Internet, which may facilitate faster access to relevant sources. New efforts similar to these can be identified by monitoring subject listservs and the major chemistry Web sites.[25]

Managing Your Own Information

Information obtained from searches can be maintained in personal computer files using database management software. The various software packages offer many advantages. Users can transfer references from a word processing document or from an electronic search directly to a personal file, can update the database easily, can add index terms and chemical structures, can search the database, and can reformat references into the format required by common journals. Software packages commonly used by chemists include EndNote (along with EndLink), ReferenceManager, ProCite, and Papyrus. Several recent articles compare these software packages.[31-32]

Obtaining Copies

Acquiring articles, patents, and books is a service provided by many organizations' libraries. Libraries have increasing options for providing copies of documents to chemists—traditional interlibrary loan between libraries, professional and commercial organizations (such as Chemical Abstracts Services' Document Detective Service, Institute for Scientific Information's Genuine Article Service, UnCover), specialized products (such as SciFinder), full-text online (for example, from STN International and Knight-Ridder Information's DIALOG Service), and increasingly via the Web. Individuals can also order articles from CAS, ISI, UnCover, or SciFinder, or via online databases or Web sources.

How to Cope with the Information Explosion

A number of aids are available:

1. The major chemical information textbooks.[1-4]
2. Major articles of interest, including:
   a. Carol Carr, "Teaching and using chemical information; an updated bibliography"; *J. Chem. Ed.* 1993, *70*, 719–726. This is a comprehensive list of articles, publications, and organizations that cover all aspects of teaching chemical information. It will be updated soon on the Web.
   b. "The Chemical Information Instructor"; quarterly column in *Journal of Chemical Education,* July 1991+. This column offers practical information to instructors of chemical information. Editor Arleen N. Somerville, University of Rochester, Carlson Library, Rochester NY 14627-0236 (ansv@dbv.cc.rochester.edu), welcomes suggestions for topics for articles and potential manuscripts.
3. Education Committee, ACS Division of Chemical Information. Members offer teaching modules, hold workshops about teaching chemical information, and sponsor programs at ACS national and regional meetings. In addition, members can be contacted as consultants and as outside speakers. Education Committee activities can be monitored via the division's Web site.
4. ACS/SLA Clearinghouse for Chemical Information Instruction Materials. The clearinghouse is a project initiated by the chemical information divisions of the American Chemical Society and the Special Libraries Association. Its purpose is to collect and distribute items for use in instructing students and scientists in the use of chemical information sources. In 1996 it included approximately 220 items. Materials include course syllabi, practice questions, lecture materials, tests, teaching tips, etc. Cost to requestor is $0.10 per page. A list of available items can be acquired via:
   a. WWW at URL: http://www.indiana.edu/~cheminfo/cciimnro.html
   b. Contacting Dr. Gary Wiggins, Chemistry Library, Chemistry Bldg. 003, Indiana University, Bloomington, IN 47405; Phone: 812/855-9452; FAX: 812/855-6611; Internet: wiggins@ucs.indiana.edu
5. Other Internet sources for chemical information:
   a. Courses: In 1997 two major courses are offered via the Internet:
      i. CICOURSE offered by Gary Wiggins, Indiana University, Chemistry; available at http://www.indiana.edu/~cheminfo/400home.html
      ii. CIStudio—a Chemical Information Course offered by Professor Joseph Warden and Instruction Librarian Colette Holmes, Rensselear Polytechnic Institute; http://www.rpi.edu/dept/chem/cheminfo/cistudio/index.html
   b. The "Chemical Information Sources (CIS-IU)" Web site, initiated by Gary Wiggins at Indiana University, describes how to use major sources.
   c. Publishers and computer companies have begun to place instructional materials on their Web pages. For example, Chemical Abstracts Service has introduced three training manuals on its Web site: (1) searching CAS databases on STN, (2) searching the CAS Registry File on STN, and (3) structure searching using the CAS Registry File on STN. More organizations will follow.

Finding Out About Databases and How to Use Them

Librarians in your academic institution or company are excellent resources, because an important part of their responsibilities includes staying current with new sources, new ways of accessing the information, changes in cost, and advantages/disadvantages of new technologies. Other information specialists in the community and information consultants are also valuable resources for chemists.

Database producers (Chemical Abstracts Service, Institute of Scientific Information, Chapman & Hall, Beilstein Information Systems, Gmelin, CRC Press, Derwent Inc., IFI/Plenum Data Corp., etc.) and vendors of chemistry databases (Knight-Ridder Information's DIALOG Service, STN International, Molecular Design Ltd., etc.) offer printed guides and workshops on how to search their databases, and offer help via toll-free help desks for phone questions and via Web sites. Some provide print or electronic materials for instructors (often providing full-text downloading of these materials), and some have booths at ACS and other conferences where staff answer questions. Each organization has a Web site and an 1-800 phone number, often in more than one country (see Tables 1 and 2). The Web sites offer ways to stay current on new features and new databases and to download instructional material. In addition, many vendors and database producers have account representatives, who are excellent resources for updated information and as lecturers. Monitoring a subject listserv (such as the Chemical Information listserv or Chemical Education listserv) or occasional checking of the major Web sites for chemistry resources, will enable you to stay current on new electronic publications and changes in databases.

Summary

This chapter summarized how computers have changed the way chemists obtain information, suggested ways to stay current on changes in sources, offered advice on how electronic forms of information can help chemists acquire information needed on the job, suggested ways to stay current on new sources, and provided information useful for teaching students about these sources. Information skills needed by current and future chemists were defined, including how to use the sources efficiently and selecting the optimum sources to answer specific questions. While this everchanging array of sources can be daunting, the increased ease of the more advanced search features will encourage frequent use of the sources.

References

1. Wiggins, G.; *Chemical Information Sources;* McGraw-Hill, New York, 1991.
2. Wolman, Y.; *Chemical Information: A Practical Guide to Utilization,* 2nd ed.; Wiley, New York, 1988.
3. Maizell, R. E.; *How to Find Chemical Information: A Guide for Practicing Chemists, Educators and Students,* 2nd ed.; Wiley, New York, 1987 (new edition in preparation).
4. Mellon, M. G.; *Chemical Publications: Their Nature and Use,* 5th ed.; McGraw-Hill, New York, 1982.
5. Neufeld, M. L.; Cornog, M.; "Database History: From Dinosaurs to Compact Discs"; *J. Amer. Soc. Infor. Sci.* 1986, *37,* 183–190.
6. Tenopir, C.; Hover, K.; "When is the Same Database Not the Same?: Database Differences Among Systems"; *Library Journal,* July 1993, *17,* 20–27.

7. Hawkins, D. T.; "Applications of Artificial Intelligence (AI) and Expert Systems for Online Searching"; *Online,* 1988, *12,* 31–42.
8. Koenig, M. E. D.; "The Convergence of Moore's/Mooers' Laws"; *Infor. Process. & Manage.* 1987, *23,* 583–592.
9. Borman, S.; "Electronic Publishing Increasingly Offered as Alternative to Print Medium"; *C & E News,* March 27, 1995, *73,* 42–49.
10. Taube, G.;"Science Journals Go Wired"; *Science* February 9, 1996, *271,* 764–766.
11. Knowledge is Power, Derwent Direct Catalog; 1995.
12. Thompson, N. J.; "Intellectual Property Materials Online/CD-Rom: What and Where"; *Database* 1992, *15,* 14–34.
13. Lambert, N.; "The Idiot's Guide to Patent Resources on the Internet"; *Searcher,* May 1995, *3,* 34–39.
14. Williams, J.; "SciFinder: Information at the Desktop for Scientists"; *Online,* July 1995, *19,* 60–66.
15. Hartwell, I.; Palma, M. A.; "Knight-Ridder Information's ScienceBase: A Preview"; *Infor. Today,* April 1995, *12,* 22–23.
16. Warr, W.; "More Chemistry on CD-ROM"; *Database,* Feb/Mar 1995, *18,* 60–64.
17. Warr, W.; "New Chemical Databases on CD-ROM"; *Database,* Feb 1993, *16,* 59–67.
18. Warr, W.; Wilkins, M. P. "Front End Software for Chemical Structure Searching: A State-of-the-Art Review"; *Online,* January 1992, *16,* 48–55.
19. "News from ISI: ISI Releases Reaction Citation Index Database"; *Information Today,* July 1995, *12,* 15.
20. Roth, W. P.; Shearer, C. J.; Taylor, G. L.; "The Times They Are A-Changin' for R&D"; *ChemTech,* June 1995, *25,* 6–12.
21. *Current Trends in Chemical Technology, Business, and Employment;* American Chemical Society: Washington, DC, 1994.
22. Carr, C.; Somerville, A. N.; "Chemical Information Instruction in Academe"; National Chemical Information Symposium, University of Vermont, June 1994; augmented by discussions 8/95.
23. Undergraduate Professional Education in Chemistry: Guidelines and Evaluation Procedures; Committee on Professional Training, American Chemical Society: Washington, DC, 1992, 12. (http://www.acs.org/cpt.guide.htm)
24. ACS Division of Chemical Information, Education Committee, conducted a survey of all chemistry departments in the United States that granted chemistry degrees, per the ACS Committee on Professional Training's 1992 Annual Report; 390 responses were received; report to be published.
25. A few major WWW URL addresses are: http://www.rpi.edu/dept/chem/chem-info/chemres.html (most comprehensive list of chemistry Internet resources, developed by G. Wiggins, Indiana University, and maintained by J. Warden, Rensselaer Polytechnic Institute); http://www.chem.ucla.edu/chempointers.html (WWW Virtual Library: Chemistry); http://www.yahoo.com/Science/chemistry (Yahoo).
26. Database of educational materials collected by the Educational Resources Information Center of the U.S. Department of Education; consists of two subfiles: Resources in Education, and Current Index to Journals in Education.
27. From CARL, Inc. Includes tables-of-contents from over 20,000 journals since 1988, approximately 50% of which are science journals.
28. Somerville, A. N.; "Information Sources for Organic Chemistry, 1: Name Reaction; Type of Reaction"; *J. Chem. Ed.* 1991, *68,* 553–561.
29. Somerville, A. N.; "Information Sources for Organic Chemistry, 2: Functional Groups"; *J. Chem. Ed.* 1991, *68,* 842–853.
30. Somerville, A. N.; "Information Sources for Organic Chemistry, 3: Reagents, Solvents"; *J. Chem. Ed.* 1992, *69,* 379–386.
31. Zaroukian, M. H.; "Managing Bibliographies"; *Med. Software Rev.* 1993, *2,* 1–3.
32. Stigleman, S.; "Bibliography Formating Software: An Updated Guide for 1994";*Database* 1994, 17, 53–65.

Part III
Fundamental Computer Skills in Modern Chemical Practice

8

Beyond the Basics: What Chemistry Students Need to Know About Computing

Peter C. Jurs

Chemical education is meant to prepare students for their future as chemists by providing them with the knowledge and skills they will need to function as professionals, and, in addition, the ability to think critically and apply their knowledge. Whatever the surroundings in which they will function as chemists, an important part of their duties will probably involve computers, computer software, and computational methods. It is essential, then, for chemical educators to provide the necessary information to our students so that they will be prepared for the chemical workplace they will enter.

It is by now a truism that chemistry students need to be computer literate, just like all college graduates, especially those in technical fields. Computer literacy is the third step in the logical progression starting with being literate, then moving to being numerate, and culminating with being computer literate. To be functionally computer literate, chemistry students need to know about word processors, spreadsheets, graphics packages, e-mail and newsgroups, personal computers, and many other topics relating to both software and hardware. Most of these topics are taught outside of chemistry classes, as they should be. Many of these topics are learned by students entirely outside of formal educational settings. They are general skills and are broadly applicable. Most well-prepared students begin their college or university education already being computer literate. Given this starting point, the relevant question for chemical educators then becomes: What more do chemistry students need in order to be prepared for graduate school or for a professional scientific career?

Table 1 presents a list of some of the additional computer-related skills needed by chemistry students. This is a long and daunting list, even though it is incomplete, and space does not permit provision of in-depth coverage of all these topics here. No single college course can hope to cover all these topics in depth in one semester. It is probably not feasible to include all these topics in a chemistry curriculum. Thus,

TABLE I Additional computer skills needed by chemistry students.

Topics	Numerical	Nonnumerical
Numerical topics	x	
Useful software packages	x	x
Workstations & UNIX		
Statistics	x	
Molecular modeling		x
Internet		x
Scaling phenomena	x	
Chemical databases		x
Simulations	x	x
Computational chemistry	x	x
Media		x
Laboratory interfacing	x	x
Molecular design		x
Artificial intelligence		x
Optimization	x	x
Programming	x	x
Chemical structure information handling		x
Graph theory		x
Molecular structure model building		x
Information retrieval		x
Computational neural networks		x

in practice it is necessary to provide a survey of what is important and provide some in-depth coverage of a few selected topics.

In the chemistry curriculum at Penn State University, many of the important computer techniques are learned as components of regular chemistry course material. For example, graphical visualization of three-dimensional molecular models is introduced in organic chemistry lectures, molecular modeling is introduced as an integral part of organic chemistry laboratories, simple statistics are used in general chemistry laboratories, and more advanced statistical techniques are introduced in physical chemistry laboratories for the analysis of data using spreadsheets or other packaged statistical software such as Minitab.[1] These computer uses focus on the students' use of software that is provided for them, and they do not include the generation of software, the analysis of algorithms, etc. Thus, while these computational topics are important components of an up-to-date chemistry education, there are additional topics that we feel are also important, especially for students who want to pursue such topics. To amplify on these basic topics and to introduce a number of additional ones, we have developed and taught a specialized course for advanced undergraduate students called Computer Applications in Chemistry.

This course is designed to be taken by advanced undergraduate and beginning graduate students majoring in chemistry and closely related disciplines. It is a one-semester, three-credit elective taken by approximately 25 students each spring semester. This is perhaps half of our chemistry majors. Students graduate from our department with a wide range of computer expertise picked up in a variety of ways, but our course is meant to provide a grounding in computational chemistry to those who seek it as an adjunct to the common core of chemistry courses.

The course does not teach programming because it is assumed that the students have already learned how to write code elsewhere. A programming course is not required of all chemistry majors, but many elect to take it. Most of the students entering our course have learned programming in a prior university course, but some have learned it as part of a job, while doing research, or on their own. The language used is FORTRAN because of its suitability for numerical computation and the enormous backlog of scientific software that is available in FORTRAN. The subject matter consists of a survey of computer software for scientific purposes. Many numerical topics are introduced, and a number of nonnumerical topics are also covered. The students are assigned programming homework problems that are done on the Penn State University Computation Center IBM 3090 mainframe or on Sun SparcStations running Unix in the physical sciences computing laboratory, which is run by the Eberly College of Science. The students are assigned readings in books put on reserve in the chemistry library, and photocopies of journal articles are made available for more focused reading assignments. Assessment is done through grading of the homework assignments, two one-hour written examinations, and a final examination.

The course begins with an introduction to numerical computation. The students have all been through the standard sequence of calculus courses that are required in chemistry and related curricula. However, very few of them have been exposed to the discrete, digital, iterative computation that forms the heart of computer algorithms. Some basics about what sets computers apart from other scientific tools, how the floating-point number system is implemented on computers (and the limitations that are therefore imposed), and some concepts on iteration and series are introduced. Since the numerical values of functions, such as trigonometric functions, must be computed via series or other approximations, this material can be used to introduce many related topics about series, the effects of truncation, comparisons of different types of series, etc. With this introductory material completed, the course is ready to move to a discussion of individual numerical methods.

The first item in the list in Table 1 is numerical methods, and the first half to two-thirds of our course is devoted to introductions to a number of numerical topics. This section of the course consists of a survey of important numerical techniques that the students may need later as chemists. The following topics are included: a description of the floating-point number system and its attributes, statistical analysis, root finding including the Newton-Raphson method, digital generation of functions, root finding including the Newton method, curve fitting including linear and nonlinear least-squares and iterative methods, numerical integration including Simpson's method and Gaussian quadrature, matrix methods and linear equations, Monte Carlo methods including random number generation and simulations, numerical solution of differential equations including the Runge-Kutta fourth-order method and systems of chemical equations, and the Fourier transform. Each of these topics has been included because they are all in areas that should be of utility to students. For example, curve fitting is a numerical technique that is used with experimental data continually, and having a sense of what lies behind curve fitting will allow the students to read the chemistry literature with a better understanding of how the results presented were obtained and, therefore, how much they should question what they read. Matrix methods and linear equations provide an introduction to questions that surround the solution of sets of equations, the errors that might be encountered in such analyses, etc. A discussion of random numbers for Monte Carlo calculations permits the introduction of many concepts and questions that the students have not encountered in science classes previously, such as: What constitutes randomness? How can one generate random numbers in a machine? How can random numbers be used to advantage? The material dealing with solutions of differential equations allows an introduction to numerical methods for the solution of linked systems of chemical equations, and a homework assignment in this area has the students dealing with a small but realistic chemical system of three linked reactions. The Fourier transform is at the heart of frequency-domain instrumentation (FTIR and FTNMR) which is ubiquitous in chemistry. A final numerical section deals with computational mathematics using Mathematica.[2]

Homework assignments are included with most of these numerical topics. The assignments focus on understanding and comparing algorithms, analyzing the desirable and undesirable attributes of algorithms, and analyzing sets of data. Software is usually provided to the students as a starting point for each assignment, rather than having them write software from scratch. Woven through the course is an emphasis on understanding the fundamentals of each technique. The properties of different iterative schemes are compared for efficiency, simplicity, speed of convergence, and other relevant properties. The software for homework problems is provided as operating FORTRAN code, which the students then modify or compare to alternative methods for the same problem. Mathematica is used as an introduction to the Unix operating system for symbolic algebraic manipulations and for the graphical output of functional forms that may be difficult to visualize, for example, plots of atomic orbitals. Once the students have become acquainted with Mathematica's capabilities, they are eager to try out many additional features and experiment with the software. Some students immediately find that Mathematica can be used as part of their undergraduate research projects or assignments in other courses, and they are quite creative in applying its capabilities to their own projects.

The second item in Table 1 is useful software packages. There are numerous software packages that have capabilities useful for scientific computing. All of the software from the preceding computer-literacy discussion is in this category, as is Mathematica. Chemistry students should be acquainted with word processors and general graphics packages, for example. In addition to the very general programs, there are software packages more narrowly focused on the needs of science students. An outstanding example is MATLAB,[3] which provides a language in which mathematical manipulations of data can be done without the need to learn as many conventions and as much background as with a general-purpose language. Entire chemistry courses have been designed around the capabilities of MATLAB.[4] Certainly, when the tasks to be accomplished are in keeping with the capabilities of such software as MATLAB, then it is efficient and practical to use these packages. A paper has been published that provides information about a very large number of commercial software packages that are suitable for chemistry curricula.[5] A recent paper discusses the implementation of a computer chemistry course based on commercially available software.[6]

Another item listed in Table 1 is molecular modeling. As mentioned previously, our curriculum introduces molecular modeling as part of the standard undergraduate organic chemistry course. However, the emphasis in that setting is on the visualization of molecular structure, comparisons of energies of related structures, the comparison of the steric effects among isomers, and similar concerns. In our computational chemistry course, the students are introduced to the computational effort that is required to develop the models shown in the pictures, the nature of the force field being used for estimation of steric energies, methods for minimization of the strain energy function, and other fundamentals of the computations supporting the development of molecular models. Our course has used the PCMODEL[7] software package running on Sun SparcStations. Many other approaches could be equally satisfactory, for example, Alchemy III, CAChe, Chem3D/Plus, Chem-X, HyperChem, MOBY, or Spartan. A listing of available molecular modeling software has appeared recently.[8] The literature contains some descriptions of the incorporation of molecular modeling into the undergraduate curriculum.[9] The availability of the physical sciences computing laboratory allows us to introduce the students to Unix workstations while they are working with molecular modeling. The graphical capabilities of these modeling packages is an especially appealing feature for the students. Many of the students have never had their hands on molecular modeling software, and they find it interesting and illuminating to have a somewhat sophisticated program at their command to manipulate structures. Once again, many of the students find that PCMODEL can be used in conjunction with their undergraduate research projects or assignments in other courses, and this enhances their interest in molecular modeling.

In dealing with molecular mechanics model building, the subject matter is partly numerical and partly nonnumerical. There are many additional nonnumerical topics that are important and should be included in a course such as ours. The bottom of Table 1 lists some of the nonnumerical topics included in our course.

Chemical structure information handling deals with the representation and manipulation of chemical structures by computer, that is, their representation by connection tables. Discussion of connection tables leads directly into a consideration of chemical structures as graphs. A section on graph theory and its connection with chemistry[10,11] follows, including isomer enumeration, substructure searching, canonical numbering of structures and the Morgan algorithm,[12] registration, and molecular connectivity indices.[13] The use of molecular connectivity indices and related graph theoretical indices for the development of equations for the prediction of properties of chemicals is discussed. Moving from the two-dimensional representation of structures as graphs to three-dimensional representation leads to discussion of molecular mechanics and model building. Other important nonnumerical topics include information retrieval, with CAS Online as the primary example,[14-16] optimization (with the simplex methods as a main example),[17,18] and computational neural networks.[19-21]

In addition to the topics discussed above, there are many other important subjects listed in Table 1. Chemistry resources on the Internet are introduced briefly, with instructions on how to retrieve information from several chemistry Gophers and World Wide Web sites. Specific, detailed instructions on how to link up with such sites are provided to the students. Many of the students in our course are already users of e-mail, and communication between the instructor and students in often done via e-mail. The existence of several international newsgroups that trade information about chemical topics among hundreds of chemists is pointed out. Laboratory interfacing and data capture and manipulation are extremely important topics but are beyond the scope of our course in its present configuration. The importance of simulation as a new partner in science along with the traditional duo of experiment and theory is pointed out, and some simulations are discussed. The existence of the field of computational chemistry is pointed out. Undergraduate students are not generally aware that this subfield of chemistry is the thriving enterprise that it is. The importance of science that exists at the complex interface of chemistry, mathematics and statistics, and computer science is shown by example. As one example, the five-volume series *Reviews in Computational Chemistry*[22] is shown, and the contents of these volumes are discussed. The application of computational chemistry to molecular design (drug design, material design) is covered, with examples of recent work.

A final section of the course challenges each student to design and write a program to solve a simplified traveling salesman problem presented in a chemistry context. Over several class meetings, lectures as well as class discussions reveal how to approach the problem, some pitfalls to avoid, some time-saving tricks to consider, how to report the results, how to graph the results obtained, etc. Each student must design an approach to solve the problem, write a program that implements that approach, run it, and demonstrate that it works properly. This assignment allows the gathering together of many topics that have been discussed earlier in the course such as graph theory, efficiency of algorithms, combinatorial explosions, etc. It also provides a natural entry into a discussion of P problems—those problems that can be solved in polynomial time—and NP problems—those problems that cannot be solved in polynomial time but that can be checked in polynomial time. The students have approximately one month to work on this project during the final part of the semester.

Our course on computer software in chemistry incorporates the idea that learning to program in a high-level, general-purpose language (here, FORTRAN) is a useful skill for chemistry students to learn. This is not to say that software packages should not be used; on the contrary, software packages that accomplish what is desired certainly should be used. However, there are many things that cannot be accomplished with packaged software, and there are also some principles that can be better approached through programming in a general-purpose language such as FORTRAN. As mentioned previously, chemistry students should learn the actual art of writing code elsewhere, not in a chemistry course. Given that the students passing through our course have learned the basics of how to write software for chemistry, then the question becomes: Why should they do so? What benefits accrue? Is this a useful expenditure of student and faculty time?

One of the major advantages of learning to write software is that the activity forces writers to focus their thoughts and think logically. Problems must be broken down into small, manageable pieces that can be tackled individually. The most effective way to understand software completely is to write it, debug it, and then test it to make sure it is working properly. The thorough understanding of software by the user is an effective way of avoiding its misuse or avoiding unwarranted certainty in the results obtained. Programming allows the student to get a sense of the limits of numerical computation—what is feasible and what is not feasible. These ideas are solidified in our course with discussions of intrinsically difficult problems, probabilistic algorithms, combinatorial explosions (the traveling salesman problem), and class P and class NP problems. Finally, if chemists do not learn how to program, then who will be the developers of the future? Who will write the software that stretches the limits of what is currently possible? Research occurs at the interface of what is currently possible and new, nonstandard approaches. Therefore, almost axiomatically, nonstandard problems require nonstandard approaches and new software to do new analyses. The ability to write appropriate software is a crucial capability in such circumstances. Our course assumes that programming is useful for many reasons, and the students do a substantial amount of code tweaking and comparing, and some code writing, as part of this course.

Students seldom acquire all these skills and capabilities in one or two university courses. However, they do learn the basics, and they are exposed to the rigor of software analysis and generation. For those whose interest is sparked, research involvement with computational chemistry is an option. Students who have taken a course such as ours are prepared to delve into this field of research.

References

1. Minitab, Minitab, Inc., 3081 Enterprise Dr., State College, PA.
2. Wolfram, Stephen; *Mathematica. A System for Doing Mathematics by Computer,* 2nd ed., Addison-Wesley Publishing Co., Inc.: Redwood City, CA, 1991.
3. *Student Edition of MATLAB;* The MathWorks, Inc., Prentice-Hall: Englewood Cliffs, NJ, 1992.
4. O'Haver, T. C.; "Teaching and Learning Chemometrics with MatLab"; *Chemometics and Intelligent Laboratory Systems* 1989, *6,* 95–103.
5. O'Haver, T. C.; "Applications of Computers and Computer Software in Teaching Analytical Chemistry"; *Anal. Chem.* 1991, *63,* 521A.
6. Bowater, I. C.; McWilliam, I. G.; Wong, M. G.; *Producing Computer-Literate Chemists; Jour. Chem. Ed.* 1995, *72,* 31–34.
7. PCMODEL, Molecular Modeling Software, Serena Software, Bloomington, IN.
8. Boyd, D. B.; "Compendium of Software for Molecular Modeling"; in K. B. Lipkowitz and D. B. Boyd (Eds.); *Reviews in Computational Chemistry,* Vol. 4; VCH Publishers: New York, 1993.
9. Casanova, J.; "Computer-Based Molecular Modeling in the Curriculum"; *Jour. Chem. Ed.* 1993, *70,* 904–909.
10. Hansen, P. J.; Jurs, P. C.; "Chemical Applications of Graph Theory. I. Fundamentals and Topological Indices"; *Jour. Chem. Ed.* 1988, *65,* 574–580.
11. Hansen, P. J.; Jurs, P. C.; "Chemical Applications of Graph Theory, II. Isomer Enumeration"; *Jour. Chem. Ed.* 1988, *65,* 661–664.
12. Wipke, W. T.; Dyott, T. M.; "Stereochemically Unique Naming Algorithm"; *Jour. Amer. Chem. Soc.* 1974, *96,* 4834–4838.
13. Kier, L. B.; Hall, L. H.; *Molecular Connectivity in Structure-Activity Analysis;* John Wiley and Sons, Inc.; New York, 1986.
14. Krumpolc, M.; Trimakas, D.; Miller, C.; "Searching Chemical Abstracts Online in Undergraduate Chemistry. Part 1. CA File, Bollean, and Proximity Operators"; *Jour. Chem. Ed.* 1987, *64,* 55–59.
15. Krumpolc, M.; Trimakas, D.; Miller, C.; "Searching Chemical Abstracts Online in Undergraduate Chemistry. Part 2. Registry (Structure) File: Molecular Formulas, Names, and Name Fragments"; *Jour. Chem. Ed.* 1989, *66,* 26–29.
16. Somerville, A. N.; "Subject Searching of Chemical Abstracts Online"; *Jour. Chem. Ed.* 1993, *70,* 200–203.
17. Berridge, J. C.; "Simplex Optimization of High-Performance Liquid Chromatographic Separations"; *Jour. Chromat.* 1989, *485,* 3–14.
18. Morgan, E.; Burton, K. W.; Nickless, G.; "Optimization Using the Modified Simplex Method"; *Chemom. Intell. Lab. Syst.* 1990, *7,* 209–222.
19. Zupan, Z.; Gasteiger, J.; *Neural Networks for Chemists;* VCH Publishers: New York, 1993.

20. Burns, J. A.; Whitesides, G. M.; "Feed-Forward Neural Networks in Chemistry: Mathematical Systems for Classification and Pattern Recognition"; *Chemical Reviews* 1993, *93*, 2583–2601.

21. Wythoff, B. J.; "Backpropagation Neural Networks. A Tutorial"; *Chemo. Intell. Lab. Syst.* 1993, *18*, 115–155.

22. Lipkowitz, K. B., Boyd, D. B., Eds.; *Reviews in Computational Chemistry*, 5 volumes; VCH Publishers: New York, 1990–1994.

9

Spreadsheets for Doing and Teaching Chemistry

Daniel E. Atkinson

Introduction

Chemical calculations span a wide range of size and complexity, from simple stoichiometric arithmetic, for which a hand calculator is suitable, to treatments of structural or kinetic systems that may require hours of time on a large computer. Most of them, however, fall in a wide middle range for which personal computers are appropriate, and for many of these a spreadsheet is the approach of choice. The transparency of a spreadsheet model, the ease of generating and modifying it, and the fact that normal chemical and algebraic concepts and formats are used directly in setting it up make such models uniquely valuable in teaching chemistry and other scientific subjects.

Spreadsheets in Teaching Science

Although they were initially developed for use in generating financial predictions, spreadsheets are ideally suited to many kinds of calculations dealing with experimental results and especially to the construction and manipulation of models to be used in teaching science or in research. One unique advantage of spreadsheets in such cases is that the model itself and all of the consequences that result from variations in experimental or conceptual inputs are displayed in transparent detail. A model written in a conventional computer language is essentially a black box. It supplies a means of getting from input to output, but the operations by which that journey is made are specified in an abstruse computer language and have no direct relationship to the chemical model.

A student setting up a spreadsheet model deals with chemistry rather than with the grammar of a computer language, and thinks, at every stage, in chemical terms. In the spreadsheet, each feature of the physical or chemical model is displayed directly. Changes in the concentrations of intermediates during the course of a reaction, for example, or the rate of an enzymic reaction as a function of the concentrations of substrates and of activators and inhibitors of various types, are directly accessible to visual inspection. Any row or column of the model can be plotted as a function of any other, and the effects across the whole model that are caused by changes in relevant parameters can be explored by entering of new values into single cells. The model may suggest relationships and possible correlations that would not otherwise have been considered by the user, and the ease of modifying or extending the model invites the student to pursue such possibilities and gain additional insight into the chemical system.

The preceding sentence may be overly optimistic with respect to most students. A few will be entranced by their ability to modify characteristics of the system that is modeled and see graphs change immediately to reflect the results of the modifications, and they will explore possibilities and combinations of possibilities on their own. Not only will they gain deeper insight into the phenomena that are modeled, but they will be well on the way to developing the habit of generating and manipulating models when they later encounter situations in which models can aid their understanding. However, most students are not so strongly motivated by intellectual curiosity, and instructors will need to urge them along the first stages of becoming comfortable with numerical modeling using spreadsheets and should try to help them develop interest in further exploration on their own.

This chapter includes for illustration a few models dealing with concepts that are encountered in elementary chemistry or biochemistry courses. For any such model, an instructor can pose questions to be answered by manipulating or extending the model. Students might be asked, for example, how the model of Figure 4 (ionization of H_3PO_4) should be modified to deal with a system that contains three ionizable systems on three discrete kinds of molecules. The graphs of mole fractions of ionic forms (Figures 4d and 4e) would clearly not be relevant, but would the titration curve (Figure 4c) be identical? The better students will see the answer intuitively without need of numerical validation, but setting up the rather different model for that case may still be worthwhile. They could be asked to design a spreadsheet to generate a graph of relative buffer strength against pH for a two- or three-buffer system and answer questions such as how widely the pK_a values can differ while retaining at least half of the peak buffer strength across the interval between them, and how wide a pH range can be buffered to that degree. Directed exploration of even such simple models will introduce students to the power of numerical modeling and the ease of using models for exploring effects of changes in inputs or in constraints of a system.

Introduction of spreadsheet models into elementary science classes thus has other goals in addition to clarifying the specific systems that are modeled. Generation of numerical models requires a student to think clearly about the system he or she deals with, and to focus on the questions that he or she wants to answer. For the intellectually curious, those answers are likely to lead to extensions of the model to answer other questions perhaps not thought of until they are suggested by the model itself. An explicit goal is that students should, as a matter of course, turn to models to solve problems or clarify situations that they encounter in later courses or after graduation, and to aim at a deeper understanding of the underlying systems. By posing appropriate questions for exploration and by pushing students to set up models to answer questions most effectively, instructors can recruit an additional group of students beyond the self-starters who take to modeling with enthusiasm on first contact.

The unique value of spreadsheets in teaching science was recognized by David Barkley, who recruited colleagues to produce three workbooks that present basic models and pose questions for further exploration by students. Spreadsheet applications that were available when the first workbook, which deals with biochemistry,[1] was written did not include graphing capabilities, so the models were limited to numerical results. Students had to resort to manual plotting if they were to see the results graphically. The same models could now, of course, be used to generate graphs automatically and instantaneously. The workbook dealing with concepts from first-year chemistry[2] includes graphing as an integral part of the models. Most of the figures in this chapter are modified from models in those books. A third workbook deals with physics.[3]

Setup Procedures

This chapter aims to demonstrate the usefulness of spreadsheets in chemistry and especially in teaching chemistry. It does not include detailed instructions for constructing spreadsheets and plotting results. Such information is supplied by the software companies that sell spreadsheets, and the techniques vary slightly between applications. An indication of what is involved in using a spreadsheet may be useful, however, to indicate to readers who have no experience with this kind of software how models are set up and manipulated. Examples are presented as illustrations of the use of spreadsheets in teaching science.

A spreadsheet is an array of cells, each of which is identified by row and column. In most applications, numbers are used for rows and letters for columns. Thus, B5 refers to the cell at the intersection of column B and row 5.

In simple applications, cells may be used in three ways, which are illustrated in Figure 1a.

1. They may contain text, such as labels (cells A1 to A4), column headings (cells A6 to C6), clarifying comments, and the like.
2. They may contain numbers (constants), as in cells B1 to B4.
3. They may contain mathematical formulas, as in row 7 and the remainder of the array.

Cells of types *a* and *b* are passive; that is, they contain whatever the user has put into them and are not changed when the program is run.

Cells of type *c* are active; that is, the operation that is specified by the formula is calculated each time the program runs.

Much of the value of spreadsheets results from the ability of cells to refer to other cells. An array contains citing cells, which refer to one or more other cells, and cited cells, which are referred to by at least one other cell. Most cells in a typical array are of both types: they refer to other cells and are referred to by other cells.

Those relationships are illustrated in Figure 1a; each formula refers to at least one other cell. When the formula is calculated, the values in those cited cells are used. For example, when the program runs, cell A10 will generate the sum of the numbers in cells A9 and B4. Note that both active cells and passive cells can be cited.

	A	B	C
1	pK=	5.3	
2	[acid]+[base]=	1	
3	[base]init =	0.01	
4	[base]increm =	0.01	
5			
6	[base]	[acid]	pH
7	=$B$3	=$B$2-$A7	=$B$1+LOG10($A7/$B7)
8	=$A7+$B$4	=$B$2-$A8	=$B$1+LOG10($A8/$B8)
9	=$A8+$B$4	=$B$2-$A9	=$B$1+LOG10($A9/$B9)
10	=$A9+$B$4	=$B$2-$A10	=$B$1+LOG10($A10/$B10)

(a)

	A	B	C
1	pK=	5.30	
2	[acid]+[base]=	1.00	
3	[base]init =	0.01	
4	[base]increm =	0.01	
5			
6	[base]	[acid]	pH
7	0.01	0.99	3.30
8	0.02	0.98	3.61
9	0.03	0.97	3.79
10	0.04	0.96	3.92

(b)

FIGURE I Titration of a weak acid, pK_a = 5.3, by a strong base; pH as a function of amount of base added. a, Spreadsheet model showing formulas; entries that were made manually are shown in boldface. b, Spreadsheet model showing values. c, Titration curve generated from the spreadsheet.

Cell B4 contains a constant and A9 contains the result of a calculation. This ability to use results of calculations in other calculations is responsible for much of the adaptability of spreadsheets.

The operation specified by the formula in an active cell can be as simple as copying a constant, as in cell A7, or it may require using numbers from several cells

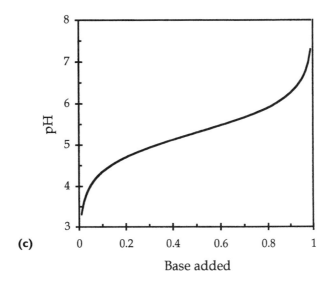

pH

Base added

FIGURE I (continued)

and carrying out a complex calculation. Formulas are written essentially in standard algebraic format, with the computer conventions that the asterisk (*) signifies multiplication and the caret (^) designates an exponent. Spreadsheet programs include a wide range of logarithmic, trigonometric, and other mathematical functions that can be used in writing operational formulas for cells. They also include logical functions (IF, AND, OR, EQUAL TO, GREATER THAN, LESS THAN, and the like), so that the result of a calculation can determine what further calculation will be done or what statement will be displayed.

In ordinary use, the formulas are not shown in the cells; Figure 1a is for illustration only. The cells on the computer monitor or printouts show the results of the calculations, as in Figure 1b. The formulas are in the background, and unless they are called up for viewing or modification, they are not visible.

A graph is often the desired output of a spreadsheet model. The curve of Figure 1c is generated by the model in Figure 1a.

Spreadsheets would not be practical if it were necessary to enter formulas in each cell of an array. Even for very small problems, the work would be too tedious. Spreadsheets can be set up quickly and easily because the program itself generates most of the cells of an array. The sheet shown in Figure 1a contains 300 active cells in three columns. The user keys in only the prototype formulas (in the illustration, cells A7, B7, C7, and A8) and specifies the size of the array. The program fills the cells of the array with the appropriate formulas.

In order to have the program generate the correct formulas, citations to other cells are of two types: absolute or relative. Those terms refer to the spatial relationship between the citing cell and the one cited. In Microsoft Excel, which was used in this illustration, citations are relative unless marked with the "$" sign, which specifies an absolute citation. The formulas in column A supply a simple illustration of the difference between the two kinds of citations. The value in each cell should be the sum of the cell immediately above and the increment, which is given in cell B4. That addition is specified in cell A8, and the formulas for the other cells in column A are generated automatically. In cell A8, the term "A7" cites the cell that is directly above the citing cell. Thus, the program enters "An" in each cell, where "n" is the row immediately above the cell. Unlike the first term, the second term should be constant in all cells of column A, since it is the constant value by which the concentration of base is to be incremented. That citation is specified by using the "$" sign to fix both the row and column, that is, to fix cell B4.

This matter of absolute and relative citations may seem complex when first encountered, but it is actually quite simple and makes it possible to set up even large and fairly complex arrays by manually entering each formula only once. Rows as well as columns can be cited absolutely, which makes it possible to generate arrays from a single formula in some cases. Figure 2a provides a simple illustration of relative and absolute citations. The calculation models the velocity of an enzymic reaction as a function of the concentration of the substrate (reactant). The enzyme in the model, which is typical of many that participate in metabolic regulation, catalyzes a modified fourth-order reaction with regard to substrate, and its affinity for substrate is modulated by the concentration of a regulatory metabolite. The behavior of most such enzymes is modeled closely by the equation

$$\text{velocity} = [S]^n/([S]^n + [A]^n)$$

where $[S]$ is concentration of substrate, $[A]$ is a parameter specifying affinity of the enzyme for its substrate (which is a function of the concentration of the regulatory metabolite), and n is the effective order of reaction. Figure 2c was generated from a spreadsheet that contained that expression in 300 cells, but it was keyed in only once. A range of substrate concentrations was generated in column A, and three values of $[A]$ were entered in row 5. Next, the formula " = $A7^$B$1/($A7^$B$1+ B$4^$B$1)" was entered into cell B7 and expanded into the other 299 cells of the array. A user who is new to spreadsheets should note how the use of absolute and relative references for both rows and columns made it possible to generate the three curves of this figure by entering the expression only once. The behavior of the enzyme can be explored by changing the values of A in row 4; the expressions in all cells and the curves of the graph will change immediately. Other spreadsheet applications use different symbols to designate relative and absolute references, but replication of a basic expression is possible in all of them, and they all can generate graphs from any selected array of cells.

Students should be encouraged to define all parameters and variables in passive cells, rather than incorporating them into formulas. In this illustration, the range of concentrations of substrate could have been generated by entering "0.01" in cell A8 and "A8*1.05" in cell A9. However, if the user later wished to modify the range to optimize the resulting graph, he or she would need to change the formulas in those cells and replicate cell A9 down the column. As the model is written, the range can be modified by merely changing values in cells B2 and B3. Similarly, a student who wishes to generate a curve like that of Figure 2c (or who is asked to do so as an exercise) might enter the formula " = $A7^4/($A7^4+B$4^4)" in cell B8. The resulting calculations and graphs would be identical with those in the illustration, but if the student later wanted to see the effect of changing the exponent (that exponent is an important parameter in modeling the regulation of metabolic processes), he would need to change "4" to the desired value and generate the array again. When the exponent is placed in a passive cell and cited in the formula cells, the dramatic effect of change in reaction order can be seen immediately in the associated graph by merely changing the value in cell B1. Ease of exploring the effects of changes in parameters or conditions is a major advantage of spreadsheet models, and ease of modification depends on having all assignable values in passive cells at the top of the model, where they are readily accessible for change.

Buffers

Numerical Differentiation: Buffer Strength

The variation of buffer strength with pH can be easily illustrated by extension of the model of Figure 1a. The added columns are shown in Figure 3a, and the resulting curve in Figure 3c. This example illustrates how easily spreadsheets can be expanded to provide additional output, often related to questions that were not considered when the initial model was written. In this case, only two additional formulas were entered into the preexisting sheet. The first (cell E7) determines pH at a value of $[B]$ slightly above that of the cells in column A, and the other (F7) uses the result of that calculation to determine the amount of base needed for a given change in pH. The resulting curve (Figure 3c) shows the importance of choosing a buffer system with a pK_a close to the pH value that is to be stabilized.

Multiple-Buffer Systems

Titrations that involve several ionizable groups are difficult or tedious to deal with algebraically, but numerical simulation is easy. When a spreadsheet is used for the simulation, the chemical relationships are displayed directly on the model (the spreadsheet) itself, rather than being hidden in a program. A titration curve for a system containing three conjugate acidic groups is generated by the spreadsheet in Figure 4a and is shown in Figure 4c. Cells B1, B2, and B3 contain the pK_a values for the three acid dissociations of H_3PO_4, but any others can, of course, be entered; for example, the pK_a values for arginine are 2.3, 9.1, and 12.5. A plot of buffer strength against pH could be generated for this model, and students could vary pK_a values to explore the overlap of buffer strength in a multiple-buffer system, varying the distance between pK_a values.

Concentrations of the different ionic forms in a phosphate (or any other) buffer can be plotted as a function of pH from the model of Figure 4a. Comparison of the resulting graph, Figure 4d, with a graph of buffer strength would illustrate the relationship between rate of change of ionic concentrations and buffer strength. If they were assigned to plot ionic concentrations against base added, students would probably be surprised by the result (Figure 4e). They might be asked to explain the difference between Figures 4d and 4e, and to relate both to a plot of buffer strength.

This model could be used to illustrate for students the advantages of thinking about the approach before setting up a model. Direct calculation of pH as a function of the amount of base added to a multiple-buffer system is moderately complicated, but when the problem is approached the other way around—calculating the amount of base addition that would result in a given value of pH—it is trivially simple. The results can then be plotted in any desired way, including that shown in Figure 4c.

(a)

	A	B	C	D
1	n =	4		
2	[S]init =	0.01		
3	[S]mult fact=	1.05		
4	[A] =	0.5	1	2
5				
6	[S]	rate	rate	rate
7	0	=$A7^$B$1/($A7^$B$1+B$4^$B$1)	=$A7^$B$1/($A7^$B$1+C$4^$B$1)	=$A7^$B$1/($A7^$B$1+D$4^$B$1)
8	=$B$2	=$A8^$B$1/($A8^$B$1+B$4^$B$1)	=$A8^$B$1/($A8^$B$1+C$4^$B$1)	=$A8^$B$1/($A8^$B$1+D$4^$B$1)
9	=$A8*$B$3	=$A9^$B$1/($A9^$B$1+B$4^$B$1)	=$A9^$B$1/($A9^$B$1+C$4^$B$1)	=$A9^$B$1/($A9^$B$1+D$4^$B$1)
10	=$A9*$B$3	=$A10^$B$1/($A10^$B$1+B$4^$B$1)	=$A10^$B$1/($A10^$B$1+C$4^$B$1)	=$A10^$B$1/($A10^$B$1+D$4^$B$1)

FIGURE 2   Rate of the reaction catalyzed by a regulatory enzyme as a function of concentration of substrate. The enzyme binds substrate cooperatively at four sites, and the affinity of binding is modulated by the concentration of the end product of the sequence. The curves show three values of affinity. a, Spreadsheet model showing formulas; entries that were made manually are shown in boldface. b, Spreadsheet model showing values. c, Curves generated from the spreadsheet.

	A	B	C	D
1	n =	**4**		
2	[S]init =	**0.01**		
3	[S]mult fact=	**1.05**		
4	[A] =	**0.5**	1	2
5				
6	[S]	rate	rate	rate
7	**0.000**	**0.000E+00**	0.000E+00	0.000E+00
8	**0.010**	1.600E-07	1.000E-08	6.250E-10
9	**0.011**	1.945E-07	1.216E-08	7.597E-10
10	0.011	2.364E-07	1.477E-08	9.234E-10

(b)

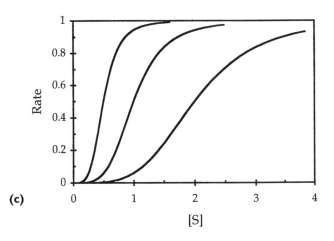

(c)

FIGURE 2 (continued)

Numerical Integration: Gibbs Energy vs. Extent of Reaction

The model of Figure 5a presents the relation between the Gibbs energy of a system and the extent of reaction in a way that may clarify the relationship for students. The model generates a plot (Figure 5c) of relative value of Gibbs energy against the concentration of B for the reaction A $\leftrightharpoons$ B.

The model is based on the equation $\Delta G = RT\ln(Q/K)$, where $Q = [B]/[A]$ and K is the equilibrium constant. In the interests of simplicity in the example, integration is carried out by multiplying the length of each concentration interval by the slope (ΔG) at the end of the interval. This shifts the curve by about half of the length of an interval, which is trivial. Students could of course be asked to generate a more accurate curve by multiplying each interval by the value of ΔG at its center. Since G has no absolute value, the curve has been normalized to a value of zero at equilibrium.

(a)

	D	E	F
1			
2			
3			
4			
5			
6		pH'	buffer strength
7		=$B$1+LOG10(($A7+0.001)/($B7-0.001))	=0.001/($E7-$C7)
8		=$B$1+LOG10(($A8+0.001)/($B8-0.001))	=0.001/($E8-$C8)
9		=$B$1+LOG10(($A9+0.001)/($B9-0.001))	=0.001/($E9-$C9)

(b)

	A	B	C	D	E	F
1	pK =	5.3				
2	[A] + [B] =	1				
3	[B]init =	0.01				
4	[B]inc =	0.01				
5						buffer
6	[B]	[A]	pH		pH'	strength
7	0.01	0.99	3.30		**3.35**	**0.02**
8	0.02	0.98	3.61		3.63	0.05
9	0.03	0.97	3.79		3.81	0.07

FIGURE 3   Relative buffer strength as a function of pH; pK_a = 5.3. a, Columns added to the spreadsheet of Figure 1a; entries that were made manually are shown in boldface. b, Complete spreadsheet model showing values. c, Curve generated from the spreadsheet.

The curve of Figure 5c graphically illustrates the fact that a reaction moves downhill with regard to the Gibbs energy of the system toward the minimum of the curve, which is equilibrium. By shifting the value of K, students can explore the relationship between K, $\Delta G°$, and ΔG. $\Delta G°$ is merely the value of ΔG (the slope of the curve) when $Q = 1$, which is indicated in the figure by the arrow. The difference between ΔG and the decrease in Gibbs energy as the system goes to equilibrium, which confuses many students (and even an occasional textbook writer in related fields) is shown clearly. ΔG is the slope of the curve at any point, and the Gibbs energy change when the system goes to equilibrium is the value of the curve at that point, as read from the vertical axis. Generating such models and then modifying

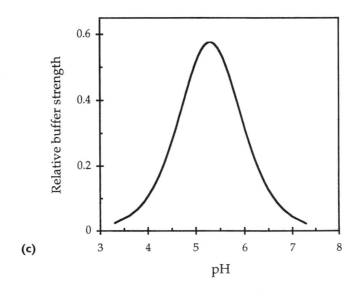

(c)

FIGURE 3 (continued)

them and thinking about the resulting curves can aid students in developing a deeper understanding of relationships that are often merely accepted. The more conventional plot of ΔG against the progress of the reaction (Figure 5d) does not contribute to understanding the relationship between Gibbs energy and equilibrium: the curve does nothing to clarify the fact that $\Delta G = 0$ at equilibrium.

The Blood Buffer System

A student knows from his or her first chemistry class that the buffer strength of an ordinary system is greatest when $pH = pK_a$. That relationship was shown graphically in Figure 3c. But he or she may then learn in a biology or biochemistry class that the pH of blood is held within a very narrow range around 7.4, and that the main buffer in blood is the $(CO_2 + H_2O)/HCO_3^-$ system, for which $pK_a = 6.1$. From the shape of the buffer strength curve in Figure 3c, it seems that the bicarbonate system would buffer very weakly in blood, 1.3 pH units above its pK_a value. The student, if he or she has much scientific curiosity, might not merely file those seemingly contradictory statements in separate mental compartments, one labeled chemistry and one labeled biology, but might wonder how they can be reconciled. A physiology textbook may say, rather cryptically, that blood is strongly buffered at pH 7.4 because regulation of breathing holds the concentration of CO_2 in the blood at 1.2 mM, within a few percent. The student may well wonder whether that fact can really affect buffer strength strongly enough to allow significant buffering by the bicarbonate system at pH 7.4. One aim of an instructor who uses spreadsheets in teaching science should be to encourage students to realize that when any question of this kind arises, it can easily be explored by making a spreadsheet model. Students should also be encouraged to feel that such exploration is interesting and worthwhile.

(a)

	A	B	C	D	E	F	G	H	I	J
1	pK1=	2.12	pH start=	0						
2	pK2=	7.28	pH incr=	0.25						
3	pK3=	12.32								
4										
5	pH	H2A/H3A	HA/H2A	A/HA	H3A	H2A	HA	A	Tot H	base added
6	0.00	0.008	0.000	0.000	0.992	0.008	0.000	0.000	2.992	0.008
7	0.25	0.013	0.000	0.000	0.987	0.013	0.000	0.000	2.987	0.013
8	0.50	0.024	0.000	0.000	0.977	0.023	0.000	0.000	2.977	0.023

(b)

Cell	Formula
A6	=$D$1
A7	=$A6+$D$2
B6	=10^($A6-$B$1)
C6	=10^($A6-$B$2)
D6	=10^($A6-$B$3)
E6	=1/(1+$B6+$C6*$B6+$D6*$C6*$B6)
F6	=$B6*$E6
G6	=$C6*$F6
H6	=$D6*$G6
I6	=$G6+2*$F6+3*$E6
J6	=3-$I6

FIGURE 4 Titration curve for H_3PO_4. a, Spreadsheet model showing values; cells to which entries were made manually are shown in boldface. b, Formulas that were entered manually into the spreadsheet. c, pH as a function of amount of base added. d, Mole fractions of ionic forms as a function of pH. e, Mole fractions of ionic forms as a function of amount of base added (HP, H_3PO_4; $P-$, $H_2PO_4^-$; $P=$, HPO_4^{2-}; $P\equiv$, PO_4^{3-}).

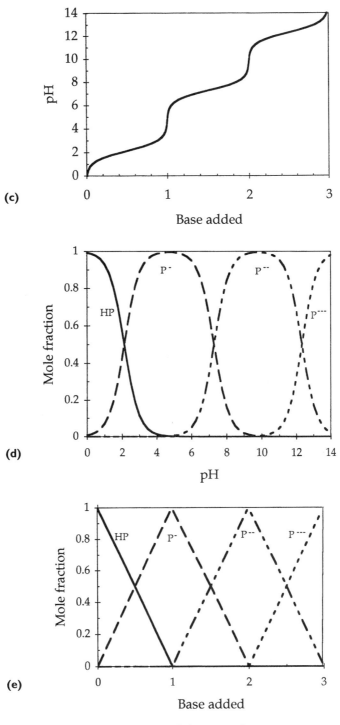

(c)

(d)

(e)

FIGURE 4 (continued)

(a)

	A	B	C	D	E	F
1	K=	**4**	R=	**0.00831**	kJ/mole	
2	B+A=	**1**	T=	**300**	K	
3	Binit=	**0.01**				
4	Bincr=	**0.01**		Min of rel G	**-3.766**	
5						
6	[B]	[A]	Q/K	ΔG	G(rel)	G(norm)
7	**0.01**	**0.99**	**0.003**	**-14.912**	**0.000**	**3.766**
8	**0.02**	0.98	0.005	-13.158	**-0.132**	3.634
9	0.03	0.97	0.008	-12.122	-0.253	3.513

(b)

Cell	Formula		Cell	Formula
A7	=$B$3		D7	=$D$1*$D$2*LN($C7)
A8	=$A7+$B$4		E4	=MIN(E7:E105)
B7	=$B$2-$A7		E8	=$E7+($B$4*$D8)
C7	=$A7/($B$1*$B7)		F7	=$E7-$E$4

FIGURE 5 Relative Gibbs energy of a system as a function of the progress of the reaction A $\rightleftharpoons$ B; K_{eq} = 4 . a, Spreadsheet model showing values; cells to which entries were made manually are shown in boldface. b, Formulas that were entered manually into the spreadsheet. c, Relative Gibbs energy as a function of [B], with [A] + [B] constant; normalized to a value of zero at equilibrium. d, ΔG as a function of [B].

The exploration can start with the model of Figure 3a. Another pH calculation is added in which [B] varies as before, but [A] is held at 1.2 mM (Figure 6a). Buffer strength is calculated as before. The resulting curves, Figures 6c and 6d, may surprise the student, and certainly will help him or her understand pH relationships in the blood. If students can be motivated to undertake such explorations on their own, they will be prepared to handle similar situations in the future, and also will probably understand and remember the answers more clearly than if they were merely presented in a textbook or a lecture.

A student might, for example, be baffled if told that the blood buffer system and a conventional bicarbonate system with a total buffer concentration of 25.2 mM are identical in composition (and of course in pH value) at pH 7.4, but that the buffer strength of the blood buffer is 21 times that of the other. That statement might seem almost impossible to accept. But if the student considers Figure 6c, and especially if he or she understands the curves because of having written the spreadsheet model that generated them, the apparent paradox will be easily resolved.

(Many students may also be surprised to learn that when they pant heavily during hard exercise, they are really panting primarily to dispose of CO_2 rather than to obtain

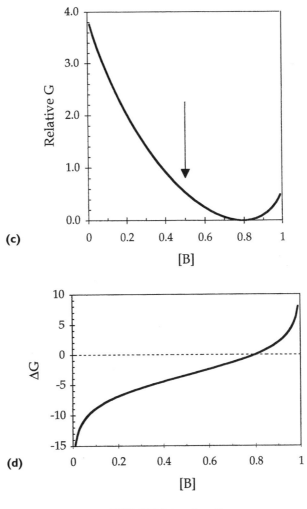

(c)

(d)

FIGURE 5 (continued)

oxygen, but that insight follows from the statement above concerning regulation of breathing, and not from the spreadsheet model.)

Spreadsheet Models In Research

The output of a model is, of course, independent of the language or procedure in which it is implemented. Spreadsheets and ordinary languages are alternative possibilities. Spreadsheets have significant advantages in many cases where the model is not large or complex. Those advantages are similar to those we have discussed in connection with small models for use in teaching. They are quick and easy to set up, they can be debugged or modified very easily because every operation in the model is clearly set forth on the face of the model itself, and little or no comment or documentation is needed to make the model easily comprehensible to other workers, or even to the author a year after it was written.

	A	B	C	D	E	F	G	H	I	J
1	pK =	6.1								
2	A+B=	25.2								
3	Binit=	5.0								
4	Binc =	0.3						[A] =	1.20	
5	base					buffer				buffer
6	added	[B]	[A]	pH	pH'	strength		pH	pH'	strength
7	0.0	5.0	20.2	5.494	5.494	9.229		**6.720**	**6.720**	**11.514**
8	0.3	5.3	19.9	5.525	5.526	9.638		6.745	6.745	12.205
9	0.6	5.6	19.6	5.556	5.556	10.030		6.769	6.769	12.896

(a)

Cell	Formula
H7	=$B$1+LOG10($B7/$I$4)
I7	=$B$1+LOG10(($B7+0.001)/$I$4)
J7	=0.001/($I7-$H7)

(b)

FIGURE 6   Comparison of the blood buffer system with a standard closed buffer. a, Spreadsheet model showing values; columns B to F are identical to columns A to E of Figure 3; cells to which additional entries were made manually are shown in boldface. b, Formulas that were entered manually into the spreadsheet. c, Titration curves for blood buffer and standard buffer. d, Relative buffer strengths of blood buffer system and standard buffer as functions of pH; the physiological blood pH value, 7.4, is indicated by the arrow.

Figure 7 illustrates the use of a spreadsheet in construction of a small research model. The model deals with regulation of a metabolic biosynthetic pathway, indicated generically by eq 1.

$$x_0 \rightarrow x_1 \rightleftharpoons x_2 \rightleftharpoons x_3 \rightarrow \text{metabolic use} \tag{1}$$

Such pathways are controlled in large part by negative feedback relationships in which the end product of the pathway (x_3 in eq 1) binds to specific regulatory sites on the enzyme that catalyzes the first step. When one or more product molecules are bound, the conformation of the enzyme changes in such a way that the affinity of the enzyme for its substrate x_0, the starting material for the pathway, is decreased. Thus, an increase in the concentration of product x_3 causes the rate of its synthesis to decrease, which tends to stabilize the concentration of the end product over a wide range of utilization rates. It is known from experiment that such control is extremely effective and that small increases in the concentration of an end product can decrease the rate at which that product is synthesized in a living cell by a factor of 100 or more.

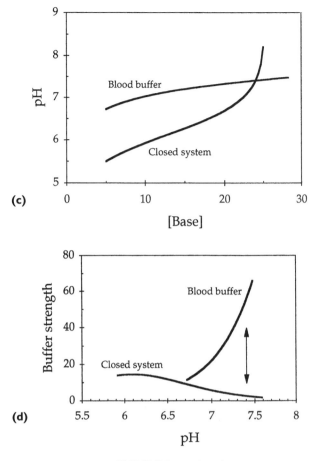

FIGURE 6 (continued)

The time course of regulation after a steady-state system is perturbed, for example by an increase or decrease in the rate of use of the end product x_3, cannot be modeled analytically, and to my knowledge no numerical model has been produced using a computer language. The spreadsheet model used in generating Figure 7 was easy to write and is transparent in the sense that all intermediate values are available for inspection and in the sense that it can be easily understood and modified by anyone familiar with spreadsheets, with minimal documentation. The model has 18 assignable parameters, among them the inherent affinity of the first enzyme for its substrate and for the regulatory product, the degrees of cooperativity of binding of those ligands, the change in free energy of binding of substrate when the end product binds to a regulatory site, the inherent maximal velocity of the first reaction, and the kinetic parameters of the simple (nonregulatory) enzymes that catalyze the other two steps. Each parameter can be caused to change by any desired amount after a steady state has been reached, and the model generates time curves for attainment of a new steady state. The rate of any reaction or the concentration of any intermediate can be plotted as a function of time.

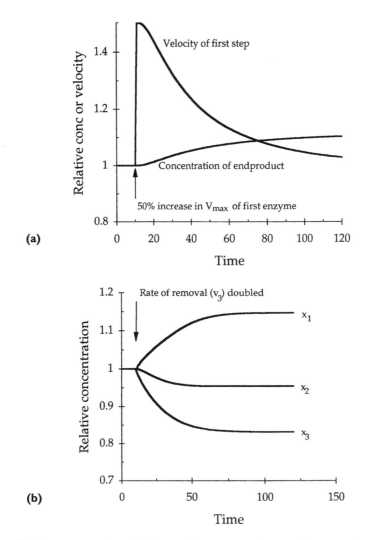

(a)

(b)

Figure 7   Time course of establishment of a new steady state after perturbation of a model biosynthetic sequence that is regulated by product negative feedback. The generic sequence is shown in the text. a, Velocity of the first (regulated) reaction and concentration of the end product as functions of time; the amount or activity (V_{max}) of the first enzyme is increased by 50% at time 10. b, Concentrations of intermediates x_1 and x_2 and of the end product x_3 as functions of time; the rate of metabolic use of the end product x_3 is doubled at time 10. (Figure 7a is modified with permission from ref. 4. Copyright 1990 Plenum Press.)

Figure 7a shows attainment of a new steady state after the activity of the first enzyme is increased by 50%. It demonstrates the strong control exerted by negative feedback and the fact that, as in technological feedback systems, the flux through the reactions of the sequence is nearly independent of the maximal capacity of the system (if enough enzyme is present to support the necessary rate, an increase in amount has very little effect). With the parameter settings that were used to generate this figure, an increase of about 8% in the concentration of the end product suffices to offset a 50% increase in inherent enzyme activity.

Figure 7b shows the changes in concentrations of intermediates when the rate of metabolic removal of the end product doubles (for example, when the rate of protein synthesis increases, amino acids are used more rapidly). When concentration of end product x_3 falls slightly, feedback control causes the rate at which the first enzyme delivers intermediate x_1 to the sequence to increase until a new steady state is reached, and the concentrations of the intermediates adjust to permit doubling of flux through the other reactions. The figure illustrates an important property of feedback-controlled systems. If the rate of removal of the end product of an unregulated sequence of reactions increased, the concentrations of all intermediates would fall. In the regulated system, the concentration of x_1 at the new steady state is higher than its initial value.

The model shows that the ability of cells to regulate precisely the rates of their metabolic pathways to meet certain needs, which has in the past seemed to some observers to be mysterious or specifically biological, results from straightforward chemical interactions that are dependent on very specific genetic design of catalytic sites, regulatory sites, and the interactions between them. It also permits users to vary experimentally known properties of such systems, such as the number of sites at which substrate and modifier are bound cooperatively, the strength of the cooperativity, and the change in free energy of binding of substrate when modifier (end product) binds to regulatory sites on the first enzyme, in order to evaluate the contribution of each to sensitivity of metabolic control. For the purposes of this book, the model is included to show that spreadsheet methodology is ideally suited to small research models.

Conclusion

A spreadsheet cannot, of course, do anything that cannot be done in FORTRAN or BASIC. However, it is not, in practice, reasonable to expect all science students, at least in elementary courses, to be proficient in computer programming. Models that, in practice, would not be generated otherwise can be made quickly and easily by use of a spreadsheet. Also, ease of use is not gained at the expense of inferiority in the product. On the contrary, the features that make spreadsheet models easy to create are the same that make them valuable in teaching and learning. In setting up models, students focus on the concepts of the system that they are modeling and deal with the relevant mathematics in familiar and conventional ways. The model illuminates the system and facilitates its exploration. A student who generates a model in a conventional language necessarily thinks in terms of the conventions of that language, instead of in terms of chemistry, when setting up and modifying the model, which limits the chemical insight that the model can provide.

In addition to being easy to make and transparent to understand, spreadsheet models allow students to generate modifications and extensions of all kinds. Thus, they provide a unique means of exploring scientific principles and interactions. "What if" questions of almost unlimited variety can be asked and answered. Given innovative interest on the part of instructors and intellectual curiosity on the part of students, the use of spreadsheet models provides an additional dimension to learning the mathematical aspects of chemistry and other sciences.

References

1. Atkinson, D. E.; Clarke, S. G.; Rees, D. C.; Barkley, D. S.; *Dynamic Models in Biochemistry;* N. Simonson: Marina del Rey, CA, 1987.
2. Atkinson, D. E.; Brower, C. D.; McClard, R. W.; Barkley, D. S.; *Dynamic Models in Chemistry;* N. Simonson: Marina del Rey, CA, 1990.
3. Potter, F.; Peck, C. W.; Barkley, D. S.; *Dynamic Models in Physics, Vol. I: Mechanics;* N. Simonson: Marina del Rey, CA, 1989.
4. Atkinson, D. E; "An Experimentalist's View of Control Analysis"; in *Control of Metabolic Processes;* Cornish-Bowden, A., Cárdenas, M. L., Eds.; NATO ASI Series A, vol. 190; Plenum Press: New York, 1990, pp. 413–427.

Software for Teaching and Using Numerical Methods in Physical Chemistry

Sidney H. Young
Jeffry D. Madura
Frank Rioux

Overview

Physical chemistry is the first course in the chemistry curriculum that extensively uses numerical methods to calculate quantities of physical and chemical interest from measurable data. The complexity of the quantitative aspect of the course is demonstrated by the technical prerequisites for the course: one year each of calculus and physics. However, there seems to be a gap in the chemistry curriculum, in particular, with respect to a course that addresses the introduction and application of numerical methods as used in chemistry. Current physical chemistry courses do not address this issue satisfactorily. Popular texts present problems involving numerical methods but do not give sufficient background material for students to apply these methods, perhaps assuming that they have been introduced to them in other courses. Laboratory manuals may provide written computer programs written in BASIC, FORTRAN, or Pascal that can be used by the student to analyze data, but the underlying algorithms are not usually presented to the student systematically, and are often treated as "black boxes".

Easy access to personal computers and software devised specifically for the implementation of numerical methods makes it possible now to teach these methods, as needed, to an undergraduate audience unfamiliar with programming languages. Spreadsheet programs such as Quattro Pro and Excel, and numerical and symbolic engines such as Mathcad, MAPLE, and MATHEMATICA, allow the student not only to use numerical methods but to learn the algorithms incorporated by those methods as well.

Following an introduction to the various software packages that are useful for implementing various computational algorithms, a survey of some of the numerical methods that are useful in both physical chemistry lecture and laboratory is presented, along with selected examples from thermodynamics, kinetics, and quantum mechanics.

Review of Relevant Software

Today, students and teachers have available to them various methods with which to solve scientific numerical problems. These methods include using a programming language such as FORTRAN, Pascal, C, or BASIC, or a spreadsheet such as Microsoft Excel, Corel Quattro Pro, or Lotus 1-2-3. Software can additionally be used to solve problems symbolically. Currently, there are several programs that fall into the numerical/symbolic manipulation category. These include, but are not limited to, programs such as Mathcad, MATHEMATICA, MAPLE, MATLAB, Derive, and Macsyma. In this section we will review some of the pros and cons of the aforementioned software as possible tools for teaching physical chemistry and assisting students in understanding the complexities of physical chemistry concepts. Several examples of how to use modern software will be given. Linear least-squares will be a common theme for some of these so that readers can choose the problem-solving method that will be "best" for their students. The formulas used in calculating the various quantities of the linear least-squares problem come from H. G. Hecht, *Mathematics in Chemistry: An Introduction to Modern Methods,* pages 267–274 (Prentice-Hall, New York, 1990).

Programming Languages

Programming languages are an "easy" way for people to communicate directly with a computer. One of the first computer languages, FORTRAN, was developed in 1957 by IBM. FORTRAN (FORmula TRANslator) was one of the first high-level languages introduced to the scientific community after assembly language. Initially, the language contained few constructs and was primarily designed to aid in the evaluation of complex arithmetic formulas. In order to evaluate a formula or set of formulas, a program had to be written, compiled, i.e., converted into machine language, and finally executed. The writing of any program requires the programmer to follow a set of strict rules known as syntax. If these rules are not followed, then the compiler will return error messages and the program will not run. Once the program is compiled and executed, it has to be checked for semantics (accuracy). If a correct result is not obtained, then one of two problems has occurred and has to be resolved. The first problem could be that the compiler generated incorrect machine code; this is usually not the case. The second more likely problem is a programmer logic error. Samples of FORTRAN code for various applications in physical chemistry can be found in many standard data analysis and laboratory texts.

Pascal, developed in 1971 by Niklaus Wirth of the Technical University of Zurich, Switzerland, was one of the first highly structured modular programming languages. Wirth's intention was to provide a language to aid in teaching a systematic approach to computer programming. For this reason it is the first programming language taught in many schools. In 1983, Phillipe Kahn introduced an inexpensive PC version of Pascal called Turbo Pascal.

C, defined in 1978 by Ritchie and Kerrigan at Bell Labs, is a general-purpose programming language very similar to Pascal in structure and modularity. An "improved" version of C, C++, was developed in 1986 by B. Stroustrup.

BASIC, invented in 1964 at Dartmouth College, stands for beginner's all-purpose symbolic instruction code. It is an interpreted language; that is, statements are executed by the computer as they are entered by the programmer. This avoids the compilation step found in the other languages. The disadvantage here is that these BASIC programs execute at a slower rate since each statement has to be interpreted each time. One of the best features of BASIC, prior to the introduction of DOS 5.0, was that it was included as part of the operating system on the PC. It should be noted that BASIC was one of the first languages that contained commands for writing graphics programs. Many programs were written in BASIC to solve a variety of numerical problems in chemistry. Unfortunately, BASIC is not a structured language; therefore, the programs were sometimes long and hard to read. Later versions of DOS (5.0, 6.X) are packaged with Q-BASIC. This language lends itself to the modular, structured paradigm of modern programming practice. Immediately obvious to the users of early BASIC was the lack of line numbers.

Even with all of the difficulties of learning a programming language, developing the algorithm, writing syntactically correct code, and then testing the code, programming is still the best method for solving large, complex problems.

Spreadsheets

A spreadsheet is composed of columns labeled with the letters of the alphabet and rows labeled numerically starting with the number 1. To use a spreadsheet, the user enters numbers or formulas into cells, each cell having a specific column and row designator. Two currently popular spreadsheets used throughout academia and industry are MS Excel and Corel Quattro Pro. The power of spreadsheets resides in the user's ability to manipulate data and generate graphical representations of the data. One good application for a spreadsheet is keeping students' grades. With a spreadsheet an instructor can instantaneously tell each student how he or she is doing in the class. They also are useful in exploring "What if" situations that can occur during a chemistry lecture or in a discussion with a student about a grade. The spreadsheet provides one important feature not found in the programming languages: an easy way to graph data. More details about spreadsheet use in the chemistry curriculum are given by D. Atkinson in Chapter 9 of this volume, Spreadsheets for Doing and Teaching Chemistry.

Numeric/Symbolic Software

In this section we will review some of the numerical/symbolic software available for use in the physical chemistry curriculum. Two products that will not be reviewed are the programs Derive and Macsyma because the authors, at this time, have had very limited experience with them. This should not be taken to mean that these programs are not useful in the physical chemistry program. Indeed, several papers using one of them, Derive, have been published in the *Journal of Chemical Education*. The reader might also be interested to learn that Derive is available on a variety of programmable calculators and palm-sized computers.

MATHEMATICA

MATHEMATICA is described by Bahder as a modern computer language designed for the needs of scientists and engineers. MATHEMATICA is able to carry out symbolic calculations as well as the numeric calculations traditionally done using FORTRAN. In addition to its computational prowess, MATHEMATICA is capable of performing simple and complex graphics manipulations. MATHEMATICA is an interpreted language. This is one way in which MATHEMATICA differs from Mathcad. In MATHEMATICA one usually enters an instruction one line at a time; the result follows immediately below. MATHEMATICA notebooks, a structured mix of text, equations, and graphics, can be created and given to students. Although MATHEMATICA has numerous numeric functions built in, some calculations may still take a long time due to the interpreted nature of the program. In addition, as with MAPLE and MATLAB, these programs can consume large amounts of memory. The drawbacks should not impede the introduction and use of MATHEMATICA in the physical chemistry curriculum. In the proper hands MATHEMATICA is a very powerful tool for solving many different kinds of problems. MATHEMATICA may be a natural choice over other software such as Mathcad because students learn to use MATHEMATICA in their calculus courses.

MAPLE

MAPLE was introduced in 1983 as a powerful and reliable way of doing mathematics by computer. MAPLE, like the other programs in this category, is able to perform numeric and symbolic calculations, and generate two- and three-dimensional graphics. MAPLE is also an interpreted program and has many features similar to MATHEMATICA. By using MAPLE one can create procedures that are analogous to subroutines found in FORTRAN. MAPLE has been used to generate complex quantum chemical expressions that were then incorporated into a FORTRAN program. MAPLE has a more relaxed syntax than MATHEMATICA. MAPLE also has built-in graphing capabilities.

MATLAB

MATLAB, which stands for MATrix LABoratory, is best described as a high-performance technical computing environment for numeric computation and visualization. The basic data element in MATLAB is a matrix that does not require dimensioning. Initially, the reader may think MATLAB may be limited in the number of different types of problems it can solve. The opposite is true considering the following: A matrix is normally envisioned as a mathematical construction of a rectangular array arranged in rows and columns. However, a matrix of pixels could represent an image, a matrix of signal fluctuations can be a sound, and a matrix can also describe the linear relationships among the components of a mathematical model. MATLAB is an expression language, which means that it interprets and evaluates expressions entered from the keyboard. MATLAB has a syntax similar to

MATHEMATICA and MAPLE. MATLAB Toolboxes can be used to enhance the strong MATLAB foundation and can be used to address specific technical computing needs such as image processing, statistics, symbolic math, chemometrics, signal processing, and optimization. Although the use of MATLAB in physical chemistry is unknown, there are three books worth mentioning. The first is *Numerical Methods for Mathematics, Science, and Engineering* by J. H. Mathews; the second is *Numerical Methods Using MATLAB* by G. Lindfield and J. Penny; and the third is *Numerical Methods for Physics* by A. Garcia.

MATHCAD

Mathcad is best described as an electronic blackboard. The program allows the user to place data, variables, constants, and equations virtually any place on the computer screen. Earlier versions of Mathcad, which were MS-DOS programs, could not perform symbolic manipulations. Mathcad version 3.0, the first Microsoft Windows version, had the capabilities of performing symbolic manipulation using MAPLE. There are several features of Mathcad that make it an attractive addition to the physical chemistry curriculum. The first feature, already mentioned, is its free-form placement of information such as text, data, variable and constant assignments, and equations. Second, equations appear as one would see them in a textbook or scientific paper. This is a big advantage over having to translate an equation into a cryptic form, as would be required with FORTRAN. Third is the ease of making graphs of equations and data. The second and third features are especially big pluses for students because they enable students to instantaneously and interactively observe equation behavior or trends in their data. The final feature is the symbolic manipulations that are available in Mathcad. With this feature a student can carry out a complex derivation, differentiation, or integration without worrying about dropping a sign or making a mistake. One may argue that with such software, students will not perfect their skills in differentiating or integrating functions. As a rebuttal we contend that one can always go to a handbook and look up the formulas necessary to perform the differentiation and integration. What is important, though, is that the student learn to recognize whether the software performed the symbolic manipulation correctly. In addition, using Mathcad will help students achieve the ultimate goal: a result that they can use, interpret, and comprehend without being left frustrated because they could not complete the derivation. Figure 1 represents a typical example of a Mathcad document. Each section of the document is clearly delineated; formulas and graphs appear as they do in a book. Like spreadsheets, Mathcad documents automatically recalculate whenever anything is changed.

Survey of Numerical Methods

Statistical Methods

Linear-Least-Squares Regression.   One of the most important algorithms that any scientist needs is the method of least-squares regression for fitting data to a straight line. There are many texts that present both the matrix formalism and explicit equations used in the least-squares algorithm, especially certain physical chemistry lab manuals.[1–3]

Multivariate Linear Least Squares

$i := 0 .. 9$

$x_i := i + rnd(1)$

This generates x data with some random error

$y_i := 0.4 \cdot (x_i)^2 + 2 \cdot x_i + 3 + rnd(0.1)$

y=0.4x^2+2x+3 with random error added

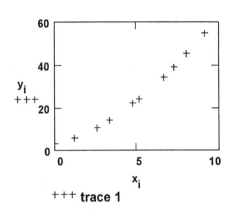

Plot of x-y data

+++ **trace 1**

$X_{i,0} := 1 \qquad X_{i,1} := x_i \qquad Y_i := y_i$ **Definitions of X and Y**

$$b := (X^T \cdot X)^{-1} \cdot X^T \cdot Y$$

**b is the vector of fitting parameters.
For a linear fit, b_0 is the intercept and b_1
is the slope.**

$$SS := \sum_i \frac{(y_i - b_0 - b_1 \cdot x_i)^2}{8}$$

**SS is the variance of the fit.
There are 8 degrees of freedom.**

$$S_0 := \sqrt{\left[(X^T \cdot X)^{-1}\right]_{0,0} \cdot SS} \qquad S_1 := \sqrt{\left[(X^T \cdot X)^{-1}\right]_{1,1} \cdot SS}$$

S_0 and S_1 give the uncertainties in the fitting parameters.

FIGURE I   This Mathcad document illustrates the implementation of multivariate least-squares fitting using data pairs with random error generated from the quadratic function $y = 0.4x^2 + 2x + 3$, where x ranged from 0 to 10. Both linear and quadratic fitting were obtained by matrix multiplication. The standard deviations of the fitting parameters are easily obtained by this method. Each term is clearly expressed and easy for students to learn and use.

Results for the linear fit:

$$fit_i := b_0 + b_1 \cdot x_i$$

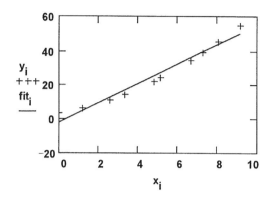

$$b = \begin{pmatrix} -2.066 \\ 5.676 \end{pmatrix}$$

$$S = \begin{pmatrix} 2.071 \\ 0.368 \end{pmatrix}$$

$$SS = 11.21$$

Now expand to a quadratic fit: $$X_{i,2} := \left(x_i \right)^2$$

$$b := \left(X^T \cdot X \right)^{-1} \cdot \left(X^T \cdot Y \right)$$

$$SS := \sum_i \frac{\left[y_i - b_0 - b_1 \cdot x_i - b_2 \cdot \left(x_i \right)^2 \right]^2}{7}$$

$$S_0 := \sqrt{\left[\left(X^T \cdot X \right)^{-1} \right]_{0,0} \cdot SS} \qquad S_1 := \sqrt{\left[\left(X^T \cdot X \right)^{-1} \right]_{1,1} \cdot SS}$$

$$S_2 := \sqrt{\left[\left(X^T \cdot X \right)^{-1} \right]_{2,2} \cdot SS}$$

$$fit2_i := b_0 + b_1 \cdot x_i + b_2 \cdot \left(x_i \right)^2$$

FIGURE 1 (continued)

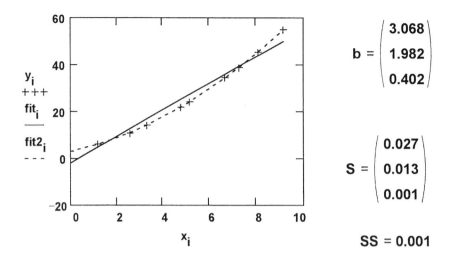

$$b = \begin{pmatrix} 3.068 \\ 1.982 \\ 0.402 \end{pmatrix}$$

$$S = \begin{pmatrix} 0.027 \\ 0.013 \\ 0.001 \end{pmatrix}$$

$$SS = 0.001$$

Note the improvement in SS

FIGURE 1 (continued)

In cases in which the data points should be weighted, the relevant formulas are more difficult to find in the literature and are presented here. In these equations, m and b are the slope and intercept of the fitted line, SS is the "sum of squares" or the variance of fit, and $S(m)$ and $S(b)$ are the uncertainties in the slope and intercept:

$$m = \frac{\sum_i w_i \sum_i w_i x_i y_i - \sum_i w_i x_i \sum_i w_i y_i}{\sum_i w_i \sum_i w_i x_i^2 - \left(\sum_i w_i x_i\right)^2} \qquad b = \frac{\sum_i w_i x_i^2 \sum_i w_i y_i \sum_i w_i x_i y_i \sum_i w_i x_i}{\sum_i w_i \sum_i x_i^2 - \left(\sum_i w_i x_i\right)^2}$$

$$SS = \frac{\sum_{i=1}^{N} \left(y_i - mx_i - b\right)^2 w_i}{N-2} \qquad S(b) = \sqrt{SS} \sqrt{\frac{\sum_i w_i x_i^2}{\sum_i w_i \sum_i w_i x_i^2 - \left(\sum_i w_i x_i\right)^2}}$$

$$S(m) = \sqrt{SS} \sqrt{\frac{N}{n \sum_i w_i x_i^2 - \left(\sum_i w_i x_i\right)^2}}$$

When is linear least squares insufficient? For unweighted linear least-squares to be sufficient for data analysis, some conditions are required: The variance must be constant for each point, and the data points must randomly lie around the fitted line with no correlation. To test these conditions, a residual analysis is often performed. For each point, define the residual as $res_i = (y_i - m*x_i - b)$ and then a plot res_i vs. x_i. Three cases are usually observed:

1. If the linear unweighted model is correct, there will be no trend observed. The points will lie in a band centered around res = 0. Under these conditions it is safe to use the model unchanged. Outliers will appear as points with unusually large values of res and may be omitted (with caution! [see Seber[4]] and the model refined.
2. If the magnitudes of res_i change smoothly with x_i, generating a "cone" either as x decreases or x increases, then a weighting factor as a function of x must be included in the analysis. As an alternative, the model may be changed through power transformations, which have the effect of removing the weighting factor.
3. If the residuals seem to be correlated, i.e., not randomly spaced around zero, then a linear model is inadequate and must be modified. The standard test for this situation is the Durbin–Watson test, in which the number of times that the residual passes through zero is recorded and compared with that expected of a random uncorrelated function. See Draper and Smith[5] for further information.

Polynomial Least-Squares Fitting.   Polynomial least-squares fitting is an extension of linear least-squares fitting that allows powers of the function to be arbitrarily added to the fitting function until convergence is met. Algorithms are easily found in the literature[6] and are supported by most statistical packages, including SAS and SPSS.

As an example of multivariate least-squares fitting, consider a set of points fit first to linear and then to quadratic functions. The set of data used in this example was generated from the function $y = 0.4*x^2 + 2x + 3$, where x ranges from 0 to 10, with random increments added to both x and y.

The results of this fit are found in Figure 1 and Mathcad template 1. (Several templates are shown explicitly in this paper as figures. Instructions for obtaining working copies of all templates mentioned are given at the end of this chapter. Other Mathcad models for physical chemistry can be found in the literature, especially the *Journal of Chemical Education.*) Identical results for the linear fit were obtained for both parameters and estimated uncertainties by using the regression package from EXCEL 4.0. Note that from a pedagogical perspective, using the Mathcad template is preferable to an EXCEL spreadsheet for teaching the fitting algorithm because all formulas in the algorithm are presented instead of being hidden in a previously constructed macro.

The improvement due to the extension to a quadratic fit is expected. The addition of a few lines to the Mathcad document is sufficient to modify the algorithm from linear to quadratic. This method could easily be extended to any order.

Nonlinear Least-Squares Fitting.   In cases where the mathematical fitting function is insufficient to fit the data, or there is theoretical justification for using a function that cannot be reduced to a polynomial equation, it is necessary to use a nonlinear least-squares regression.

The method of choice for nonlinear least-squares curve fitting is the Levenberg-Marquardt algorithm.[7] This method, a combination of two other standard nonlinear techniques, the inverse-Hessian method, and the method of steepest descent, is discussed later in this paper. The algorithm is fully described in Press et al.[8]

Template 2 presents a Mathcad formulation for a nonlinear curve fit. After generating an artificial Gaussian line with noise, the Minerr function is used to fit the data to the parameters, which in this case relate to the intensity, line width, and wavelength of a spectral line with Gaussian line shape.

Under certain conditions, nonlinear functions may be linearized by appropriate coordinate transformations, which allow linear curve-fitting methods to be used. These must be applied with caution, since round-off error of the variables may occur with the linearization, leading to larger uncertainties than the data would indicate.

Calculus Methods

Integration

Analytical integrations of functions with known antiderivatives are usually calculated using either integral tables or symbolic processors (MAPLE, Mathcad) with known lookup tables. For analytical functions without known antiderivatives, there are many procedures, some of which date from precomputer days, that have been implemented with more modern methods.

The most commonly used method of numeric integration of smooth functions is to break the region of integration into small blocks and use the trapezoidal rule. The integral then has the form

$$\int_a^b F(x)\,dx = \frac{1}{2}x_0 + \sum_{i=1}^N x_i + \frac{1}{2}x_N$$

where the region $[a,b]$ is split into N portions with end points ranging from x_0 to x_N, with an error in the order of the second derivative of F. There are other "point integral" formulas that expand upon this theme; see Stroud[9] for further details.

For integrals of x-y data points for which the analytical function is unknown, there are at least two recommended procedures: using a trapezoid rule as described above on neighboring data points, or curve-fitting to an analytical function that then can be used to produce an analytical formula for the integral.

Template 3 (Figure 2), contains a Mathcad document that performs the numerical integration of heat capacity data to determine the absolute entropy of a molecule. Different techniques, including the trapezoid rule and curve-fitting approaches, are compared.

Differentiation.   Derivatives of simple analytical functions can be performed systematically using the rules of calculus and will not be discussed here. For derivatives of complicated functions, the symmetric form of the derivative is useful. It has an error that is proportional to the square of the step size h:

$$f'(x) = \frac{f(x+h) - f(x-h)}{2h}$$

The formula given in calculus books,

$$f'(x) = \frac{f(x+h) - f(x)}{h}$$

has an error proportional to h and converges much more slowly.

Determine the absolute entropy of anhydrous potassium hexacyanoferrate(III), based upon heat capacity data. This is a Mathcad solution to problem #4.19 found in Physical Chemistry, 5th Edition, by Peter Atkins.

Assume no residual entropy. Obtain the the entropy from 0 K to the first experimental point by applying the Debye rule. T and C_p are read from ASCII files.   There are 17 data pairs in this problem.

$T := \text{READPRN}(\text{integ\_x})$ $N := \text{length}(T)$ $N = 17$

$i := 0..(N-1)$

$Cp := \text{READPRN}(\text{integ\_y})$ $M := \text{length}(Cp)$ $M = 17$

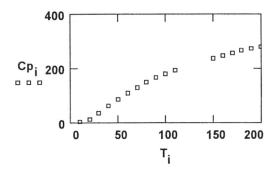

Either integrate Cp/T vs. T or Cp vs. ln(T) from the first observable point to the final point. Add $C_{p0}/3$ to give the integral from 0 to the first data point. Both curves are shown here.

FIGURE 2 Integration and curve fitting can be accomplished efficiently in Mathcad by using the trapezoid rule or by fitting a cubic polynomial to C_p/T data. The polynomial is then used to demonstrate how Mathcad is used to evaluate a definite integral.

$$CpT_i := \frac{Cp_i}{T_i} \qquad\qquad InT_i := \ln\left(T_i\right)$$

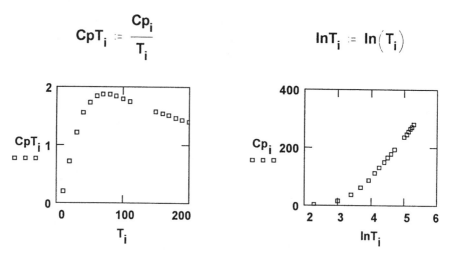

This is the preferable function for integration

Integrals

1. Trapezoid Rule

$$Int1 := \left[\sum_{i=1}^{N-1}\left[\left(\frac{CpT_i + CpT_{i-1}}{2}\right)\cdot\left(T_i - T_{i-1}\right)\right]\right] + \frac{Cp_0}{3}$$

Int1 = 296.486

$$Int2 := \left[\sum_{i=1}^{N-1}\frac{Cp_i + Cp_{i-1}}{2}\cdot\left[\ln\left(\left(T_i\right)\right) - \ln\left(\left(T_{i-1}\right)\right)\right]\right] + \frac{Cp_0}{3}$$

Int2 = 298.904

FIGURE 2 (continued)

2. Curve fitting to the polynomial: Cp=a+bT+cT^2+dT^3. After the curve fitting is complete the integral is straightforward.

$$T2 := \overrightarrow{\left(T^2\right)} \qquad i := 0 .. N - 1$$

$$T3 := \overrightarrow{\left(T^3\right)}$$

$$X_{i,0} := 1 \qquad X^{<1>} := T \qquad X^{<2>} := T2 \qquad X^{<3>} := T3$$

$$b := \left(X^T \cdot X\right)^{-1} \cdot \left(X^T \cdot Cp\right) \qquad b = \begin{bmatrix} -34.601 \\ 2.832 \\ -0.008 \\ 6.838 \cdot 10^{-6} \end{bmatrix}$$

The b vector contains the set of coefficients for the polynomial fit to the heat capacity data.

$$Int3 := \int_{T_0}^{T_{N-1}} \left(\frac{b_0}{z} + b_1 + b_2 \cdot z + b_3 \cdot z^2 \right) dz + \frac{Cp_0}{3}$$

Int3 = 299.674   Joules/degree

FIGURE 2 (continued)

For derivatives of *x-y* data pairs, Chebyshev polynomials should be used to find a polynomial representation of the function, which can then be used to find the derivative.

Chebychev Polynomials.   Chebychev polynomials are defined as functions of x:

$$T_0(x) = 1$$
$$T_1(x) = x$$
$$T_2(x) = 2x^2 - 1$$

$$T_{n+1}(x) = 2xT_n(x) - T_{n-1}(x)$$

The polynomials have the property that in the range $[-1,1]$, the values of T range between -1 and 1, with maxima that are spread evenly throughout the range. It can be shown[10] that by curve-fitting an arbitrary function to a series of N Chebychev polynomials, the error is approximated by a term proportional to $T_N(x)$, which has

N+1 maxima spread evenly throughout the interval. Thus, since the error in fit is spread evenly throughout the domain, it is also useful in spreading out errors in derivatives or integrals of functions.

For a domain ranging from arbitrary limits *a* to *b*, a change in variable,

$$y = -1 + 2 * \frac{(x-a)}{(b-a)}$$

allows a Chebychev approximation to be made to *y*, which now ranges from −1 to +1.

An example can be found in template 4, consisting of a Mathcad template in which the derivative of the EMF of an electrochemical cell with respect to temperature is calculated, using both the trapezoid rule and curve fit, to Chebychev polynomials.

Matrix Methods

A matrix is an array of numbers that can be manipulated through a set of rules. Applications of matrices in chemistry are numerous; examples range from the solution of simultaneous equations, including finding the concentrations of components in a mixture, and matrix diagonalization to determination of energies and wave functions of molecules. The most important matrix manipulations for chemists are matrix addition, multiplication, determination of the inverse, and matrix diagonalization.

Template 5 gives an example of Mathcad implementation of matrix inversion and diagonalization using built-in routines. In Mathcad, the inverse of a matrix, *A*, is calculated by evaluating A^{-1}. There also exist built-in functions of matrix diagonalization, which produce the eigenvalues and eigenvectors of a square matrix that are easy to use.

Advanced Examples Using Numerical Methods

Function Minimization Using Mathcad

In performing a molecular mechanics calculation, one tries to locate a minimum on the potential energy surface that is described by a function commonly called the potential energy function. This potential energy function describes the relationship between the atoms in the system, and this description usually consists of intramolecular and intermolecular terms. How one locates the minimum of this potential energy function is the topic of this section. Template 6, Figure 3, illustrates how the method known as steepest descent is used to transverse model functions. The steepest descent method is one of the simplest ways in which to locate a minimum of a function. The method calculates a gradient and based on the gradient makes a step in the downhill direction. The step size is taken as a scaled gradient value. The scaling depends on how large (steep) the gradient is; if the gradient is large, then a big step is taken.

Function Minimization Methods

Start with defining a function to minimize.
There are two that we will explore here.

Function #1 $f1(x,y) := x^2 + 5 \cdot y^2$

$N := 10$ $i := 0 .. N$ $xmin := -2.0$ $xmax := 2.0$ $xdel := \dfrac{xmax - xmin}{N}$

$M := 10$ $j := 0 .. M$ $ymin := -2.0$ $ymax := 2.0$ $ydel := \dfrac{ymax - ymin}{M}$

$x_i := xmin + i \cdot xdel$ $y_j := ymin + j \cdot ydel$ $z_{i,j} := f1(x_i, y_j)$

This is the contour map for the function f1(x,y) defined above

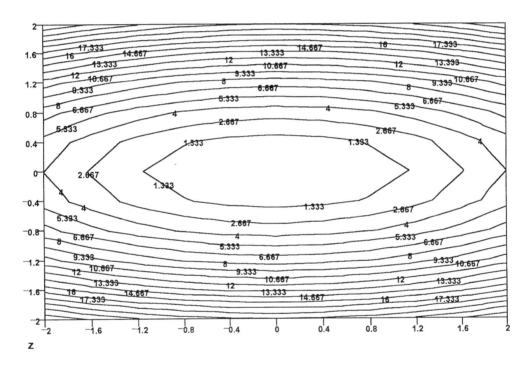

z

FIGURE 3   This Mathcad sample document illustrates how the software can be used to locate the minimum of a function using the method of steepest descent.

Function #2 $f2(x,y) := (1 - x)^2 + 100 \cdot (y - x^2)^2$

$N := 10 \quad i := 0..N \quad xmin := -1.0 \quad xmax := 1.0 \quad xdel := \dfrac{xmax - xmin}{N}$

$M := 10 \quad j := 0..M \quad ymin := -.5 \quad ymax := 1.5 \quad ydel := \dfrac{ymax - ymin}{M}$

$x_i := xmin + i \cdot xdel \qquad y_j := ymin + j \cdot ydel \qquad z_{i,j} := f2(x_i, y_j)$

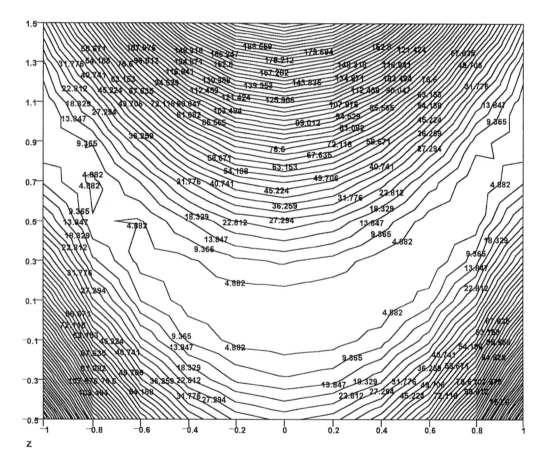

z

FIGURE 3 (continued)

steps := 20   **Number of steps to be taken**

$i := 0 .. steps$ $j := 0 .. 1$ $k := 1 .. steps$

$$r1_{i,j} :=$$

$g1_0 := 0.75$ $g1_1 := g1_0$ $g1_2 := g1_0$ $\begin{bmatrix} 1.5 \\ 1.5 \end{bmatrix}$ **Initial starting point**

Steepest descent:

$$gradx(v,w) := \frac{d}{dv}\left[(1-v)^2 + 100 \cdot (w - v^2)^2\right]$$

$$grady(v,w) := \frac{d}{dw}\left[(1-v)^2 + 100 \cdot (w - v^2)^2\right]$$

$$grad1(v,w,j) := if(j=0, gradx(v,w), grady(v,w))$$

$$\begin{pmatrix} g1_k \\ r1_{k,j} \end{pmatrix} := \begin{bmatrix} if\left[k>2, if\left[\left(f2\left(r1_{k-1,0},r1_{k-1,1}\right) - f2\left(r1_{k-2,0},r1_{k-2,1}\right)>0\right), 0.3 \cdot g1_{k-1}, 1.2 \cdot g1_{k-1}\right], g1_0\right] \\ r1_{k-1,j} - if\left(k>2, g1_{k-1}, g1_k\right) \cdot \dfrac{grad1\left(r1_{k-1,0},r1_{k-1,1},j\right)}{\sqrt{\displaystyle\sum_{l=0}^{1} grad1\left(r1_{k-1,0},r1_{k-1,1},l\right)^2}} \end{bmatrix}$$

The resulting data, shown below, can be plotted on the contour plot for f2(x,y).

	0	1
0	1.5	1.5
1	0.788	1.737
2	1.422	1.335
3	0.714	1.584
4	1.452	1.068
5	1.197	1.156
6	0.898	1.281
7	1.237	1.091
8	1.129	1.135
9	1.001	1.192
10	1.151	1.117
11	1.105	1.137
12	1.091	1.143
13	1.075	1.15
14	1.055	1.159

$r1 =$

$$grad1\left(r1_{0,0}, r1_{0,1}, 0\right) = 451$$

$$grad1\left(r1_{0,0}, r1_{0,1}, 1\right) = -150$$

$$f2\left(r1_{0,0}, r1_{0,1}\right) = 56.5$$

$$grad1\left(r1_{steps,0}, r1_{steps,1}, 0\right) = -3.71$$

$$grad1\left(r1_{steps,0}, r1_{steps,1}, 1\right) = 1.8$$

$$f2\left(r1_{steps,0}, r1_{steps,1}\right) = 0.013$$

$$grad1(1,1,0) = -7.772 \cdot 10^{-14}$$

$$grad1(1,1,1) = 0$$

$$f2(1,1) = 0$$

FIGURE 3 (continued)

Molecular Dynamics Using Mathcad

Molecular dynamics is the simulation of the motion of atoms and molecules. In particular, one can simulate pure liquids, small molecules in solution, proteins in vacuum, or proteins in solution. Not only does molecular dynamics determine the motions of atoms and molecules, but the results of a molecular dynamics simulation can be used to calculate thermodynamic variables such as ΔH, C_p, etc. There are three basic concepts one needs to know in order to perform a molecular dynamics simulation. They are (1) the potential function, (2) calculation of the forces from the potential function, and (3) the relationship between the forces and displacement (Newton's second law)—i.e., the equations of motion. Template 7 allows the student to see and understand each of the components of a molecular dynamics simulation for two argon atoms in a periodic box. In the document the student can explore the potential energy function by changing the Lennard–Jones parameters sigma and epsilon. The student learns about terms such as periodic images, minimum image, and cutoffs. In addition, the document brings to the student's attention the relationship between atom velocity and temperature as well as the knowledge that experimentally measured temperature is an average of instantaneous temperatures.

Fitting Ab Initio Data to a Potential Function

As mentioned previously the potential function describes the interactions between the atoms and molecules of a system. The potential function contains several parameters that must be defined before it can be used. Some of these parameters, such as equilibrium bond lengths and angles as well as the corresponding force constants, can be obtained from experimental data. Other parameters, such as nonbonded interactions and charges, do not have corresponding experimental data from which they can easily be determined. Therefore, some effort has been made to use ab initio calculations to determine the values for the nonbonded interactions and charges. Template 8 demonstrates how one uses ab initio data to calculate the sigma and epsilon of the Lennard–Jones potential function for an argon dimer.

Numerical Methods in Quantum Chemistry

Solving Schrödinger's equation is the primary goal in the field of quantum chemistry. However, Schrödinger's equation can only be solved exactly for a very small number of simple systems. Therefore, the purpose of this section is to illustrate three computational methods used to obtain approximate solutions for Schrödinger's equation: numerical integration of Schrödinger's equation, an LCAO-MO variational calculation, and a calculation based on the perturbation method. A prior study of simple applications, such as the ones presented here, enables one to appreciate more fully the more complicated electronic structure calculations being done routinely today on fairly large molecular systems.[11] The material presented in this section will draw upon some of the numerical techniques described earlier and also introduce several methods that have not been presented previously. Additional examples of the use of Mathcad in the area of quantum chemistry can be found in the literature.[12–15]

Numerical Integration of Schrödinger's Equation

Template 9 shows a Mathcad worksheet for solving Schrödinger's equation numerically for a particle in an infinite potential well with an internal barrier. It employs a reasonably accurate finite difference algorithm reported in the physics literature by Bolemon.[16] One of the appealing features of doing a calculation of this type in the Mathcad programming environment is that the integration algorithm is clearly seen by the user: Everything of significance is out in the open and looks like it does in a textbook or a set of lecture notes.

The algorithm works in the following way. The grid size is determined by dividing the difference between the limits of integration into small segments. The left boundary value provides a value for $\Psi(x\text{-}\delta)$, and $\Psi(x)$ is given an arbitrary positive value. Next, a guess is made for the energy, and the algorithm generates a value of Ψ for each value of x on the grid. If the wave function so-generated satisfies the right boundary condition, then a solution to Schrödinger's equation has been found. If the wave function does not satisfy the right boundary condition, another guess for the energy is made. In general it is easy to bracket the correct energy eigenvalue and zero in on a solution.

A special feature of this Mathcad worksheet is that it generates the momentum-space distribution function by numerical Fourier transformation of the position-space wave function. It is also important to note that this worksheet is a template for all other one-dimensional problems. The main difference between one quantum mechanical problem and another is in the potential energy term. Thus, while other minor parameters will also have to be changed, the major change in moving from one problem to the next is to edit the potential energy term. Of course, it is also quite easy to prepare Mathcad templates for the numeric integration of the radial part of Schrödinger's equation for two- and three-dimensional problems.[13,14]

Variational LCAO-MO Calculation on H_2^+

The variation principle, which is generally combined with the SCF procedure in atomic and molecular applications,[11] is introduced with a very simple molecular orbital calculation on the hydrogen molecule ion shown in template 10. In the variation method the expectation value of the energy,

$$\langle E \rangle = \int \Psi^* \hat{H} \Psi d\tau$$

is minimized with respect to any variational parameters contained in the trial wave function, and the minimized energy is an upper bound to the true energy of the system under study. Therefore, a fundamental tenet of the variational procedure is that the best trial wave function is the one that yields the lowest energy, assuming, of course, that the Hamiltonian energy operator is correct.

Today, high-level ab initio calculations on large molecules have become routine due to the tremendous advances in computer hardware technology and the development of very efficient electronic structure codes.[11] However, to fully appreciate a complex calculation it is necessary to understand the simplest of ab initio molecular orbital calculations. To this end, template 10 illustrates this calculation on H_2^+ using Mathcad.[12,13] A calculation of this type is a standard illustrative exercise in all introductory quantum chemistry courses, and many years ago Robiette[17] published a version of this calculation using BASIC. A novel feature of the calculation presented here is that it incorporates part of Ruedenberg's well-known[18] analysis of the physical nature of the chemical bond. As the exercise indicates, it is very easy to extend this

Numerical Solutions for the Particle in the 1-D Box with an Internal Barrier

Set the parameters:

Grid: $n := 300$ Mass: $\mu := 1$ V_o: $V_0 := 100$

Left boundary: $lb := .45$ Right boundary: $rb := .55$

Integration limits: $xmin := 0$ $xmax := 1$

$$\Delta := \frac{xmax - xmin}{n}$$

Initial values for the wave function: $\Psi_0 := 0$ $\Psi_1 := .001$

Calculate the potential energy:

$i := 1 .. n$ $x_i := xmin + i \cdot \Delta$ $V_i := if\left[\left(x_i \geq lb \right) \cdot \left(x_i \leq rb \right), V_0, 0 \right]$

$f_i := 2 \cdot \mu \cdot \left(V_i - energy \right)$ **The energy is defined by a global equality given below.**

The integration algorithm: $i := 2, 3 .. n$

$$\Psi_i := \frac{\left(2 \cdot \Psi_{i-1} - \Psi_{i-2} \right) + \dfrac{\Delta^2}{12} \cdot \left(f_{i-2} \Psi_{i-2} + 10 \cdot f_{i-1} \Psi_{i-1} \right)}{\left(1 - f_i \cdot \dfrac{\Delta^2}{12} \right)}$$

Make energy guess: $energy \equiv 15.64$ $i := 0 .. n$

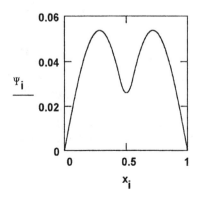
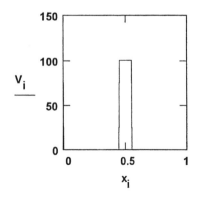

FIGURE 4   This sample document shows how Mathcad software can be used to numerically solve differential equations of interest to physical chemists. This case presents the application of the technique to the problem of the particle in a 1-D box with an internal barrier.

Transform the coordinate-space wave function to a momentum space wave function.

Define i: $i := \sqrt{-1}$

Define the range for the momentum: $p := -20, -19.5 .. 20$

Evaluate the Fourier transform numerically:

$$\Phi(p) := \frac{i}{\sqrt{n}} \cdot \sum_{j=0}^{n} \Psi_j \cdot \exp\left(-i \cdot p \cdot x_j\right)$$

Display the wave function:

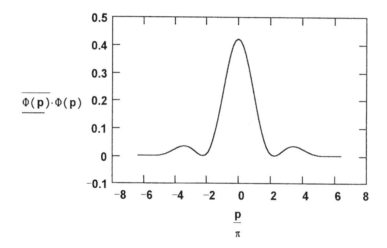

Sample Exercise:

Find the first six energy levels for the particle in the box with a 100-Hartree internal barrier. Sketch the wave function and plot the energies on a vertical scale. Change the barrier height to 50 hartrees and repeat the exercise.

FIGURE 4 (continued)

calculation to include a molecular orbital formed from a linear combination of scaled hydrogen 1s orbitals. A comparison of the results of this calculation with the one using unscaled orbitals helps clarify the basic idea behind the variational theorem. The molecular orbital calculation presented in template 10 might be considered the simplest example of a geometry optimization.

Perturbation Theory

Another important way of finding approximate solutions to Schrödinger's equation is perturbation theory. The hydrogen atom has an exact quantum mechanical solution yielding energy eigenvalues and associated eigenfunctions. However, Schrödinger's equation for the hydrogen atom in an external electric field does not have an exact analytical solution. Fortunately, an approximate solution can be obtained by assuming that the electric field is a small perturbation on the total state of the hydrogen atom. In the perturbation method, as shown in Template 11, the eigenfunctions for the unperturbed hydrogen atom are used to calculate a correction to the energy due to the presence of the electric field. This particular exercise is an example of degenerate perturbation theory.

Templates

The material contained in all of the templates, plus additional Mathcad documents, can be obtained using the following Internet site:

http://milo.chem.usouthal.edu/Mathcad

The files can be downloaded using Netscape or a similar WWW browser. Use Mathcad 6.0 to view these files. The Mathcad documents for the templates have the form

templatX.mcd

where X is the template number. For example, to look at the template 1 document, one would enter the following into Mathcad:
http://milo.chem.usouthal.edu/Mathcad/templat1.mcd

Additional Mathcad materials can be found at http://www.niagara.edu/~tjz

References

1. Matthews, P.; *Experimental Physical Chemistry;* Clarendon Press: Oxford, 1985; p xxv.
2. Shoemaker, D.; Garland, C.; Nibler, J.; *Experiments in Physical Chemistry,* 5th ed.; McGraw-Hill: New York, 1989, p 801
3. Syme, R.; *Physical Chemistry: Methods, Techniques, and Experiments;* Saunders: Philadelphia, 1990; p 72.
4. Seber, G.; *Multivariate Observations;* Wiley: New York, 1984; p 169.
5. Draper, N.; Smith, H.; *Applied Regression Analysis,* 2nd ed.; Wiley: New York, 1981.
6. Bevington, P.; *Data Reduction and Error Analysis for the Physical Sciences,* 2nd ed.; McGraw-Hill: New York, 1992.

7. Marquardt, D.; *Journal for the Society of Industrial and Applied Mathematics* 1963, *11*, 431.

8. Press, W.; Teukolsky, S.; Vetterling, W.; Flannery, B.; *Numerical Recipes in Fortran: The Art of Scientific Computing,* 2nd. ed.; Cambridge University Press: Cambridge, 1992; p 678.

9. Stroud, A.; *Approximate Calculation of Multiple Integrals;* Prentice-Hall: Englewood Cliffs, NJ, 1971.

10. Press, W. op. cit., p 184.

11. Hehre, W.; Radom, L.; Schleyer, P. v. R.; Pople, J.; *Ab Initio Molecular Orbital Theory;* Wiley: New York, 1986.

12. Coleman, W. J.; *Chem. Educ.* 1990, *67,* A203.

13. Rioux, F.; *J. Chem. Educ.* 1992, *69,* A240.

14. Rioux, F.; *J. Chem. Educ.: Soft.* 1993, Volume 1D, Number 2.

15. Rioux, F.; *J. Chem. Educ.: Soft.* 1995, Volume 3D, Number 1.

16. Bolemon, J. S.; *Am. J. Phys.* 1972, *40,* 1511.

17. Robiette, A. G.; *J. Chem. Educ.* 1975, *52,* 95.

18. Ruedenberg, K.; *Rev. Mod. Phys.* 1962, *34,* 326.

A Framework for the Teaching of Computer–Instrument Interfacing

Kenneth L. Ratzlaff

Introduction

Computer interfacing (or, more precisely, computer–instrument interfacing) is the task of making possible the transmission of information between the computer and an external device for purposes of data acquisition and control.

Because the range of measurable phenomena (the whole world of material existence) is unlimited, the topic of computer–instrument interfacing is as broad as the range of scientific experimental endeavor. As we will find, the subjects involved in interfacing can include computer architecture, programming, electronics, and sensors and actuators; the latter topics can bring in mechanical or electronic engineering or bring us back to the chemistry of the measurement itself.

At the same time, chemistry students (or computer science students for that matter) seldom receive any exposure to even the fundamentals of the interface outside the chemistry curriculum. Consequently, in addressing the question of "what every chemist should know...", we must begin by defining not only terms but also our goal.

Goal

The goal of a chemistry student's study of interfacing can be that the student understand interfacing sufficiently well that he or she might function effectively as a research chemist. That goal must be attained without expecting a background in electronics beyond a knowledge of Ohm's Law and without teaching moderately complex topics such as transmission lines, circuit analysis, or even the solid-state physics of electronic devices.

FIGURE 1   Representation of the off-line mode.

In order to reach that target, we will establish the following objectives:

- An understanding of the information/data flow between real-world phenomena and the computer. As Megargle points out in Chapter 3, most chemists will not have the task of interfacing an instrument; however, to use that instrument effectively, the concepts of data flow are necessary.
- Some understanding of possibilities and limits inherent in computer-based data acquisition and control.
- Acquisition of sufficient concepts and jargon that the chemist might collaborate with the engineer or other interfacing professional.

Modes of Laboratory Computing

A look at how the computer interacts with the experiment can help define the scope of this study.

Before smaller computers that could be dedicated to an experiment became available, laboratory data could only be computer-processed on a shared mainframe. The user was a direct link between the experiment and the computer, and this mode of computing is termed off-line (Figure 1).[1] Clearly, the quality of interchange between computer and experiment is dependent on the human link.

Stimulated to a significant extent by NASA's requirements, computers were developed in the 1960s that were small enough to dedicate to single instruments. The mode of operation changed because the computer and experiment could exchange data directly and, as a consequence, at high speed. This mode is termed on-line (Figure 2).

Computers continued to decline in price so that more and more experiments could be operated in the on-line mode. After undergoing a cost and size revolution, the ubiquitous microprocessor came to be found in almost every type of instrument.

However, the microprocessor enabled a new mode of interaction in which the computer also controlled the user's interaction. The term in-line, and Figure 3, can describe this mode.[2,3] Virtually every commercial instrument and electronic tool in today's laboratory follows the in-line mode.[2,3] An operational comparison of these modes is found elsewhere.[1]

Modern experimental methods and instrumentation cannot be understood without understanding how data flows between computer and experiment, and new instruments for new experiments cannot be developed in the laboratory without understanding the techniques. Disciplines outside science and engineering seldom require this understanding, and computing courses outside the sciences seldom

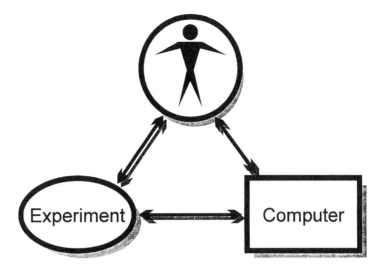

FIGURE 2   Representation of the on-line mode.

present these techniques. Consequently, chemists often provide their students' only exposure to computer–instrument interfacing. An approach is presented in this chapter.

Some Basic Terminology

Interface.   An interface is the point of exchange across a boundary. Thus, in chemistry, for example, we speak of an interface between phases. The computer–instrument interface facilitates the exchange between the experiment and the computer, enabling exchange in both directions for data acquisition and experimental control.

Real-Time.   The effective use of a computer in an experiment closes a loop: The data are acquired and processed during the experiment so that experimental control is based on the data of that experiment. When this loop is closed, the computer is operating in real time, as opposed to processing data after the experiment is complete, in preparation for a new experiment.

Transducers: Sensors and Actuators.   Sensors are devices that, for purposes of electronic data acquisition, change an electrical parameter in response to a change in a physical or chemical phenomenon. An actuator operates in reverse, changing a chemical or physical parameter in response to an electrical signal. For practical purposes, transducers are sensors and actuators.

Two important examples of sensors for chemists are the pH electrode and the photomultiplier tube. The former is an active transducer: it generates a signal (potential) independent of any energy source external to the sample being studied. The photomultiplier tube is a passive transducer: its signal (current) depends not only on the light intensity but also on the applied voltage.

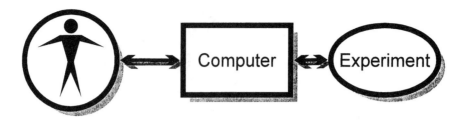

FIGURE 3   Representation of the in-line mode.

Data Domains

Scientific study requires the measurement of information that is physical or chemical in nature: temperature, pH, position, flow, intensity, and so on. The target information to be captured lies in the chemical or physical domain.[4] The task will be to accurately transfer that information to the computer and ultimately, in many cases, to the user.

These phenomena that must be controlled and measured in the physical and chemical domains are, for the most part, analog. That is, for practical purposes the values are continuous. Analog information is subject to the influence of noise—any unwanted component of the information—and because the noise is also analog, it is measured and transmitted together with the desired information.

In some cases the physical phenomenon may not be continuous but is commonly treated as such; for example, if photon flux is much greater than 10^7-sec^{-1}, it is measured as a continuous parameter, even though light-intensity photons are discrete.

If the computer is to process the data, the information must arrive in the digital domain, encoded in a format suitable for the digital computer. That encoding method is binary. The numerical and logical information is represented by a series of bits that can only take the logical values 1 or 0, HIGH or LOW (frequently written HI or LO), ON or OFF; with no allowed values in between the two states, the data should no longer be susceptible to the addition of noise during processing.

The Electronic Representation of Binary Values.

A somewhat arbitrary standard exists for representing a binary value (a bit) electronically. At the "point of entry" into the computer, the datum must conform to the computer's standard. Consequently, that standard must be clear.

The logic levels 1 and 0 are defined in TTL (transistor–transistor logic).[5] First, logic 1 is +5 V and a 0 is ground (zero volts). In practice, to account for the non-idealities in the system, the voltage range for logic 0 extends to approximately 0.6 V (depending somewhat on the transistor technology in use), and logic 1 is recognized by inputs as voltages from about 2.4 V to 5 V. In any case, there is a significant gap between 1 and 0.

The voltage does not completely define the system. TTL is designed so that outputs can "sink" current but not be a source for current to power another device; that means that a logic 1 will do a poor job of being a +5-V power source, but a logic 0 acts as a good ground. Furthermore, in true TTL technology, if an input pin is unconnected, the circuit will respond as if that pin were connected to a logic 1, a nonintuitive phenomenon for many of us.

These facts will come into play in interfacing methods for sensing and for control.

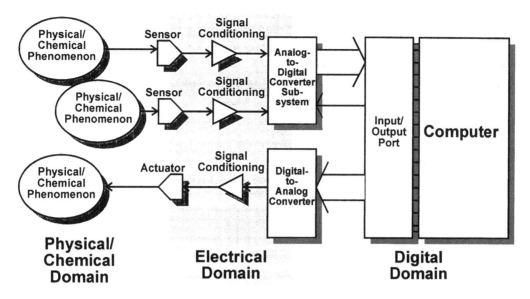

FIGURE 4   Information transfer from the chemical/physical domain to the computer and back.

The Interdomain Transfers in Interfacing.

The process of acquisition of information is outlined in Figure 4. Information travels between the analog physical/chemical domain and the binary digital domain of the digital computer.

For input, a sensor generates an analog electrical signal representative of the property measured. That signal is electrically conditioned to match the input of the analog-to-digital converter subsystem, which produces a digital representation that can be read by the computer through the input port.

Similarly for output, the output port feeds a value to a digital-to-analog converter whose output is usually voltage. That voltage is conditioned (usually amplified in voltage or current) for control of an output device or actuator that can affect the physical or chemical property.

For each type of conversion, the transfer or conversion must preserve the integrity of the information. In the following sections, various components of this system will be reviewed.

Conversions To and From the Digital Domain

Conversion from the Electrical Analog to the Digital Domain[6]

A properly conditioned electrical signal representing a chemical or physical property is converted from the continuous electrical domain to the digital domain with an analog-to-digital converter (ADC), a device that is usually a single integrated circuit but is part of an ADC subsystem. Over the past two decades, the techniques of conversion have changed little, but the size has shrunk and the price has dropped about three orders of magnitude, while performance and convenience continue to improve.

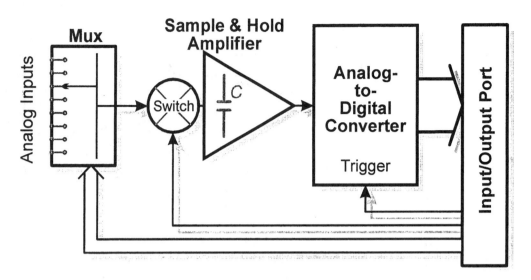

FIGURE 5   The typical ADC subsystem.

The configuration for a basic ADC subsystem board is shown in block diagram in Figure 5. As a unit, it usually has 1–8 inputs, for voltage typically in the 0 to 10 V range. The output to the computer port is typically 8–12 bits of binary data. Some knowledge of the operation is necessary for selection and optimal operation of the full system.

The Multiplexer.   In the past, the ADC was a very expensive and bulky component and, consequently, was shared between a group of signal inputs. In standard design, it still is usually shared among signal inputs with a solid-state switch called a multiplexer (MUX). The technology of these switches can be studied elsewhere,[4] but two key properties are:

1. Switching is not instantaneous and may require as much as tens of microseconds;
2. Because only one channel is selected at a time, the configuration of Figure 5 requires that the multiple input signals be measured sequentially.

Precision Conversion from the Electrical Domain.[1,7]   Many conversion technologies exist for analog-to-digital conversion, but three are important, presenting speed and signal-to-noise-ratio (SNR) trade-offs. Each is a comparison method in which the input signal is compared with a reference signal.

Successive Approximation.   General-purpose systems use the method of successive approximation, and the conversion requires time, typically 2–20 µs. This implies that the conversion must be started at a particular instant, and the result cannot be transferred to the computer until the conversion is complete.

Furthermore, if the device makes successive approximations, it cannot have a moving target. The sample-and-hold amplifier (S&H) closes a switch to sample the signal and opens it to hold the signal constant during the conversion process. The signal held by the S&H amplifier is representative of less than a microsecond of the input signal.

In summary, the successive-approximation technique has moderate speed and samples the signal only briefly. Commercial devices typically convert to 1 part in 4,096 (12 bits) but are available from 8 bits (1:256) to 16 bits (1:65,536).

Integrating ADCs.   Some devices integrate the signal over the period of the measurement. Several integrating technologies exist, but they share these features: no S&H amplifier is necessary, and the conversion is relatively slow, typically on the order of 10 ms. By integrating the sources of noise, these converters are capable of very high resolution, up to 1 part in 10^6 (20 bits). Chromatographic data stations and other high-precision but low-speed applications (on the computer's time scale) typically use this technology.

High-Speed ADCs.   For conversion rates of 1–100 MHz, flash converters are required. Although scientists frequently need to make measurements at these rates, the biggest market is for digitizing video signals; consequently, these are often called video ADCs. The technology can be expensive (an 8-bit converter requires 256 high-quality amplifiers) and the resolution is typically 6–10 bits. The technique differs from the previous in that no timed sequence of events takes place; instead, the speed is determined primarily by the settling time of the analog amplifiers.

Sampling Constraints.   The concept of sampling the signal is very important. While the traditional strip-chart recorder, having a relatively long response time, performs some averaging or filtering, the result of a conversion by a general-purpose ADC represents less than a microsecond of the signal.

An important consequence can be aliasing.[1,5] When a wave form is not sampled at a sufficient frequency, the result, when the points are connected, is a wave form whose apparent frequency is lower than the actual frequency. An example of the resultant wave form for an undersampled signal is shown in Figure 6.

Two conclusions must be emphasized:

1. The wave form must be acquired at a frequency that is greater than twice the highest frequency in the signal. Unfortunately, that frequency is not obvious for nonsinusoidal signals, and students are easily lulled into expecting that for repeating signals the highest frequency is the reciprocal of the repeat period. Fourier analysis should be invoked, not from a mathematical point of view but from a graphical intuitive view. The illustrations in the book by Brigham[8] can be very helpful.

2. Noise whose frequency is greater than half the acquisition frequency must be filtered in the analog domain before digitization. If high-frequency noise is present, that signal component will also be aliased so that it then appears at the same frequency as the desired information. Noise at a frequency higher than that of the desired signal can be removed with analog filters, but if it is aliased to the same range of frequencies as the signal, it cannot easily be removed by digital techniques. Consequently, an analog filter should be employed to remove frequencies higher than half the acquisition frequency; this filter is often described as an antialiasing filter.

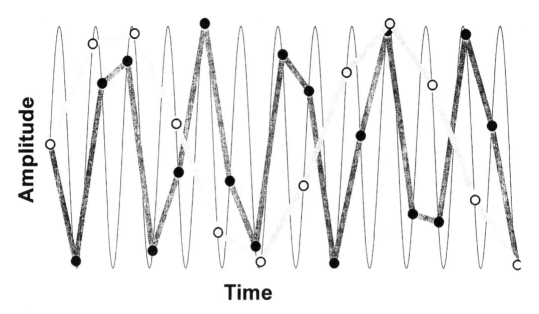

Amplitude

Time

FIGURE 6   Examples of aliased signals. The gray traces represent aliased images of the original trace. The closed circles illustrate acquisitions at less than one point per cycle, and the open circles illustrate acquisition at between one and two points per cycle.

An advantage to integrating ADCs is that if they convert continuously, the integration process provides the antialiasing remedy. If sampling ADCs are employed, oversampling by an order of magnitude or more may be possible; the results can be block-averaged for noise reduction.

Single-Bit Conversion.   In many cases, only on/off information is needed. The measurement may involve setting a threshold for a continuous parameter, or the parameter may be inherently digital.

A contact closure is inherently digital: it is either closed or open. With a single resistor and a +5 V power source (Figure 7), that switch will conform to a TTL input (see "The Electronic Representation of Binary Values", p. 190) and produce the logic levels of +5 V and ground, compatible with standard digital input ports. If the input conforms to standard TTL technology, the resistor R_{pu} will actually be unnecessary. This switch circuit can be used for detecting when a door is shut or when a button is pressed by a user.

The signal from a phototransistor that detects when a light beam is blocked may be converted to computer logic levels with the addition of only a transistor, since the two states, light and dark, are clearly separated.

An important ADC converts continuous voltage levels to only one bit of precision: the comparator. It simply delivers a logic 1 or 0 in response to the magnitude of the signal relative to a threshold. The operational amplifier technology that is used is described below.

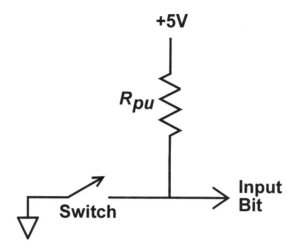

FIGURE 7 Interface for a contact closure. R_{pu} represents a "pull-up" resistor of about 1 kW.

Conversion from the Digital Domain to the Electrical Domain.

The complement to analog data acquisition is the analog control of external devices. Applications could include setting the potential on an electrode or controlling a high-power device. We will take a brief look at several categories of output.

Conversion to the Analog Electrical Domain. A digital-to-analog converter (DAC) accepts a binary input from an I/O port and produces a current or voltage in response; the devices on commercial analog I/O cards nearly universally output voltage in a 10 V range. Response is relatively fast, typically microseconds or less. DACs typically deliver only up to a few milliamps; higher currents present heat-dissipation problems in the integrated circuits.

Many high-capacity power supplies have external control inputs; an input signal of 0 to 5 V from the DAC may be amplified by the power supply to tens of amps or thousands of volts.

Single-Bit (On–Off) Control. An output bit from a computer port can switch power for many types of applications, and many have relatively simple solutions involving single transistors or modules.

Switching DC power is uncomplicated with power-FET transistors switching the ground side of the load. However, the concept of switching the low side rather than the high side, as illustrated in Figure 8, is nonintuitive and difficult for many students. To place the switch at the +5 V side of the load is possible but would require greater complexity since the voltage at the control input would have to be greater.

Many or most devices powered from 120-V ac can be switched with a solid-state relay,[24] which is controlled directly from a standard output bit of a computer port. These are available as self-contained modules, allowing even rather large devices to be controlled simply and safely.

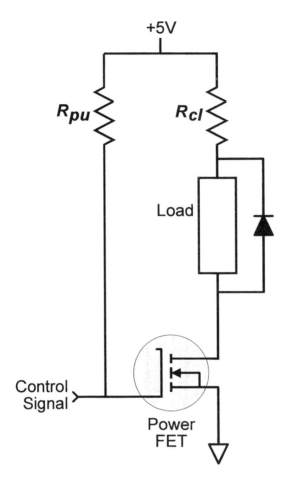

+5V

R_{pu}

R_{cl}

Load

Control Signal

Power FET

FIGURE 8   An example of switching power to a dc load. The load could be an incandescent bulb, a motor, etc. R_{pu} is a pull-up resistor, and R_{cl} limits the current to the load.

Pulse-Width Modulation for Power Control.   Instead of controlling the level of the power delivered to a load, the average power can be controlled by the technique of pulse-width modulation (PWM). The control signal is a square wave whose ratio of time on to the period (the duty cycle) is controlled.

PWM makes much more efficient use of power than the alternative of placing a source of resistance in series with the load to be powered; wasted power in that resistance must be dissipated as heat ($P = i^2 R$ where P is power in watts, i is current in amps, and R is resistance in ohms). For a switch, however, when the power is switched on, R is near zero, and when the power is switched off, i is zero.

Since the power is delivered in pulses, the load must have inertia: a motor that will keep turning and an incandescent lamp that will keep glowing between pulses which occur typically at 20 KHz.

Sensors and Actuators

A good sensor has the characteristics of a good analytical method: good sensitivity, limit of detection, range, selectivity, linearity, and rapid response, plus small size and price.[1,5] The output can be any type of electrical signal, which will be converted and conditioned for the ADC input.

Useful Sensor Examples

Temperature sensors provide examples of devices that vary dramatically in range and sensitivity. The outputs may be resistance (thermistor), voltage (thermocouple), or current (integrated-circuit transducer).

The ubiquitous pH electrode provides a valuable lesson in output impedance and is often interfaced in student experiments.[9,10]

The photomultiplier tube generates a current, and when employed in a closed loop, it illustrates extended dynamic range with computer control of applied voltage.[1]

Of course, many people simply use the entire instrument as a sensor, as in an example in which the NMR instrument is treated simply as a sensor for the instrument created in the computer.[11]

Useful Actuator Examples

Typically, the computer will be used to turn a device such as a motor or a light source on and off.

Light-emitting diodes (LEDs) typically require 5 to 40 ma dc. It may be possible to switch an LED directly from a port's output bit by connecting the output bit to the LED and a current-limiting resistor to +5 V. (Estimate resistance R from Ohm's law: $R = V/I = +5$ V$/0.02$ A $= 250\ \Omega$).

Small motors are easily interfaced with the circuitry of Figure 8. Stepping motors are very powerful tools for positioning under computer control in the laboratory. Four windings are interfaced, and some programming is required; the motor that steps the head of a 5.25 in. diskette drive can be controlled from the computer with only a 5 V power source and four power FET transistors. The built-in printer port can control the FET switches.[12]

Analog Signal Conditioning

At this point we are ready to bridge the electronic gap between the requirements of the sensor and those of the ADC input. Each type of sensor brings its own set of electrical characteristics: signal range, output impedance, response time, bandwidth, etc. This must be matched with an appropriate amplifier or series of amplifiers. The output of the amplifier(s) must then match the input range (0–10 V, typically) of the ADC.

Introduction to Operational Amplifiers

A revolution was enabled by the introduction of operational amplifiers (OAs) so that chemists were able to design amplifier systems whose characteristics were defined

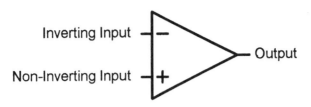

FIGURE 9   The symbol for the operational amplifier.

by the resistors and capacitors in the circuit.[13] Today's OAs are easier than ever to use and require fewer performance trade-offs than ever.[4,14,15]

The OA, shown in schematic in Figure 9, has three signal pins: the output, and the inverting and noninverting inputs. Without external components in the feedback loop, it has a very simple transfer function: The difference of the two inputs is multiplied by infinity (ideally); that function is limited by the voltage of the power supply (not shown).

Most OA circuits use negative feedback, a concept that is important to entire instrumentation systems. When the feedback loop is properly configured, a beginning student can understand the circuit by applying the combination of Ohm's law and the principle that "the OA generates the output that is necessary to keep the two input voltages equal."

Some Representative Circuits

The operation of the amplifier depends on the configuration of circuit components in a feedback loop. Without these components, the amplifier goes "out of control" in one direction or the other, depending on which input is greater.

In fact, this "out-of-control" amplifier is a useful function on its own—a sensitive comparator whereby a signal can be compared with a reference voltage (see Single-Bit Conversion). Integrated circuits sold as comparators are simply OAs with an open loop and an output adapted to TTL.

Important instrumentation examples can be studied after understanding a few OA circuit configurations:

- The follower amplifier, as in Figure 10a, is a buffer for sensors with high-impedance outputs such as the pH electrode. The output is fed back to the inverting input so that the output (and inverting input) voltage will be equal to the noninverting input voltage; however, the inputs of properly chosen OAs draw so little current that they can be used with pH electrodes.
- The current-to-voltage amplifier, as in Figure 10b, is used for current measurements from the working electrode of a potentiostat and from photomultiplier tubes.
- The inverter with gain, as in Figure 10c, is used to amplify voltage to match the sensor output to the ADC input.
- Addition of a capacitor to the inverter with gain, as in Figure 10d, creates a low-pass filter that can perform as an antialiasing filter.

OAs were once the building blocks of the analog computers that preceded digital computers, and ingenious designers invented hundreds of other functionalities: many sorts of filters, integrators, differentiators, logarithmic conversion amplifiers, and

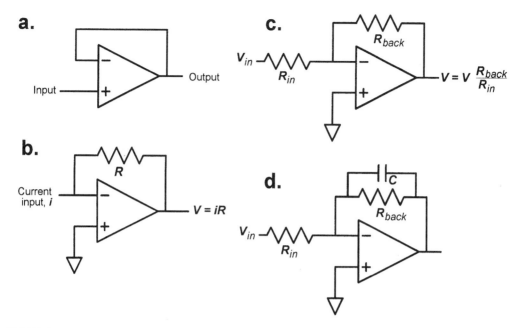

FIGURE 10 Some basic operational amplifier configurations. a. follower; b. current-to-voltage converter; c. inverter with gain; d. as in c, but with low-pass filter.

more. Most of these functions are better turned over to the digital computer in today's instrumentation, but those shown in Figure 10 are still necessary.

Timing and Interrupts

The acquisition of data for many experiments must be controlled with precision timing, and furthermore, acquisition and control must continue even while many other processes coexist in the computing environment. Most computing applications outside the laboratory are not dependent on precision timing; if the computer fails to respond for 200 ms, it is not noticeable. However, for data acquisition and control, prompt response is required. Part of the response capability in the real-time environment comes from the use of interrupts; precision timers generate the time base.

Interrupts

Modern computers have inputs that respond to a pulse from the interface so that the current task is suspended and a different task is performed. The software for the response task is the interrupt service routine, and in data acquisition and control, that routine is "a subroutine called by hardware" rather than by software.

If a computer or a program is to both be able to carry on computing-intensive tasks and be immediately responsive, interrupt-handling is required. (It's like being able to get your taxes done, but when your child needs a shoe tied, you pause and do it.)

Interrupts are occurring continuously already in the computer, even while your programmed task continues: clock updates occur at about 18-s^{-1}; position updates are sensed when you move your mouse, etc. We can program another clock in the computer to interrupt at rates in excess of 1 KHz to execute a task that we define. Within an experiment, events in the experiment such as a heartbeat can be interfaced to trigger an interrupt; then heart rate can be monitored continuously in the background while other activities (such as plotting) continue in the foreground.

For timed data acquisition, either a precision clock can trigger the ADC and the pulse signaling end-of-conversion can trigger an interrupt, or the clock can trigger an interrupt and the resulting interrupt service routine can contain code to trigger the ADC and retrieve the result. In the latter method there will be some delay between the clock pulse and the trigger for which one may need to account.

Employment of interrupts makes processing nonlinear; the computer appears to do several things at the same time. This is different from the linear program flow concepts of our beginning programming courses, and it requires careful thought.

There is, however, an important caveat. Newer operating systems, such as Windows, OS2, and MacOS, are designed to "juggle" multiple tasks, and to accomplish that, they take control of the interrupt processing. Use of interrupts and the computer's internal clocks to control precision experiment timing has become much more complex and beyond the scope of a beginning treatment.

Timing

If a time-dependent phenomenon is being measured, accuracy in the time base is as important as accuracy in the acquisition of magnitude. The source of this accuracy is a crystal-based oscillator that drives a counter; the counter counts down to zero, delivers a pulse, and starts over, generating a train of pulses at precise intervals. The divider might be on the laboratory I/O board, or it may be the clock chip on the computer.

Some users may remember the technique used often in Apple II environments in which timing was handled by a software loop; in modern operating systems, there are simply too many other activities of an indeterminate nature underway to use such a timer.

Some new analog input boards designed for easier use with Windows place the timing control on the board rather than in the operating system. Some of the data and operations are buffered on the board in order to synchronize the experiment with the operating system.

Input/Output Ports

General I/O Ports

At one time it was necessary for scientists to design input/output ports for any interface.[16] These ports are parallel ports because the bits in the data byte are presented together, in parallel. The commercially available ports of today can be configured for either input or output and have some level of protection against abuse. Nevertheless, in most cases the user will have to program its features, and the procedures can be somewhat arcane.

Applications of the "Printer Port"

One simple input/output port on the PC is the so-called "parallel printer port", named not because it is used with parallel printers, but because it is typically used with printers that are interfaced to a parallel port. On PCs, additional ports (usually up to three or four) can be added inexpensively, so this port can be a versatile port for many instructive interfacing tasks.[11]

Reference 11 gives a good description of the parallel printer port, which has changed over the years. With proper configuration, it has at least eight output bits and eight input bits; some can be bidirectional, and one input pin is easily used for interrupt inputs. In the author's lab, it is used for many projects in order to avoid placing additional cards inside the computer. Some tasks are controlling stepping motors, sensing switch positions, and monitoring shaft encoders. The interrupt input can be used to enable timing and frequency measurements with a resolution of 0.1 to 1 millisecond.

An Introduction to the Serial Port

The serial port is more complicated and somewhat specialized. However, it presents a serious headache to many users who must connect a poorly documented instrument to the computer's serial port.

If the serial port is to be discussed, it presents an environment for the discussion of handshaking which must occur in the data-transfer scheme on several levels.[1]

Software

The interface is incomplete without the software to drive it. Although advertisements claim data acquisition and control without programming, the user still must provide information both about sequences and about the algorithm. This may be expressed graphically, but it is nonetheless programming.

Many tools are now available both for handling the input/output for data acquisition and control and for display, on a graphical panel that simulates an instrument panel. Increasingly, this combination is called a virtual instrument.

Virtual instrument software development places significant demands on the programming environment. For example, depending on the rate of operation, the data acquisition/control operations may tax the CPU; in some systems it may be difficult to acquire data at rates faster than 100 Hz, while a carefully crafted low-level program may achieve rates over 50 kHz. At the same time, a graphical user interface (GUI) must be maintained. It is clear that as the power and complexity of operating systems increase, the difficulty in maintaining precise control will become more challenging.

The following section provides a modest framework for understanding software.

Writing Real-Time Software in Traditional General-Purpose Languages

Traditionally, the user with the interfacing task wrote the support software in the language used for all other tasks. Originally with microcomputers, the language was BASIC, followed by Pascal (also FORTH, but thankfully that died). The next step was to follow the dictum of "computer jocks" that "real programmers use C", which was followed by C++; it is the view (prejudice?) of this author that the learning curve for C++ is too steep to require that chemistry majors become conversant. The requirement to use the latest "in vogue" language was always an attempt to apply a general-purpose tool to the chemical problem, and debates over the merits of different programming languages continue.

Programming for real-time applications requires the ability to respond to events, and early BASIC seldom had precision timing or interrupt capability. Later versions of Pascal[17] added some of that capability and became a popular language for developing data acquisition systems.

Without going into technical detail, it is important to remember that the operating system can add a layer of control that makes it more difficult to control timing and interrupts. Asynchronous activities such as disk-caching and clock updates compete with the data acquisition software for high-performance applications. Additional complexity is added in multitasking environments such as Windows.

Event-Driven Programming

One newer approach to programming is event-driven. That is, subroutines are called by events to which a "control" is subjected; the event is an actual or simulated interrupt. For example, one of the objects to which events are related is the timer; specific actions may be programmed to take place at each timer interval concurrent with additional activities.

The environment is also graphical, using the paradigm of forms. Updated versions of BASIC and Visual BASIC, and of C and Visual C++,[18] are now major platforms for real-time software application.

Visual BASIC and Visual C++[18] can be extended, and products have been developed to serve the needs of scientific and engineering data acquisition and control.[19]

Science and Engineering Development Environments

Several environments were developed specifically for real-time data acquisition/control requirements. Both the input/output functions for data acquisition and the display functions simulating an instrument panel are included. The various products display a distinct trade-off between ease-of-use and power.[20]

One example is ASYST and its variant ASYSTANT,[21] which contain many prepared modules for common scientific activities including data acquisition from standard data acquisition boards. In one example, it was used for acquisition from a CW NMR instrument.[10]

Among the most successful is LabView.[22] LabView is "icon-based"; that is, the sequence of operations is defined not by writing lines of code in text, but by connecting icons and entering operating parameters for those icons. HP VEE[23] is also icon-based.

The icons and the connections provide a more intuitive approach to programming. It can be a rapid approach to solving problems as long as the required algorithms are those anticipated by the developer of the environment.

LabWindows CVI supports programming in a conventional language such as C, providing the extensions in the form of libraries.

A "Wedge" for Familiar Software

Regardless of whether or not data are acquired with the computer, many scientists use spreadsheets to process the data. The software package LIMSport[9,24] (Laboratory Information Management System port) is a wedge for controlling data acquisition and importing data directly into a spreadsheet. It has the advantage of low cost, being inexpensively available directly from the LIMSport author; it requires spreadsheet software, but that may well be an existing component of the laboratory.

Conclusions

A student should be able to substantially understand the process of communication between the real world of the experiment and the virtual world of the computer without having to be an electronics or software engineer. With coverage of the topics introduced here, a student can be literate in this area and have the basis for pursuing projects that can give a great deal of satisfaction.

References

1. Perone, S. P.; Jones, D. O.; *Digital Computers in Scientific Instrumentation*; McGraw-Hill: New York, 1963.
2. Ratzlaff, Kenneth L.; *Introduction to Computer-Assisted Experimentation*; Wiley-Interscience: New York, 1988.
3. Willard, H. H.; Merritt, L. L.; Dean, J. A.; Settle, F. A.; *Instrumental Methods of Analysis*, 6th ed.; VanNostrand: New York, 1981.
4. Malmstadt, H. V.; Enke, C. G.; Crouch, S. R.; *Electronic Measurements for Scientists*, W. A. Benjamin: Menlo Park, CA, 1974.
5. Horowitz, P.; Hill, W.; *The Art of Electronics*, 2nd ed.; Cambridge University Press: New York, 1989.
6. Malmstadt, H. V.; Enke, C. G.; Crouch, S. R.; *Microcomputers and Electronic Instrumentation: Making the Right Connections*, American Chemical Society: Washington, DC, 1994.
7. Vassos, B. H.; Ewing, G. W.; *Analog and Computer Electronics for Scientists*, 4th ed.; Wiley-Interscience: New York, 1993.
8. Brigham, E. O.; *The Fast Fourier Transform*; Prentice-Hall: Englewood Cliffs, NJ, 1974.
9. Ratzlaff, K. L.; Engh, S.; "The Use of Computer-Assisted Instruction in the Teaching Laboratory"; *J. Chem. Educ.* 1980, *57*, 1207.
10. Vitz, E.; Betts, T. A.; "LIMSport(V): pH Data Acquisition: An Inexpensive Probe and Calibration Software"; *J. Chem. Educ.* 1994, *71*, 412.
11. Prais, M. G.; "NMR Spectroscopy: An Introduction to the Operation of Modern Instrumentation"; *J. Chem. Educ.* 1993, *70*, 381.
12. Bergsman, Paul; *Controlling the World with Your PC*; HighText Publications: Solana Beach, CA, 1994.
13. Kalvoda, Robert; *Operational Amplifiers in Chemical Instrumentation*; Halstead Press: New York, 1975.
14. Peyton, A. J.; *Analog Electronics with Op Amps: A Source Book of Practical Circuits*, Cambridge University Press: New York, 1993.
15. Schweber, B.; "Choosing an Op Amp"; *EDN* 1995, *40*(11), 38.
16. Ratzlaff, K. L.; "Interfacing Microcomputers: A Guide to the Scientific User of S-100, TRS-80, PET, and Apple Microcomputers"; *J. Chem. Educ.* 1981, *58*, 503.
17. *Turbo Pascal*, Borland International, Scotts Valley, California.
18. *Visual BASIC*, Microsoft Corporation, Redmond, Washington.
19. *VisuaLab*, IOtech, 25971 Cannon Rd., Cleveland, OH 44146; *Real-Time Graphics Tools*, Quinn-Curtis, 35 Highland Circle, Needham, MA 02194.
20. Clarkson, M.; "Tools on the Fly: Virtual Instruments Take Off"; *Desktop Engineering* 1995, *1*(1), 20.
21. Keithley ASYST, 440 Myles Standish Blvd., Tauton, Massachusetts.
22. National Instruments, 6504 Bridge Point Pky., Austin, Texas.
23. Hewlett-Packard Measurement Systems Div., P.O. Box 310, Englewood, Colorado.
24. Vitz, E.; "LIMSport: Spreadsheet Laboratory Information Management System for General Chemistry"; *J. Chem. Educ.* 1992, *69*, 744; Vitz, E.; "LIMSport(II): Use of the Interfaced Balance for Pressure Measurements, Streamlined Syntheses, and Titrations"; *J. Chem. Educ.* 1993, *70*, 63; Vitz, E.; Reinhard, S.; "LIMSport(III): High-Resolution Input and Output Using the IBM Gameport and Lotus '@-Functions'"; *J. Chem. Educ.* 1993, *70*, 245.

Statistical Methods in Chemistry: Why and How They Should Be Integrated Into the Curriculum

Karen D. Rappaport

Introduction

In preparing chemists for the environment in which they can expect to work, the role of the computer is significant. The computer enables the chemist to collect and store large amounts of data. What do we do with the data that we collect? How do we analyze it and make it into useful information? How can we present this information so that we can prove our hypothesis? How do we sort out all the measurable variables and focus on those that are most important? There are various statistical methods that can enable us to sort through data and develop it into meaningful information. It is the goal of this chapter to present some of these methods and to show how they can be integrated into the training of young chemists.

Background

The industrial chemical environment today has become very team-oriented and cost-conscious. Companies that market a product early in its life cycle will be able to obtain market share and price advantages. This makes the time required to develop a new product a significant factor in its life cycle. In addition, limited resources in staff and money increase the pressure of developing a product within the time available. Given these constraints many chemical companies are looking to apply the methods of statistical experimental design early in the product development cycle.

While attempting to integrate statistical techniques in chemical processes, industry has encountered several roadblocks. Many of these are a result of either a lack of knowledge or limited communications within the ranks of scientists and statisticians. More specifically, some of these roadblocks are the following:

1. Lack of knowledge of statistics by chemists, primarily in the areas of experimental design and optimization.
2. Lack of knowledge of chemistry by statisticians, especially the failure to fully understand the scientific process and what chemists do.
3. Language differences between the two fields: The terminology used in statistics is a barrier for the chemist and is different from the way a chemist would describe an experiment.
4. Communications issues: In addition to the barrier of language, the statistician is not always part of the scientific team and does not always understand the process under study.
5. Limited knowledge of areas of application: The chemist has little idea of how statistical methods could help in his or her work.
6. Unwillingness to spend preparatory time in design vs. "tinkering" in the lab: The chemist often sees statistics as limiting her or his ability to do experiments and not as part of the scientific exploration process.

The inability of scientists and statisticians to understand and communicate with each other has caused inefficiencies and limited productivity in the scientific process. Efforts are underway in statistical education to better train the statistical consultant in chemistry.[1,2,3] However, our experience has shown us that there are advantages to having scientists more knowledgeable in statistics. The chemist better understands the process—what to look for and what to expect. This gives them the advantage in selecting factors to study. The scientist can recognize unusual behavior more quickly. Knowing what to look for gives the scientist the edge in starting to apply Design of Experiments. Having some statistical knowledge also better enables the scientist to know when the statistician's help is needed.

Knowledge of statistics alone cannot provide successful results in chemical experiments. However, statistics integrated together with chemistry have proven to be very effective in optimizing the results of experimentation. If we can train chemists to understand statistical concepts early in their careers, they will be more effective in performing industrial research and become more desirable candidates for a position in the chemical industry, or in other careers that are based on a chemistry background.

How Statistics Are Used in Industrial Chemistry

There are several areas where statistics have been applied in the industrial environment. In their R&D efforts many companies employ Design of Experiments to reduce the cycle time required to develop a new product. The methodology of experimental design provides for the selection of specific combinations of experiments that will cover a possible set of results. At various stages during product development, statistical techniques are useful to measure and analyze customer reaction to new products. In the manufacturing process, various techniques are applied to optimize a process and to make it more efficient and cost-effective. As a component of customer technical support, statistical methods are used to help sell products. Statistical methods are also used to calibrate instruments and to validate and troubleshoot equipment. The following are examples of several types of applications that demonstrate shortened discovery time, lower manufacturing cost, and customer impact.

1. The goal was to find and replace an expensive ingredient in a major product line. We also wanted to determine and optimize the yield, solubility, and strength. A design of thirty experiments was developed with factors such as percent water, catalyst concentration, total amount of chemical AA, concentration of ingredient, reaction time, and reaction temperature. After the experimental data were analyzed, the major factor controlling the results was found to be reaction temperature. The optimal temperature for the reaction to maximize yield was determined. By using experimental design in the synthesis process, we were able to design, run, and analyze the experiments in 10 weeks. We were thus able to answer our initial question in a time frame that was significantly shorter than the normal nine-month period of experimentation and analysis. In addition, we were able to find the controlling factors in the synthesis, which enabled control of the scale-up process. As an added benefit, we were able to provide the research team with a methodology and an approach for future experimentation.

2. The goal was to lower the manufacturing cost of a commercially available polyester blend. We wanted to reformulate the product while keeping or improving material properties and reducing the production costs. The polyester blend was composed of different percentages of recycled starting material, polyester, fillers, additives, and types of glass. The cost could be lowered by changing the quality of the recycled starting material, changing the filler, or using glass fibers or additives. All of these options were selected as factors in an experimental design. The factors were all the contributing materials and the percentage range of each. The material properties to be measured, known as responses, included tensile strength and elongation, flexural strength and modulus, melt viscosity, and warpage. An experimental design, using these factors and responses, enabled a limited number of experiments to be run in order to find the proper blend of ingredients. The optimal blend was found in 10 weeks. The evaluation of the results was done with multivariate analysis. These methods, principal component analysis (PCA) and projection to least squares (PLS), were applied to the measured results to explain the variation. The results showed that warpage and elongation properties were explained by the same factors, while strength properties were explained by different factors. The resulting analysis provided the blend of materials that gave the optimum properties. These statistical techniques enabled us to find the combination of properties and processing conditions that would reduce costs. In this case the costs were reduced by 32%, better than the original target.

3. Experimental design and multivariate analysis were used to determine the resolution and color properties of a photographic film. This is normally a difficult problem because of the complex interaction between the film and pigment composition, the process variables, and the development factors. The first prototype of the processor was rejected because it could not prepare an acceptable four-color negative pressmatch proof that met the requirements of the films specified in the product profile. As a result, a designed experiment was proposed to develop contour plots that would enable the determination of the optimal operating settings of the processor. The mechanical and physical properties of the processor, which would affect the processing of the film, were determined. These included developer temperature, developer dosage, transport speed, brush speed, and tilt angle. These five factors were tested at a high and a low level. A

designed experiment of 36 cases was developed. After the experiments were run, measurements of properties, describing the film image, were made. Analysis showed that the tilt angle was not an important factor. A graphical analysis was made using contour plots. These plots suggested operating conditions at which acceptable image resolution could be found. The result of the designed experiment was an improved film with less variability.

4. The goal of an experiment in a lab-scale organic synthesis was to optimize the product yield. There were two qualitative factors to be studied: solvent types and reactor types. Three solvents and three factors were studied. Other factors included pressure, stir rate, temperature, starting material, and percent catalyst. An experimental design was developed that included 18 runs. The product yield was measured for each experiment. The regression equation and coefficients were calculated, and a contour plot was developed. The regression coefficients showed that the catalyst percent, pressure, and starting material had the greatest effect on the yield. The plot showed the region of activity that would give the best yield. As a result of the design, product yield was improved from 45% to 71%. Using the mathematical models, the conditions needed for 90% yield were extrapolated.

5. A study was developed, for one of our plants, to determine the major factors impacting the yield of our chemicals. The primary goal of the study was to determine the yield and flowability of chemical N. The plant produced data for 217 lots of material, and there were 37 variables, of which 23 related to the process and 14 were measured responses. Because of the number of variables collected, a multivariate analysis was suggested. PCA and PLS were used to study the variation. These analyses showed that the concentration of one of the feed chemicals had a strong negative effect on the yield. They also showed a significantly lower yield for one of the equipment locations. This unusual result caused the engineers to inspect the equipment further, and in the process, a partial equipment failure was discovered. Further evaluation of the problem pinpointed the failure, and when the failure was corrected, the production yields were significantly increased.

6. Design and multivariate analysis were used to improve the yield of a lab-scale product and to maximize the product selectivity when it was scaled up to production. The factors that were studied were the amount of oxidant, amount of acid, time, and temperature. There were several constraints for the experiment. High oxidant and low acid had to go together, and few experiments could be run because each experiment lasted two days. The responses to be measured included product yield and selectivity and impurity yields for two impurities. Nine experiments were run using a D-Optimal design. Using the measured results, regression equations were calculated for all the factors. Plots were also generated so that the optimal conditions for getting the highest yield and selectivity could be seen. The results of this analysis showed the conditions needed to increase the yield from 76% to 84% and the product selectivity from 80% to 96%. In addition, we developed a better understanding of the mechanism of the reaction. The resulting mathematical model (regression equations) was used to scale up the product.

The preceding examples demonstrate how experimental design has been used to improve the manufacturing of chemicals and materials and to minimize the time spent in industrial experimentation. Some additional examples of industrial applications can be found in the July 1995 issue of the Journal, *Chemometrics and Intelligent Laboratory Systems*.[4] More detailed examples that can be used in the classroom will be found in later parts of this chapter.

Statistical Topics Relevant to Chemical Education

The following is an outline of the basic statistical topics that have practical relevance to chemists:

- basic concepts of data analysis and applications: organization of data, presentation of data, analysis techniques, data modeling;
- experimental design: organization of experimentation;
- response surface methodology;
- mixture designs;
- multivariate analysis.

The last three topics are usually covered on the graduate level and will not be addressed in any detail in this chapter. A more detailed discussion will follow on the other topics.

Organization and Presentation of Data

Any process that involves making an observation or taking a measurement can be considered an experiment to which we can apply statistical techniques. Therefore, in any laboratory work for which measurements are made, collected, and tabulated, we can apply statistics. There are many opportunities to discuss the organization and handling of the data, and techniques to calculate and determine the quality of the numbers collected.

The use of tables and charts is helpful in organizing the data for presentation of the work and for observing patterns and quality. When discussing data it is important to distinguish between quantitative and qualitative data. Quantitative data are those variables that can be measured. They may be either continuous (connected) or discrete (countable but separate). Examples of quantitative variables are pH, stirring rate, concentration, and temperature. Qualitative data are values that we can observe but cannot measure. Attributes such as color, shape, texture, type of equipment, and type of solvent fall into this class. These may be grouped into categories or ranked by some system. Either type of data may be tabulated in charts of rows, columns, or both. If tables are used extensively, spreadsheets are an excellent tool to help organize them. The student should be able to translate rows and columns in the spreadsheet to a chart. Students should know how to label rows and columns and how to make titles. They should also be able to look at the layout of data and see which variables should be rows or columns. Although tables are useful for collecting data, the presentation of data is made more accessible with the use of pictures.

The use of graphs to display data should also be introduced early in a chemist's career. It is important to keep the presentation of data simple in order to explain an idea. Many times, a simple bar or line graph will demonstrate a point. Emphasis on

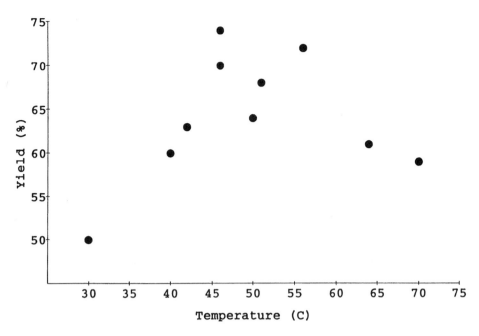

FIGURE 1   Scatter plot of yield.

simple graphs and charts may help prevent the misuse of other more technical statistics later on. Graphics should include bar graphs, histograms, scatter diagrams, and line graphs. These are all two-dimensional representations of data. They each have an X and Y axis. A scattergram uses a point to mark the position of the data on the scale. A line graph will connect (if possible) the points of location. Bar graphs are best used with qualitative data. In this case, the bar represents a category while the height represents the frequency of the category. A histogram is merely a bar graph in which the category is a numerical range. All of these presentation methods are basic tools, and they should be understood as the first choice to display data. Any spreadsheet will allow for these types of graphics and make it easy for the student to try a variety of plots easily. Using the spreadsheet to make graphs will enable the integration of both the computer and statistics into early laboratory courses.

As an example of integrating statistics into the lab class, consider an experiment in which the student is collecting data on the temperature and yield of a reaction. The results can be charted on the spreadsheet, where the rows are the temperatures and the columns are the yields. After the data are collected, they could be plotted as a graph with the X axis having the temperature range and the Y axis having the yield range. A scattergram would show the points such as T = 46° and Yield = 70%. If the data are continuous, the points could be connected as a curve. Although a simple example, this case demonstrates the first step of organizing data and then displaying it in an appropriate form. Figure 1 is an example of a scattergram, showing temperature and yield. Another example of a scattergram is shown in Figure 2. It shows the relationship of the cooling time to the sweetness of cookies. This is used

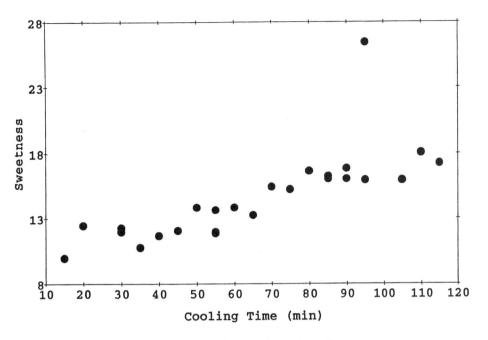

FIGURE 2   Scatter plot — Lot #1.

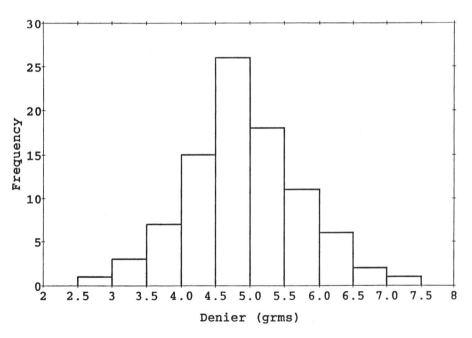

FIGURE 3   Histogram: Material Z by operator b.

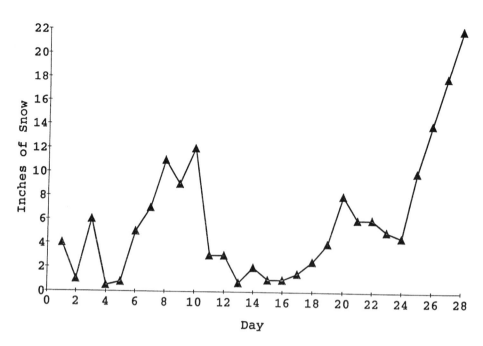

FIGURE 4   Snowfall in New Jersey, 1993.

to determine the recommended times to be given in a recipe printed on the box of cookie batter. An example of a histogram is shown in Figure 3. This is a graph of testing results that was made by a single operator for a single material. This was used to determine the properties of a new material. An example of a line graph is given in Figure 4. This case shows the snowfall for February 1993 in New Jersey.

Other statistical concepts that are important to introduce early include basic terminology such as independent and dependent variables, sample, and population. An independent variable is a measure that will have an influence on the experiment and can be controlled by the experimenter. A dependent variable is a function that depends on the value of another variable. A sample is a subset or portion of a population. These can be introduced into any and all courses involving measurement and data collection. When doing calculations, discussions should also be made about statistical concepts such as accuracy (i.e., how many decimal places are meaningful), measurement (tools and units), noise (outside contributions to the data, unmeasured variables), variation (expectation of numeric results, how much exactness is important), bias (the role of the experimenter), and outliers (the importance of data points that do not fit expectations). Discussion of data should include ideas about what happens to the results if the experiments are performed on different days and at different locations. Are the conditions of the lab the same from day to day? What, if anything, is the influence of the temperature and humidity of the room on the experiment? Does the equipment behave in the same way every day? Do people operate the experiments or measure the results exactly the same way every day? Discussion of questions such as these will help the students understand concepts such as noise, bias, and experimental error. These statistical concepts will also help in understanding the analysis of data. The use of these terms early in the chemist's

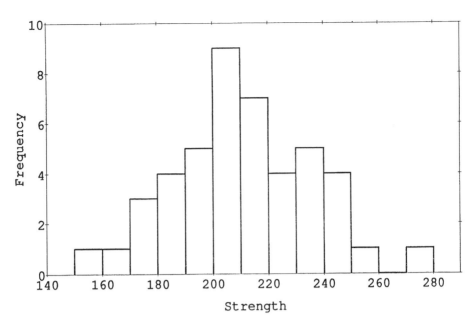

FIGURE 5   Histogram: Strength Lots #1 & #2.

education will contribute to the understanding of data collection and make it easier to apply more advanced statistical techniques in later courses.

The use of basic descriptive statistics to help review and analyze experimental data is beneficial in early chemistry classes. Measures of the middle, such as means, medians, and modes, and basic variation, such as variance and standard deviation, can also be used to describe laboratory data. The calculation of these measures is less important than the understanding of what they show and when to use them. For example, the mean or numerical average is seriously affected by the size of the numbers being summed. Therefore, if you have one piece of data that is significantly larger than 20 others, the mean may not be the best measure with which to describe your data. On the other hand, the median, the value with the same number of data points above and below it, shows relative position and is independent of the numerical values. This measure is best used when position or ranking is important. The mode, which simply shows the largest category, is best used with qualitative data. Graphical observations can be used to display these concepts. As an example, consider Figure 5. This is a histogram of the strength measurements of a material. The mode, which is the category with the highest frequency, is easily observed. It is the bar where the strength lies between 200 and 210. The median can be computed by summing the frequencies and finding the middle value. It is also found in the bar for strength between 200 and 210. The mean is a numerical average that is computed and is not as easily observed on the graph. The discussion of these measures with simple experiments can help the chemist understand the advantage of using each method. Likewise, the discussion of variation in less complex examples is extremely important. Variation is one of the most important concepts in understanding experimental data. The many statistical measures used to study this concept are often misused and

TABLE I   Initial LOC × 1,000.

Run	Sam 1	Sam 2	Sam 3	Mean	Std. Dev.
1	14	14	16	14.7	1.15
2	5	5	6	5.3	0.58
3	18	17	17	17.3	0.58
4	18	14	11	14.3	3.51
5	14	12	11	12.3	1.53
6	18	21	18	19.0	1.73
7	18	15	17	16.7	1.53
8	8	9	8	8.3	0.58
9	4	6	7	5.7	1.53
10	6	7	7	6.7	0.58

misunderstood. Using and understanding this idea in basic laboratory courses will provide the groundwork for future experimental data analysis.

Variation is a measure of how the observable results of the experiment differ from the average or expected results. Some variation is a result of noise or external variables. Some variation is due to accuracy of measurement. On the other hand, some variation can be explained by the experiment itself. The ability to differentiate between types of variation is important and should be introduced early. The variance and standard deviation measures are distances from the center or average data and can be introduced early with a computational tool such as a spreadsheet or calculator. With the availability of electronic tools, computation is less important than understanding the concepts.

As an example of the concepts of variation and measures of the center of data, consider the following problem. We have taken measurements of the "limits of color" for 10 lots of a material (see Table 1). Three samples were taken from each lot and measured. In order to compare the lots, the mean and standard deviation were calculated.

The formulas used were:

$$\text{Sample Mean } (\overline{X}) = \frac{\text{Sum of observations } \left(\sum_{i=1}^{n} x_i\right)}{\text{Number of observations } (n)}$$

$$\text{Standard deviation } (S) = \sqrt{\sum_{i=1}^{n} \frac{(X_i - \overline{X})^2}{(n-1)}}$$

To test the use of these formulas, a student could finish Table 2.

TABLE 2 LOC × 1,000 after 2 weeks.

Run	Sam 1	Sam 2	Sam 3	Mean	Std. Dev.
1	17	17	8		
2	12	6	5		
3	23	19	18		
4	88	93	84		
5	38	41	40		
6	18	18	16		
7	22	30	27		
8	38	33	37		
9	39	35	38		
10	41	41	40		

All of these topics should be integrated into any basic chemistry course as part of the collection, analysis, and display of data. Repeated use of these topics throughout the chemistry curriculum will enable the new chemist to better understand and use these concepts. They will also contribute to the understanding of more advanced statistical concepts that arise at a later stage in the chemist's education.

Software

There are several basic computer software packages that can help simplify and ease the development of data collection and presentation. Any major spreadsheet, such as Lotus or MS Excel, is an excellent tool for organizing data. In addition, these also include the basic descriptive statistics mentioned previously as well as a variety of graphs and charts. The use of software packages can significantly simplify data presentation because they make it easy to try different graphs for the same data. The student can try both a line graph and a bar graph easily and determine which better describes the data. The spreadsheet will also have built-in statistical functions and can simplify computational work. A student can pick the X and Y variables from his or her table, decide which axis is best, and quickly display a plot. Any row or column can be summed and averaged, and standard deviations for any row, column, or combination can also be calculated by the computer.

To illustrate the use of the computer to display data, let's consider the relationship between sweetness and cooling time for cookies, shown in the Figure 2 scatterplot. Two different lots were tested. Figure 6 shows a scatterplot of results; Figure 7 shows a trend line. Which figure provides more information? The scatterplot shows specific data points. The line illustrates the trend and should be used when you do not wish to reveal the actual data. Note that the information in the scatterplot about the outlier data point at 100 minutes is lost in the line estimate.

In actual applications the interpretation of data can be very complex. For example, a company may have five different plants at one location. The waste material from each plant is collected at a wastewater treatment facility. After treatment, it is released

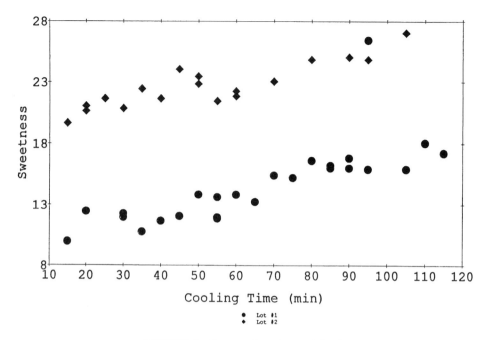

FIGURE 6   Scatter plot—Lots #1 & #2.

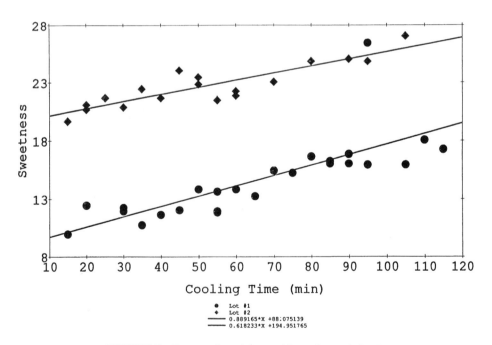

FIGURE 7   Scatter plot with trend line—Lots #1 & #2.

to the city sewer system. When the wastes exceed the limits set by the city regulations, the company will be fined. A project was organized to identify the problem materials and to recommend methods to reduce their waste.

In order to attack this problem, data were gathered from the five plants. Every day for almost a year, data were collected and analyzed. For each day a total of 38 different variables was collected. The resulting spreadsheet contained 38 columns and 347 rows. Various statistical techniques were used to define the important factors in the wastewater production. The first step involved sorting out the measured from the calculated values. Since the calculated values were not critical, they were removed from the data analysis. This reduced the variables to 22. Graphical analysis of these factors pointed toward the most important ones, and the parameters were further reduced to 12. This number was more manageable, and regression coefficients were determined for the remaining responses. The regression technique will be demonstrated in the data analysis section of this chapter.

Data Analysis Techniques

As the student progresses through the core curriculum and laboratory courses become more specialized, there are numerous statistical analysis techniques that should be integrated into the curriculum. The concepts of statistical regression, analysis of variance, and modeling of data will all fit into more advanced laboratory courses.

The analysis of variance is critical to the understanding of the source of errors in experimentation. Analysis of variance (ANOVA) is a method by which we can analyze experimental data. It is used to determine whether the error is due to chance or to one of the factors in the experiment. It enables the study of sources of variation and can be integrated into more advanced lab courses. The value resulting from an ANOVA calculation is called an F statistic. The chemistry student will need to know how to read (ANOVA) tables and graphs and the F statistic. In order to do this, some understanding of a normal distribution is preferable. It is recommended that chemistry students take a basic statistics course by their second year. If this is not the case, the concepts can still be introduced as part of analysis in the laboratory.

An example of how to use ANOVA in the laboratory follows. Suppose we are studying the performance of three different catalysts, with four different amounts of each. The yield percentage is recorded for each case. To test for differences in catalyst type, a one-way ANOVA can be performed. The ANOVA compares several sets of samples. For a one-way ANOVA, one particular treatment (factor) is varied in each sample set. In this case, the treatment being varied is catalyst type. Once the ANOVA is computed, which can be done with the spreadsheet, the resulting F statistic is used to determine whether there are any significant differences between treatments (catalyst type). The student will need to look up a table for the F statistic cutoff point. This table can be found in any basic statistics book. Discussion should focus on whether the F statistic is significant. Is it less than or greater than the cutoff point? If it is less than the cutoff, then the differences evidenced in the data are not significant. In this case, we would know if the differences in yield produced by the different catalyst types were significant. The data table for this experiment is shown in Table 3.

TABLE 3   Catalyst data table.

Run #	Catalyst Type	Catalyst Amount	Yield %
1	V-Catalyst	0	61.9
2	V-Catalyst	5	65.5
3	V-Catalyst	15	69.2
4	V-Catalyst	25	69.95
5	X-Catalyst	0	61.9
6	X-Catalyst	5	65.7
7	X-Catalyst	15	69.55
8	X-Catalyst	25	70.15
9	Y-Catalyst	0	61.9
10	Y-Catalyst	5	66.05
11	Y-Catalyst	15	69.4
12	Y-Catalyst	25	69.95

TABLE 4   One-way ANOVA table, catalyst type.

Source of Variation	df	SSq	MS	F statistic
Catalyst Type	2	0.09	0.09	0.003
Error	9	126.53	14.06	
Total	11	126.62		
F = 0.003				

First test for differences in performance due to catalyst type. To do this, we can compute the ANOVA using the spreadsheet. The results are given below in Table 4. The following definitions are used in the table:

Error   = unexplained variation.

df   = degrees of freedom.

SSq   = sum of squares, a measure of the total variation of the response Y.

MS   = mean square, the computed estimate of the variance due to the factor being tested. In this case the factor is catalyst type or amount.

F Statistic = the ratio of two variances or mean squares. It is used to test the relative significance of the factor.

From the calculation, F is found to be 0.003. This value is then compared to the F statistic in a table. At the significance or cutoff level of 0.05 for the given degrees of freedom, there are no statistical differences if F is less than 4.26. Therefore, in this case there is no significant difference due to catalyst type.

We can look at the same catalyst data table, Table 3 and ask a different question. Is there any difference in the result if we change the catalyst amount? In this case, we

TABLE 5 One-way ANOVA table, catalyst amount.

Source of Variation	df	SSq	MS	F-statistic
Catalyst Amount	3	126.38	42.13	1,404.33
Error	8	0.24	0.03	
Total	11	126.62		
F = 1,404.33				

compute the ANOVA table using the mean square for catalyst amount. The result is given in Table 5. The F statistic is 1404.33. To see if this is significant, we look up the F value in a statistical table for the 0.05 significance level and degrees of freedom. In this case there will be a difference if the F value is greater than 4.07. In our example this is the case, implying that the difference in catalyst amount will impact our yield.

Another tool used to analyze and obtain a mathematical model for experimental responses is regression analysis. Regression analysis is used to help understand the experimental system. This analysis will result in a mathematical equation and a graphical representation of the system. Once the system is understood, a model can be made to predict the responses of a system. Both of these ideas can be fit into advanced laboratory courses, and the spreadsheet can be used to both compute and plot the regression equation. Once these values are found, the interpretation of the equation and the residual plots will be the most important part.

As an example of the application of regression techniques, consider the following case. Power companies are regularly monitored with respect to environmental impact. Those still using fossil fuels will cause pollution. This pollution can be found in plants and animals. For some coal burning operations, the yearly coal consumption and pollutant concentration in the streams near the power plant are measured. We can ask the question: Is the pollutant concentration in the streams greater near plants with higher coal consumption? The overall relation between pollutant concentration and coal consumption is determined with the use of regression analysis. In this case an equation modeling the relationship can be developed. The graphical representation of this equation, called a regression line or line of best fit, can be fitted to the data. The data relating coal consumption and pollutant concentration are given in Table 6.

The scatterplot for this example can be seen in Figure 8. The regression line for this example can be seen in Figure 9.

The equation for this line can be computed using the following formulas:

$$\text{Linear regression, linear model}: \quad \hat{Y} = b_0 + b_1 X$$

$$\text{Regression coefficients}: \quad b_1 = \frac{n(\Sigma xy) - (\Sigma x)(\Sigma y)}{n(\Sigma x^2) - (\Sigma x)^2}$$

$$b_0 = \frac{(\Sigma x)(\Sigma x^2) - (\Sigma x)(\Sigma xy)}{n(\Sigma x^2) - (\Sigma x)^2}$$

TABLE 6   Title?

Coal	Pollutant
17	3.9
32	7.8
25	6.6
23	3.4
18	2.5
6	1.1
4	0.9
30	6.0
33	9.1
25	4.6
11	2.0
36	8.1
39	8.7
9	2.1
12	2.6

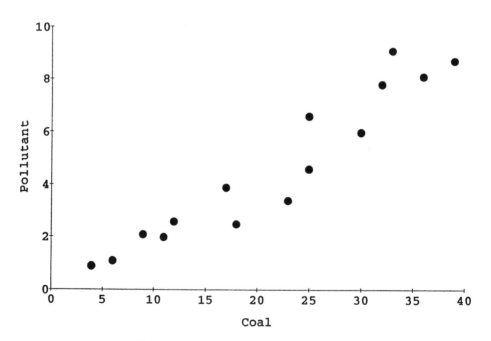

FIGURE 8   Scatter plot—Coal vs. Pollutant.

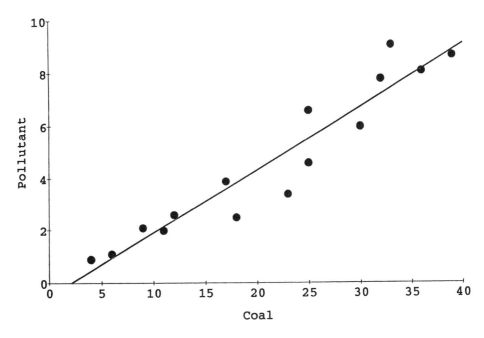

FIGURE 9   Coal line fit plot.

The equation for the line in this example is computed to be:

$$\hat{Y} = -0.51316 + 0.240929 X$$

R^2, the coefficient of determination can also be computed. R^2 is a measure of how well the regression equation fits the data and what proportion of the variation is explained by the independent variable. The closer R^2 is to 1, the more the equation explains the fit and the relationship. $R^2 = 1$ means a perfect fit, while $R^2 = 0$ means no fit at all. If we find that R^2 is a good predictor for the coal regression line, then a conclusion can be made about the effect of pollutant concentration on the streams based on the plot. The formula for the coefficient of determination R^2 is given as:

$$\text{Coefficient of determination}: \quad R^2 = \frac{SS_{Regression}}{SS_{Total}}$$

$$SS_{Regression} = \Sigma\left(\hat{Y} - \bar{Y}\right)^2$$

$$SS_{Total} = \Sigma\left(Y - \bar{Y}\right)^2$$

R^2 can be increased by adding extra terms without necessarily improving the predictive ability of the model. Therefore, the value R^2 is adjusted where the sum of squares are divided by the number of terms, i.e., the degrees of freedom (df).

TABLE 7

Catalyst A Temp., °C	Yield, %	Catalyst B Temp., °C	Yield,%
135	313	145	486
140	337	150	567
145	462	150	576
145	597	155	618
150	685	160	592
155	752	165	521
160	837		
165	811		

Large differences between R^2 and R^2 adjusted indicate the presence of unnecessary terms in the model. Adjusted R^2 is given by:

$$R^2 \text{ adjusted} = \frac{df_{\text{Total}}\left(R^2\right) - df_{\text{Regression}}}{df_{\text{Residual}}} = \frac{(n-1)R^2 - 1}{n - 2}$$

but it will be calculated by the spreadsheet. For this example, $R^2 = 0.906857$ and adjusted $R^2 = 0.899693$. In this case the regression line is a good predictor of behavior since both R^2 and R^2 adjusted are close to 1. We can therefore conclude that the pollutant concentration is greater in streams near plants with a higher coal consumption.

As another example we will look at a case where the data may be nonlinear. If the data do not fit a straight line, they may instead fit a polynomial equation of the form:

$$Y = b_0 + b_1 * X + b_2 * X^2 + b_3 * X^3 + \ldots\ldots b_n X^n$$

where n is the order of the polynomial. A polynomial model that fits is not necessarily the only one or the best one. The more parameters you add, the less meaningful the model will be. When the number of parameters equals the number of observations, you have a perfect fit but a worthless model. It is also important to remember that the equation only applies to the data that you have gathered, and you should always assume that the equations are not valid outside the range of data. Caution is required when you attempt to apply your results to other situations. When looking at polynomial models, it is important to remember that polynomials can obscure simpler relationships. You should also consider whether you have ignored some important variables. If you need to include additional variables, you may have to consider using multiple regression techniques.

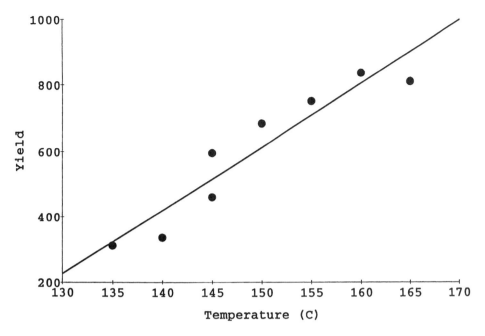

FIGURE 10   Effect of temperature on yield—Catalyst A.

As an example of a polynomial regression, consider the evaluation of two different catalysts (A and B). When temperature (X) was plotted against yield (Y), a quadratic function was needed for catalyst B, where a linear function was adequate for catalyst A.

Data are given in Table 7.

In order to determine the model (equation) that best describes the relationship between catalyst and yield, we will compute the equation and R^2 adjusted values for both a linear and quadratic case.

For catalyst A the values are:

Linear fit – Yield = –2283.5 + 19.3.Temp R^2 adj. = 89%
Quadratic fit – Yield = –12178.85 + 151.6.Temp –0.44.Temp^2 R^2 adj. = 93%

For catalyst B the values are:

Linear fit – Yield = 335.15 + 1.46.Temp R^2 adj. = 5%
Quadratic fit – Yield = –25906.85 + 340.3.Temp –1.1.Temp^2 R^2 adj. = 99%

The R^2 adjusted for catalyst A is not improved significantly by adding a quadratic term, so in this case the linear equation or model is adequate. For catalyst B the R^2 adjusted for the linear equation is only 5%, so this is not adequate and the quadratic term is needed. A graphical represention of the resulting equations is given in Figures 10 and 11. The preceding cases described, where comparisons are made or differences are being studied, are examples of problems that can be effectively understood by using either ANOVA or regression techniques.

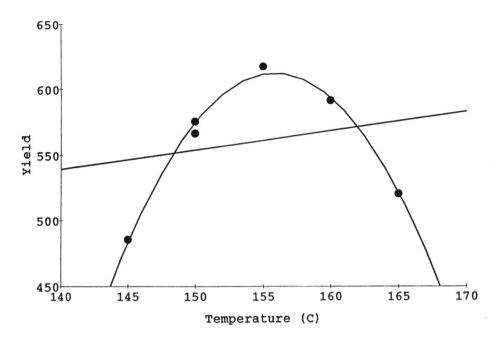

FIGURE 11   Effect of temperature on yield—Catalyst B.

Design of Experiments

Although it may be possible to predict how a reaction might occur by theoretical reasoning, in most cases we must make inferences from experimental observations. It is therefore important that chemists learn how to approach and set up experiments to solve problems. They may need to answer questions about the scope of a reaction. Can the optimal conditions in which a process will run be found? How will the optimal experimental conditions change if any part of the reaction system, such as solvents or reagents, is changed? Will a change in reaction conditions have any impact on yield or purity of the product? With these typical kinds of problems always available, the chemistry student should become familiar with experimental design concepts in her or his junior or senior year. It is preferable to have a chemistry student take a course in the design of experiments, but if that is not possible, the process of experimental design can be introduced in laboratory courses without involving the mathematical techniques.

Experimental design is a planned program of experimentation. It makes use of the chemist's knowledge and ideally combines this with statistical techniques. Experimental design helps make better use of resources and materials and enables the chemist to discover information more efficiently. The first step in the experimental process is to become familiar with the problem. The second step is to run screening designs to enhance understanding. The chemist will then try to optimize the results and find a model for the system under study.

The first step in designing an experiment is to define the problem. What is the objective? What does the student want to find? If he or she wants to obtain a certain product purity, then this would be the goal of the experiment. To solve the problem, questions should be asked about what information is known and what is not, and what we need to know in order to attack the problem.

In order to analyze the known components, the concepts of factors and responses should be defined. Factors, which are the input variables, can be objects such as temperature, concentration, pH, and catalyst. A synthesis is made with these input factors and some results are obtained. These measured outcomes are called responses. They could be variables such as yield, tensile strength, or biological activity. In the next part of the discovery process, questions can be asked about the factors. Are they quantitative or qualitative? What are the values that can or should be measured? Are there any uncontrolled factors that will influence the result, such as ambient temperature or humidity? The same questions will also need to be asked about the response values. It is important for the teacher to allow the students to think about and discuss these questions before giving an a priori answer. It is not the answer that is important here, but the process of learning to ask the right questions.

After the initial phase described above is completed, the experiment can be planned. It is a very common practice that many chemists look at one variable at a time. For instance, the influence of catalyst type and reaction temperature on the yield of a reaction is studied. In the first set of experiments the reaction temperature is fixed at a value and the catalyst type is varied. In the second set of experiments one catalyst type is chosen and the temperature is varied. A maximum yield is found for both cases. The student will report the catalyst type and reaction temperature that gave this yield as the optimal conditions. However, they will not be correct if there are any interactions between catalyst type and temperature. Using one variable at a time will often give a false optimum.

At this point in their chemistry education, students should be introduced to some form of experimental design. If a course is available in the design of experiments, it is highly recommended. If no course is available, then at a minimum the idea of planning experimentation should be discussed. For those educators with statistics experience, a reference such as *Statistics for Experimenters* by Box, Hunter, and Hunter[5] could be used for examples in chemistry classes and enable the use of some beginning designs.

Throughout the chemistry curriculum, various methods of statistical design and analysis can be applied in all courses. In organic synthesis the statistics can be used to minimize side products. In biochemistry they can be used to help optimize enzyme activity. In inorganic courses these methods can be used to study crystal growth. To optimize the understanding of interactions in physical chemistry, statistics can be used. In analytical courses statistics can be used to optimize chromatography or to study gas separations.

In the course of a graduate education, statistical design and analysis can be extensively used. A course in experimental design is particularly helpful to the graduate chemistry student. Traditional screening designs, that is, designs of relatively small experiments, are done to better understand research. These can include such designs as factorial and fractional factorials and response surface methodolgy. Multivariate techniques[6] are especially useful in the analytical area of study, as they are used for instrument calibration. Some training in these techniques would be helpful in the development of the graduate chemist.

Statistical Software Packages

There is computer software available that is effective for teaching and will help students to calculate descriptive statistics, display and analyze their data, make experimental designs and model their results.

Three types of software packages are needed.

1. Those that provide organization of data and basic statistics
2. Those that provide experimental design
3. Those that provide advanced multivariate design

When selecting statistical packages, there are several factors that should be considered. They should be easy to understand and simple to use. For statistical analyses they should have good graphics and all analyses should be available both numerically and graphically. You should not rely on one statistical software package for all your needs since there is no ONE package that does everything equally well. Some excel in descriptive statistics, others in analysis or design. For descriptive statistics spreadsheets such as MS Excel or Lotus are adequate. The large software packages such as SAS and RS1 can also be used. However, these packages are not user-friendly and require extensive statistical knowledge. We recommend them only to sophisticated statistical users. The popular PC packages—DesignEase and Design-Expert, are not easy for the chemist to interpret. Although they are seemingly easy to use for entering and calculating data, the results are often not clear or easy for the chemist to understand. If the chemist feels a need to have a general statistical PC package; we recommend Statgraphics from STSC Inc. However, since we feel that MS Excel or Lotus provides adequate descriptive statistics, we do not recommend any general PC package. For experimental design, MODDE by Umetri AB has excellent graphical presentations and is available for the Windows operating system. For advanced students a package in multivariate design such as PLS and PCA is also useful.

Conclusion

The chemist is looking to obtain the best information on how the response, i.e., yield, is influenced by experimental factors, such as temperature and pH. By experimentation chemists will proceed through stages of investigation, comparison, optimization, and modeling or prediction. Using reagents, solvents, and catalysts, they will develop an understanding of reactivity and properties. In order to accomplish this goal, statistical techniques have been shown to make the process more efficient. All of the statistical methods described in this chapter can be used with the aid of the computer to enhance the learning of chemistry.

References

1. Iman, R. L.; "New Paradigms for the Statistics Profession; Presidential Address, *Journal of the American Statistical Association* 1995, *90*(429), 1–6.
2. Snee, R. D.; "What's Missing in Statistical Education?"; *The American Statistician*, 1993, *47*, 149–154.
3. Hoerl, R. W.; Hooper, J. H.; Jacobs, P. J.; Lucas, J. M.; "Skills for Industrial Statisticians to Survive and Prosper in the Emerging Quality Environment"; *The American Statistician* 1993, *47*(4), 280–292.
4. Chemometrics and Intelligent Laboratory Systems; Rappaport, K. D., Ed.; 1995, *29*(1), 111–147.
5. Box, G. E. P.; Hunter, W. G.; Hunter, J. S.; *Statistics for Experimenters*, John Wiley & Sons: New York, 1978
6. Carlson, R.; *Design & Optimization in Organic Synthesis*, Data Handling in Science & Technology Volume 8; Elsevier: Amsterdam, 1992.

13

Visualization for Chemists

John P. Ranck

Introduction

As chemists, our work and thinking are guided by visions of interacting atoms and molecules. Through two centuries of practice, we have developed and refined line formula notation as our symbolism to explore and communicate molecular ideas. This symbolism, which represents molecules as static, two-dimensional, solid, classical objects, is inadequate for today's practice of chemistry.

The modern electronic computer provides new representations and new symbolism for our chemical thinking and communication. We may easily represent and visualize chemical reactions as dynamic collections of three-dimensional molecules. The molecules may have additional properties graphically encoded in their images. The electronic screen also supports composite representations presenting many non-conformable properties of chemical reactions simultaneously.

Data Visualization and Molecular Visualization

Any object or definable portion of the universe chosen for study is known only through its observable properties. Experimental science, in general, consists of determining the changes in some observable properties of a system as other properties are changed in a controllable or well-understood fashion. Frequently, regularities and relationships among the variables are discovered and are codified as scientific "laws".

Sometimes the relationship is simple and immediately apparent. A table of numerical data may disclose, for example, that "the volume of a sample of a gas increases with the temperature, the pressure being held constant." A graphical presentation of the same data, however, may display two additional pieces of information: that the volume increases *linearly* with the temperature and that the volume increases in *direct proportion* to the temperature measured from some point other than the arbitrary zero of the Celsius scale. By examining the volume versus temperature data on the linear scales of a graph, we are able to discern relationships not apparent in tables of numbers.

Today's scientists, chemists included, are confronted daily with vast collections of data, most of it in numeric form. The relationships or patterns in the data may not be apparent, even after careful scrutiny of the numbers. Usually, there are more than two variables, and a simple plot of dependent variable against independent variable seldom suffices. An additional variable can be plotted along a third spatial dimension and the entire three-dimensional plot projected down to two dimensions using perspective geometry. This can be accomplished easily and automatically with electronic computers. Computers also allow us to map or to code other data on the visual field as colors, textures, vectors, motion, etc., so that it is easily possible to present six or seven variables in the same data field. The human eye, with its edge detectors, motion detectors, foreground/background discrimination, etc., has an uncanny ability to discriminate small or unusual features in rich visual fields, and the techniques, tools, and conventions of electronic data visualization are rapidly being adopted by scientists in all fields.

Molecular visualization is rather different from data visualization. In data visualization, although variables are presented in two- or three-dimensional space, the variables themselves may be unrelated to the space in which they are presented. Indeed, decoupling variable space from real space is one of the major difficulties beginners experience with data visualization. Molecular visualization is somewhat easier because we are dealing with real three-dimensional objects that exist in real space; they are simply too small to be seen or experienced directly. To visualize molecules, we merely need to construct greatly magnified visual representations of objects in the familiar three-dimensional real space.

But what do molecules "look like"? Molecules do not have surfaces, edges, textures, colors, or the other familiar properties of ordinary objects. Molecules possess properties such as charge density (a scalar), electrostatic field (a vector), and polarizability (a tensor), which vary throughout the three-dimensional space that molecules occupy. To visualize molecules and their interactions, we must display the distribution of one or more of these properties (fields) throughout real space. Molecular visualization, thus, presents an ideal context in which to learn many of the elements of data visualization and coding without the "dimensional decoupling" problem.

Representation and Symbolism in Chemistry

In chemistry, as in all fields of knowledge, we utilize symbols to constitute a representation of some real system, and much of our work is performed using the symbolic representations rather than real systems. The strengths and weaknesses of our symbolic representation both guide and limit our understanding of, and our ability to manipulate, the real systems.

In two centuries of practice, we have developed a terse, tightly coded symbolism for chemistry, which we call line formula notation. This symbolism is easily represented and practiced on almost any flat surface with only a simple writing stylus. Like handwriting itself, it is a robust symbolism in that it is insensitive to a wide range of individual and local stylistic embellishments and deformations. Also, like handwriting, printing conventions provide a standard stylistic model to which we more or less adhere.

However, the line formula notation, representing Daltonian atoms (early 19th century) and Lewis bonds (early 20th century), has not been versatile enough to represent the new truths of chemistry as we have discovered them. In line formula symbolism, we are sometimes unable to represent the structure of a single molecule except as an intermediate between two acknowledged incorrect representations. We add little curved arrows and partial charges to indicate that the electrons are not exactly where the Lewis bond symbolism implies. We superimpose stylistic conventions (e.g., wedge bonds) and projection rules to imply three-dimensional features that cannot be directly represented on a plane surface. In addition, of course, the line formulas are forever static. In two centuries of practice, we have learned to "see" many things in our line formula symbolism that are not explicitly represented nor representable.

Students approaching chemistry, like any field, for the first time must learn to read and write an abstract symbolism. They can hardly learn the representable symbolism (the rules) until they have learned the unrepresentable (the exceptions). Line formula symbolism simply is not capable of representing the features of molecules that are important for understanding structure and reactivity, and like the symbolism of the alchemists, it must be abandoned in favor of a symbolism that can more adequately represent current knowledge and practice. As we approach the beginning of the 21st century, it is surely time to develop and use a richer symbolic representation for chemistry than that developed at the opening of the 19th century.

We already have one alternative symbolism for chemistry in molecular models. By combining one red and two white balls to represent a molecule of water, for example, we make explicit (1) that the atoms are real objects occupying space, (2) that different atoms have different relative size, (3) that certain pairs of atoms are linked through chemical bonds while others are not, and (4) that atoms in a molecule are joined in a particular three-dimensional arrangement.

Electronic versions of the molecular model kit further enable us (1) to construct larger models, (2) to manipulate models more quickly and easily, and (3) to represent relative sizes and geometries more accurately. However, by progressing from a physical to an electronic representation, we have lost true three-dimensionality. We must simulate the third dimension on the electronic screen by perspective geometry, stereopsis, shading and lighting models, hidden line/surface computations, occlusion, parallax, immersion (virtual reality), and other electronic and perceptual "tricks". Although some consider the loss of the physical third dimension to be so great a loss that the disadvantages of these "electronic tinker toys" outweigh their advantages, most consider the electronic representation to contain riches that can be exploited to our great advantage.

Electronic Representations and Visualizations for Chemistry

Dynamic, multimolecular, classical representations

Let us consider electronic representations we might use in the near future and how they might strengthen our understanding and practice of chemistry. Our "textbooks"—in whatever guise their electronic equivalent will take—and our symbolic workspaces will present molecules through dynamically vibrating, rotating, moving, and colliding representations with which we may interact. We are so accustomed to seeing stationary representations on a printed page—indeed, we rarely see any other—that we forget, for example, that the phases and energies of molecular vibrations greatly determine the outcomes of encounters with other molecules.* We might choose, for convenience, to "freeze" molecular motion for the sake of clarity of some detail. In doing so, we are reminded on each occasion that we have introduced this constraint artificially and that motion is the norm.

Our electronic textbook and workspace molecules will also appear not singly, but in collections of similar and/or dissimilar molecules moving in random directions with random velocities. Intermolecular distances and mean free paths will be proportioned to the same scale as the molecular dimensions, and we will gain an increased appreciation of size and distance. We may choose one molecule or one encounter for further examination, but in doing so, we are reminded that our "sample" is one among many with one particular set of dynamic properties and that the "prototype" molecule we are accustomed to considering emerges only as a statistical average among a very, very large collection of rather individual molecules.

The dynamical molecular representations will not be prescripted animations, but will be real-time graphical representations of molecular dynamical computations in which the motion of each atom during the next time interval is determined (by computation) solely from the coordinates, velocities, masses, and forces at the beginning of the interval in accord with Newton's laws of motion. We are free to modify atomic/molecular identity, molecular density, temperature, and other "initial conditions" and follow the dynamic development of the system properties.

Although we may slow down or speed up the dynamical representations of molecules, we will become accustomed to seeing molecular dynamics presented in some standard time scale and thus gain an enhanced appreciation of relative times involved in molecular reactions. Through working with dynamic collections of molecules, we will begin to gauge (and think of) molecular dynamics, not in terms of nanoseconds or femtoseconds, but in natural molecular time units such as number of molecular vibrations,[1] number of collision intervals, etc.

The important substructures in molecules are the nuclei and the electronic periphery. The nuclei, although small, contain essentially all the mass of the molecule, and their behavior is (approximately) classical. The dynamic properties of the molecule (vibration, rotation, translation) depend almost totally upon the masses of the nuclei. It is useful to represent the nuclei as small bright solid spheres at the centers of atoms. It will be necessary to exaggerate their true dimensions to make them visible, but we can nevertheless convey the information that they are truly small compared to the molecule as a whole. If we vary the diameters of the nuclear spheres according to the cube root of their masses, the largest atomic nuclei will have diameters approximately six times that of the smallest, and dynamics of masses scaled in this way will appear entirely realistic.

The second and only other important chemical property of the nucleus, the number of elemental positive charges, could be represented by simple coding, with the symbol 8, for example, for oxygen. However, because it is the effective nuclear charge experienced by the valence shell electrons that primarily determines the chemical reactivity of the molecule, it might make more sense to use the number 6

---

* See reference 1 for a dynamic simulation of the prototype bimolecular nucleophilic substitution of OH^- for Br^- in CH_3Br, in which the vibrational frequency of the C–Br bond in the reactant can be seen to be lower than that of the C–O bond in the product, the entire molecular encounter and reaction can be seen to take place in three to four "umbrella" vibrations of the hydrogens on the methyl group, and the reaction can be seen to progress as initial encounter, rebound, recapture, and finally separation as product.

for oxygen. More effectively, we could construct a scale by which we convert nuclear charge to color (e.g., 1 = violet, 2 = blue, 3 = green, 4 = yellow, 5 = orange, 6 = red, 7 = white—and gradations in between to represent effective nuclear charge). This convention must be learned, but it conforms to the common usage that progression from cooler colors to warmer colors represents increase. Such a convention also results in use of the same color for atoms in the same family of the periodic table (e.g., red for both oxygen and sulfur), emphasizing their chemical similarities. Differences between elements of the same family would need to be distinguished by other indices such as relative size of the periphery.

Enriching Molecular Representations using Data Visualization Techniques

The dynamical, multimolecular, electronic screen representations I have described so far reflect only an increase in the *complexity* of our symbolism. The computer has served only to generate and present more complex representations than we are accustomed to working with, but the assumed properties, interactions, and images of molecules are entirely classical and familiar. Such representations are only marginally beyond the capabilities of today's PCs[2] and require no special training to understand. The greatest changes in our chemical practice and our chemical thinking, however, will come from enhanced representations of the electronic peripheries of atoms and molecules where they contact and interact with each other. Simply put, the electrons in their quantum mechanical states around the charged nuclei entirely determine the structure and reactivity of atoms and molecules. To visualize chemistry, we must visualize electrons. It is here that chemists will need introduction to and practice in data visualization.

The most important property in the periphery of a molecule is the total electronic charge. Whether computed—e.g., by self-consistent field calculations—or measured by scattering or diffraction experiments, it is appropriate to interpret this function either as a *probability distribution function* or as a *measurement of the spatial distribution of the charge.*[3] The charge density has a single numeric value at each point in space—and hence is a scalar field. Electronic representation of a three-dimensional scalar field requires that we code and visualize a fourth variable at each point in three spaces—a technique called volumetric rendering. We might, for example, code the charge density in each volume element by a dot of light whose brightness is proportional to the charge density, as is shown for the water molecule in Figure 1.* This representation, though simple in concept, reveals little to the viewer. First, there is no way to discern at what depth into the molecule a particular dot exists. Front dots and rear dots appear identical. Presumably, this problem could be remedied by viewing stereo pairs of images where each dot is properly placed for viewing by each eye.

Beyond the front/rear problem, however, there is a more difficult problem that would remain even if we could see a perfect three-dimensional brightness (density) field. The charge density, varying continuously and smoothly throughout three-dimensional space, presents no surfaces or boundaries for demarcation of the field. What we see is a smudge or a blob. The essential problem is that of visualizing the internal structure of a cloud or a fog bank. While it may be possible to say that the

---

* Presenting dynamic, color representations as printed black-and-white figures is about as frustrating as presenting music in a sculpture studio. Although both statues and sonatas are art forms and although the sounds in a sculpture studio do have pitch, intensity, and harmonic structure, a hammer and chisel are poor substitutes for a piano. Electronic representations of the figures in this chapter are accessible via the WWW at URL http://albert.etown.edu/ranck/acs_book/figures.html.

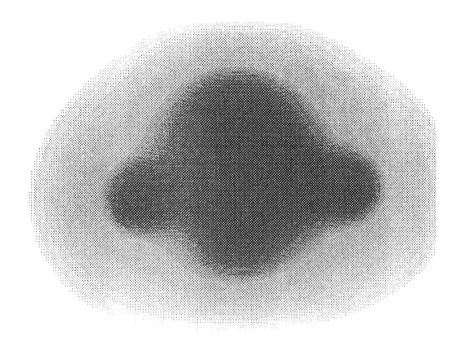

FIGURE I   Volumetric rendering of the total electron density in the water molecule.

density is greater at one point than at another distant point, it is difficult to discern local changes or at what point one enters or leaves the featureless field.

A sometimes useful technique is to map scalar field values, not as intensity, but as colors along a defined scale. The human eye is readily able to discern small changes in color, and much experimentation is carried out to develop just the right color map for displaying significant features of scalar fields. A feeling for color mapping develops only with practice, and chemistry students should surely be given scalar fields (preferably in molecules) with which to practice color mapping.

Reducing the dimensionality of the field by sectioning is another way of clarifying a volume rendering of a field. Figure 2 displays the total electron density of the water molecule in the nuclear plane. Here, we gain some clarity of detail in the region of the plane, but we have sacrificed knowledge of the field except in the plane. An interactive display in which the cross-sectional plane can be grabbed and dragged through the molecule to show adjacent surfaces in rapid succession is extremely helpful.

There are two common ways to enhance definition in a two-dimensional density plot. One is to connect all points having the same value of the density with a contour line and to generate several contour lines, usually at equal intervals (in field value), as in Figure 3. Contour maps excellently represent the "shape" of the field (in the chosen plane!), and the closeness of the contour lines represents the gradient. Because contour lines and gradients are always perpendicular, the gradient direction is also inherent in contour maps.

A variation on the contour map is the surface or relief plot in which each point in the plane is raised to an "elevation" above the plane proportional to its numeric

FIGURE 2   Cross-sectional density plot of the total electron density in the nuclear plane of the water molecule.

FIGURE 3   Contour map superimposed upon a cross-sectional density plot of the total electron density in the nuclear plane of the water molecule.

value. Here, one data dimension is plotted against two spatial dimensions. This is a common data representation and is most easily understood by comparison with contour maps, as for the three planes in the ethylene molecule shown in Figure 4.

Instead of dragging one sectional plot through space and observing its change, we might display the three-dimensional structure of the field by stacking transparent contour maps as in Figure 5. This representation is commonly used to display electron density as derived experimentally in x-ray diffraction studies.

Connecting a contour line in one plane with the lines in all planes representing the same values to create contour surfaces (Figure 6) seems an obvious extension. The outer contour surface well represents the shape of the field at that value, but the surface occludes inner surfaces whose shapes may be different. This problem can be partially solved by rendering the contour shells partially transparent (Figure 7).

The electrostatic field strength and direction in the immediate neighborhood of the molecule control the rate and direction by which ions or polar molecules will approach. The electrostatic field is a vector quantity, having both magnitude and direction. All chemists should be familiar with "hedgehog" plots (Figure 8), in which the field vectors become "flow lines".

Alternatively, one might be interested in the electrostatic potential (magnitude and sign, but not direction) at the reactive surface of a molecule. A common technique is to choose a constant charge density isosurface as a reasonable surface for the molecule and "paint" the electrostatic potential on that surface, mapped by a multicolor or a two-color and intensity scale as in Figure 9.

As an example of enhanced representation begetting enhanced communication and understanding, Bader[3] has shown convincingly that it is appropriate to speak of atoms in molecules as those regions separated from each other by gradient vector field lines in the charge density field which converge at infinity. A two-dimensional representation of the vector field in the nuclear plane of the ethylene molecule, showing clearly the regions belonging to carbon and hydrogen atoms, is reproduced in Figure 10. Chemical bonds, by extension, are represented by the unique gradient vector field lines that terminate not at infinity, but at another nucleus.

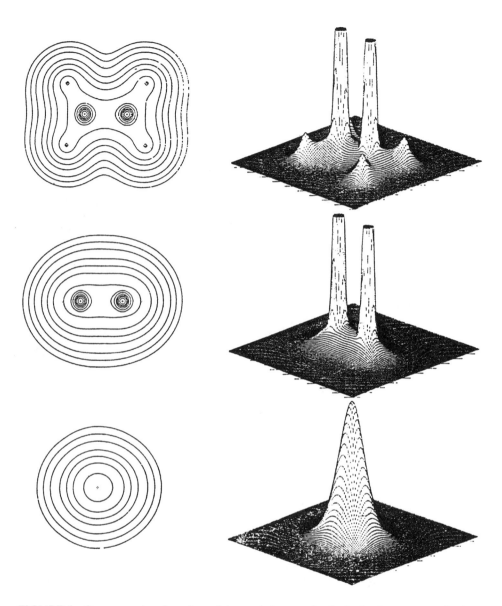

FIGURE 4   Contour and surface plots of the total electron density in the three symmetric planes of the ethylene molecule. (Reproduced with permission from ref. 3. Copyright 1990 Oxford University Press.)

Bader has also shown that positive values of the (negative of the) Laplacian of the charge density (a scalar) represent local excesses of charge, and negative values represent local deficiencies. Visualizing this function for the water molecule, for example, will display local excesses of electron density in the vicinity of the oxygen atom above and below the molecular plane and opposite the hydrogens (Figure 11). These local excesses are in the total electron density and cannot be directly interpreted as lone pairs (this is, not an orbital display), but the excess can be experimentally

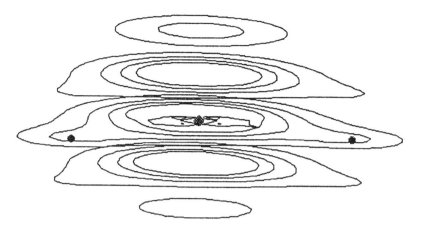

FIGURE 5   Stacked contour maps of total electron density in the water molecule. All planes are parallel to the nuclear plane.

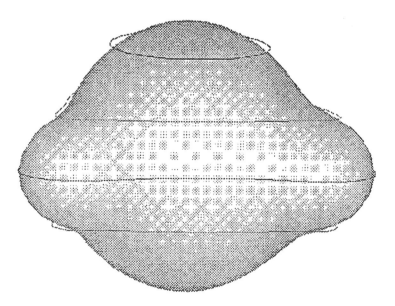

FIGURE 6   Isodensity surface representation of the water molecule. The value of the constant electron density at all points on the surface is slightly greater than the value of the outer contour line in all of the plane sections.

measured and will influence and direct reactions with other species (e.g., hydrogen bonding). Bader has presented his Laplacian maps as surface plots in one plane. Intensity and color scale volumetric rendering together with draggable sectioning planes (impossible to represent in a static black-and-white representation) make these features easier to see directly and enhance the utility of this important concept.

Using the techniques described above, it is also possible to represent the scalar fields of atomic and molecular orbitals and their squares. Many theoretical simplifications have

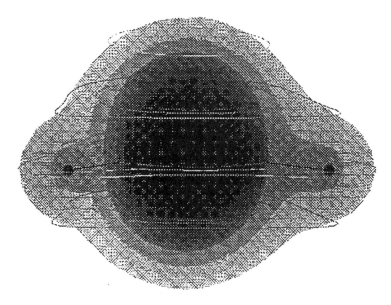

FIGURE 7   Partially transparent isoelectronic density surfaces in the water molecule revealing inner isodensity surfaces at larger field values.

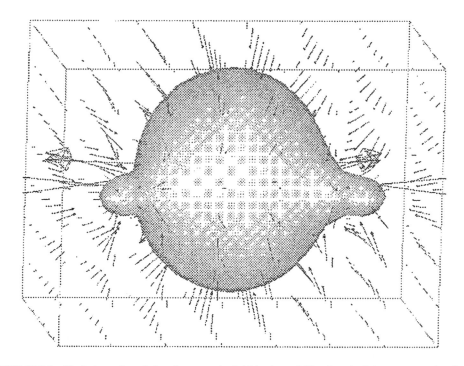

FIGURE 8   Hedgehog plot of the electrostatic field in the neighborhood of the water molecule.

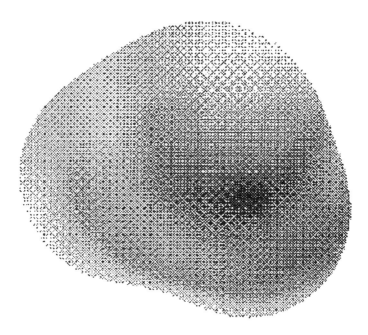

FIGURE 9 Electrostatic potential mapped on a constant electron density surface for the water molecule.

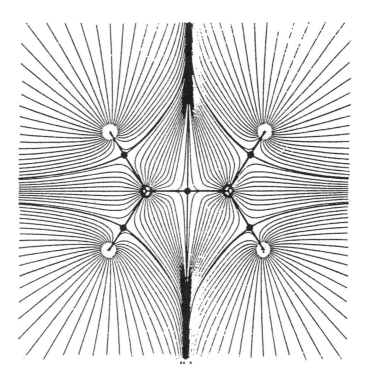

FIGURE 10 Total electron density gradient vector field in the nuclear plane of the ethylene molecule. The boundaries between atoms are along the gradient lines converging at infinity. Bonds between atoms exist where the gradient lines begin and end at nuclei. (Reproduced with permission from ref. 3. Copyright 1990 Oxford University Press.)

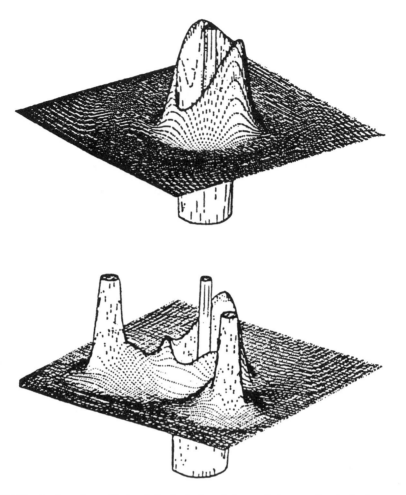

FIGURE 11   Surface plots of (minus) the Laplacian of the total electron density in the two mirror planes in the water molecule. Positive values represent local excesses of electron density; negative values represent local deficiencies. (Reproduced with permission from ref. 3. Copyright 1990 Oxford University Press.)

been applied to reach the approximation that each electron (or electron pair) is described by an orbital, and one must be careful about interpreting orbitals in molecules. Nevertheless, electronic representations of molecular orbitals have proved useful. Figure 12, for example is one frame from a dynamic representation of the electron density in the highest occupied molecular orbital during a nucleophilic substitution reaction.

We are just beginning to explore composite visualizations in which multiple representations are joined and/or different variables are plotted along the same axis. Figure 13, for example, is one frame from a time series representing the reaction H_2 + $H^- \rightarrow H^-$ + H_2.[4] The atoms and molecules, together with the reaction coordinates, are presented pictorially in the right field. The total energy reaction surface is presented at the bottom of the left field. Above this plot are two more energy surfaces (not to the same scale) representing the energies of the highest occupied and the

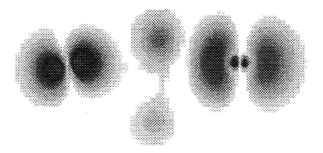

FIGURE 12   Volumetric rendering of the HOMO during the fundamental bimolecular nucleophilic reaction $CH_3Br + OH^- \rightarrow CH_3OH + Br^-$.

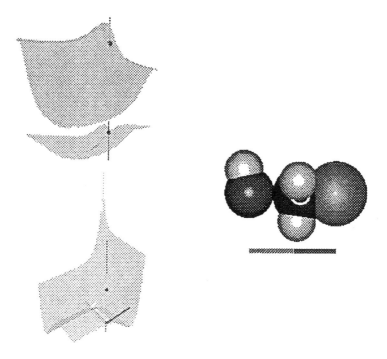

FIGURE 13   Multidimensional representation of the reaction $H_2 + H^- \rightarrow H^- + H_2$. (Adapted from ref. 4.)

lowest unoccupied molecular orbitals. As the reaction proceeds (molecular dynamics calculations), the atomic spheres in the right field move appropriately, and a point on the total energy surface moves to trace out the "reaction trajectory"—not a straight line along the reaction coordinate but the actual dynamical pathway. The current point in the trajectory is also projected upward to points on the highest occupied and lowest unoccupied molecular orbital energy surfaces. In addition, a three-dimensional rendering of one of the molecular orbitals could be superimposed over the atomic sphere representations.

Some Software for Learning Molecular Visualization

The journals and presentations at professional meetings bear witness to the quality and variety of chemistry that is practiced and presented using high-quality, versatile, and often expensive molecular graphics. I shall restrict myself to noting a few low-cost packages that run on personal computers and/or low-end workstations and are suitable for training chemists and chemistry students.

During the last few years, molecular modeling packages such as HyperChem,[5] CAChe,[6] and Spartan[7] have become part of most chemistry curricula. All of these packages compute and display orbitals, total electron density, spin density, and electrostatic potential in a variety of contour slice and isosurface representations. The CAChe system supports painting these properties upon an iso (electronic) density surface, and Spartan supports painting any property on an isosurface of any other property. Gaussian[8] does not have a graphical display interface, but provides molecular orbitals, the electrostatic potential, the electron density, density gradient, the norm of the density gradient, and the Laplacian of the density as volumetric field data for visual rendering in other graphics programs.

The volumetric data fields from Gaussian (and similar data fields exported from CAChe or Spartan) may be color-mapped and displayed in a variety of ways using Slicer,[9] an inexpensive volumetric rendering and sectioning program for PCs (Windows), or Dicer[10] for the Macintosh. For more versatility in displaying the volumetric field, data visualization packages such as IRIS Explorer,[11] IBM Visualization Data Explorer,[12] and AVS[13] are available. These packages carry a list price of several thousand dollars but can often be bundled with the purchase of a minimal workstation for much less. These general-purpose data visualization packages contain extensive libraries of modules that are connected into a data-flow network. Many chemistry modules are contained within the package or are available as extra-cost options or through module exchange sites on the Internet.[14]

References

1. Ranck, J. P.; "Visualizing Chemical Reactions"; CHEMCONF '93: Applications of Technology in Teaching Chemistry, Summer 1993, paper #3, available via WWW at URL ftp://info.umd.edu/info/Teaching/ChemConference/paper3.

2. Shirts, R. B.; "Boltzmann: Computer Software for Demonstrating Principles of Kinetic Molecular Theory in the Classroom, Laboratory, or Exploratorium"; 210th National Meeting of the American Chemical Society, Chicago, IL, August 1995, CHED261.

3. Bader, Richard F. W.; *Atoms in Molecules: A Quantum Theory;* Clarendon Press: Oxford, 1990.

4. Belsky, Alec; "Visualization of Chemical Reactions," Bachelor's Thesis, Elizabethtown College, 1994.

5. Hypercube, Inc., 419 Phillip Street, Waterloo, Ontario, Canada N2L 3X2.

6. CAChe Scientific, P.O. Box 00, Beaverton, OR 97077.

7. Wavefunction, Inc., 18401 Von Karman, Suite 370, Irvine, CA 92715.

8. Gaussian, Inc., Carnegie Office Park, Building 6, Pittsburgh, PA 15106.

9. Slicer. Spyglass, Inc., 1800 Woodfield Drive, Savoy, IL 61874.

10. Dicer. Spyglass, Inc., 1800 Woodfield Drive, Savoy, IL 61874.

11. Silicon Graphics, Inc., 2011 N. Shoreline Blvd., Mountain View, CA 94039-7311.

12. IBM Corporation, Thomas J. Watson Research Center, P.O. Box 704, Yorktown Heights, NY 10598-0704.

13. Molecular Simulations Inc., 16 New England Executive Park, Burlington, MA 01803.

14. International AVS Center, at URL http://iac.ncsc.edu.

Part IV
Computational Chemistry in the Curriculum

14

Ab Initio Techniques in Chemistry: Interpretation and Visualization

James B. Foresman

Recent improvements in the power of affordable computers and in the design of faster and more efficient algorithms have placed powerful molecular modeling tools in the hands of chemists. Ab initio molecular orbital calculations are now routinely performed by nonspecialists in the field. Because these methods do not depend on empirical parameters, they can be applied to a wide range of chemical systems and are useful to all types of chemists. These calculations often assist experimental studies by accurately predicting chemical behavior. It is widely recognized that some blend of theory and experiment must be effectively employed to solve many of our most difficult problems in chemistry.

Advances in the technology of ab initio calculations occur so fast that it is difficult to write a discussion of these techniques without becoming instantly outdated. That which defines a useful approximate model today may well be considered a crude and trivial calculation only months in the future. For this reason, I will attempt to present only the background information required to appreciate the general features of ab initio molecular orbital calculations. For the most part, emphasis will not be placed on the details of specific theoretical models or particular computer programs. Instead, I will present a broad overview of the methods, the types of chemical problems that can be reasonably solved by them, and the utility of graphical programs that can assist in the visualization of the data.

Recent Advances in Methodology

Computers that can perform quantitative molecular orbital calculations have become as useful to the bench chemist as spectrometers and vacuum lines. This situation has come about not only because of the increased speed of the low-end desktop machines, but also because of the development of algorithms that can deal

TABLE 1   List of computer codes that perform ab initio calculations.

Program	Contact
Aces II	Quantum Theory Project, Univ. of Florida (R. Bartlett)
Cache	Oxford Molecular
CADPAC	Cambridge University, U.K. (R. Amos)
Columbus	Argonne National Laboratory (R. Shepard)
DGauss	Cray Research
DMol	Biosym/Molecular Simulations
GAMES	Ames National Laboratory (M. Schmidt)
Gaussian	Gaussian, Inc.
GradSCF	PolyAtomics Research Institute (A. Komornicki)
HONDO	IBM (M. Dupuis)
HyperChem	HyperCube, Inc.
MolPro	University of Stuttgart, FRG (H.-J. Werner), and University of Birmingham, U.K. (P. J. Knowles)
Mulliken	IBM/Cache Scientific
PS-GVB	Schrödinger, Inc.
PSI	PSITECH, Inc.
Spartan	Wavefunction, Inc.
SuperMolecule	Cray Research (M. Feyereisen)
TurboMole	Biosym/Molecular Simulations

efficiently with the great number of interactions present in a system containing a large number of electrons. Table 1 lists a number of the programs that are currently available. The most popular of these (based on numbers of citations in the research literature) is the Gaussian suite of codes. It also has the greatest number of features. Standard graphical packages can be combined with these programs to visualize the results and input the data. Two of these (HyperChem and Spartan) have the graphical interface built in.

Appreciating the nature of these complex computer codes requires a review of the basic scheme for computing the energy and wave function of a molecular system. For a more detailed explanation of the equations that follow, please consult the references.[1,2]

Hartree–Fock Theory

First, a given molecular orbital (MO), ψ, can be expressed as a linear combination of mathematical functions, ϕ, centered at each atom:

$$\Psi_i = \sum_{\mu} c_{i\mu} \phi_{\mu}$$

The c's are called the MO coefficients, and the ϕ functions constitute the basis set for the calculation. These ϕ functions can be thought of as atomic orbitals (s, p, d, etc.) from the rigorous solution of Schrödinger's equation for the hydrogen atom. In practice, these functions are often fixed linear combinations of Gaussians whose exponents have been judiciously chosen by minimizing atomic energies, matching orbitals from Slater's rules, or by reproducing experimentally known molecular properties. There are many basis sets tabulated and tested in the literature for almost every element of the periodic table, and still more are developed each year. Which basis set to use is one of the first dilemmas facing a chemist who is trying to perform an ab initio calculation. That choice will depend on the problem being addressed, the level of accuracy desired, and the feasibility of the job. In general, chemists should use several different basis sets of increasing size to test what influence this choice has on the results of an investigation. In what follows, N will represent the total number of basis functions in the calculation.

The coefficients for a given MO can be thought of as arrays that, when squared and added together, will produce a density matrix, P:

$$P_{\mu\nu} = 2\sum_{i}^{occ} c_{i\mu}c_{i\nu}$$

Here, the sum runs over the occupied orbitals of the system. For the moment, we are restricting ourselves to the case of a closed-shell singlet system, so each occupied orbital will contain two electrons. The density matrix and its associated orbital basis set are all that is needed to compute electronic properties (dipole moment, charge distributions, electrostatic potential, etc.) of the molecular system. While the MO coefficients themselves do not have any connection to physical reality, it is the density matrix that can lead to the calculation of observable quantities. Since different sets of orbitals can lead to the same density matrix, examination of the orbitals in a calculation is of limited utility (unless the purpose is to establish a connection with early qualitative theories concerning bonding).

In order to determine the density matrix, the energy of the system must be minimized with respect to the unknown MO coefficients. In Hartree–Fock (HF) theory, the simplest and least accurate ab initio method, the energy of the system can be written as:

$$E_{HF} = \sum_{\mu\nu} P_{\mu\nu}H_{\mu\nu} + \sum_{\mu\nu\lambda\sigma} P_{\mu\nu}P_{\lambda\sigma}\left(J_{\mu\nu\lambda\sigma} - X_{\mu\nu\lambda\sigma}\right) + V_{nuc}$$

Here, H is a list of terms arising from the kinetic energy of the electrons and the potential energy of attraction between the electrons and the nuclei of the system. J is a list of terms due to the potential energy of repulsion between pairs of electrons. X is a list of terms that have no physical interpretation and arise solely because quantum mechanics requires that the wave function be antisymmetric with respect to exchange of any two electrons (switch electron 1 with electron 2 and this must produce the negative of what you had before). The final term, V_{nuc}, is the potential energy from the repulsion of pairs of nuclei. In order to determine the unknown MO coefficients, the energy in the expression above is minimized with respect to them. This leads to a set of N equations that need to be solved iteratively. A guess is made at the MO coefficients, and their values are varied until the system of equations is solved. In the end, a total of N molecular orbitals is generated. The lowest $n/2$ of these in energy will be occupied (n is the number of electrons), while the remaining are unoccupied and are called virtual orbitals.

Evaluating the four suffixed quantities (J and X) is the most difficult and time-consuming part of the HF calculation. These interactions are integrals that can be solved analytically if the basis set is constructed of functions such as Gaussians. In conventional calculations (those which were in practice before the mid-1980s), these values had to be computed at the start of the job and stored on a mass storage device (hard disk or magnetic tape). Taking into consideration the symmetry of these terms, there are on the order of $N^4/8$ unique integrals in all, and so calculations were limited by both the amount of storage available and the speed with which these quantities could be read from disk. For instance, a calculation with 300 basis functions (typical of a system with 20 atoms) would have required 8 gigabytes of storage! Today, the same-size calculation is done routinely on an ordinary personal computer in just a matter of hours without the need for a large disk. The truth is that with the current generation of computers and some very clever mathematical tricks involving recurrence relationships, it is actually faster to recompute the integral quantities as they are needed in the iterative solution of the equations than it is to read them from disk. It is also true that if one screens out the integrals that are small and negligible (those arising from orbitals far removed from one another in a large molecule), the time it takes to perform an HF calculation scales only as $N^{2.7}$, not N^4 as was earlier imagined. That is to say, by doubling the size of the molecule you will only increase the work by a factor of 6.5, not 16. Much effort is going into the design of new algorithms that make it possible to do ab initio calculations on even larger molecules. Because of these developments, it is now possible to use ab initio techniques to model a wide range of systems that are of interest to experimental chemists.

The energy and density that emerge from HF theory are only approximations of the true energy and density of the system. This is because certain interactions have been left out. First, the kinetic energies of the nuclei were not included. This is the so-called Born–Oppenheimer approximation and is a valid one to make since the movement of the nuclei is small compared to the electrons. Second, relativistic effects are ignored. A full treatment beginning from quantum mechanics would have to include such terms. They only are significant for the heaviest of elements (beyond the fourth row of the periodic table). Third, the J terms are computed by assuming that as a given electron moves, it experiences an average potential energy of repulsion caused by the remaining electrons of the system, as if they were stationary for that moment in time. The fact is that electrons attempt to correlate their motions (and keep out of each other's way). This leads to an additional energy term that is only partially accounted for by the exchange interactions in X. Of all the approximations in HF theory, this is by far the most severe. Calculations have shown that the neglect of electron correlation can cause properties of certain molecules and energetics of certain reactions to be inaccurate when compared to experimentation. This is especially true if the molecule exhibits unusual bonding characteristics (strained rings, transition metal complexes) or when the system being studied is far from the ground-state equilibrium geometry (transition structures, excited states).

As an example of this, consider one of the many classic problem cases for ab initio theorists: ozone. Table 2 lists the predicted and observed equilibrium geometry and vibrational frequencies for this molecule. The basis set used for each oxygen

TABLE 2 Equilibrium properties[a] of the ozone (O_3) molecule.

Method	R_{O-O}	θ_{O-O-O}	$\omega(a1)$	$\omega(b2)$	$\omega(a1)$
HF	1.193	119.4	868	1421	1541
B3LYP	1.253	118.4	749	1210	1260
CCSD(T)	1.275	116.9	717	1057	1152
Expt.	1.272	116.8	716	1089	1135

[a] Bond distance in angstroms, angle in degrees, and frequencies in cm^{-1}.

contains five s functions, four p functions, two d functions, and one f function, so it is quite a large basis set. HF theory even at this large basis set underestimates the bond length by 6%. Accurate thermochemical predictions are also difficult to obtain from HF theory. The experimental atomization energy (energy difference between a molecule and its component atoms) of triplet oxygen (O_2) is 118.0 kcal/mol. HF theory using a modest basis set yields a value of only 30 kcal/mol.

Density Functional Theory

One way to improve the HF description without adding substantially to the time or complexity of the calculation is to replace the exchange matrix with two empirically derived functions of the total electron density, ρ. These are called density functionals. The first is for the exchange term and the other is for the missing correlation term. This describes a family of ab initio methods known collectively as density functional theory (DFT). The energy in this framework can be expressed:

$$E_{DFT} = \sum_{\mu\nu} P_{\mu\nu}H_{\mu\nu} + \sum_{\mu\nu\lambda\sigma} P_{\mu\nu}P_{\lambda\sigma}J_{\mu\nu\lambda\sigma} + E_X(\rho) + E_C(\rho) + V_{nuc}$$

Using these functionals is quite different from the traditional means of evaluating integral quantities. These are mathematical functions that have the value of the electron density as one parameter. This density needs to be formally evaluated at specified points in the three-dimensional space of the molecule. Instead of integrals being evaluated with analytical expressions, a grid of points must be specified, and the required terms are computed numerically. Chemists who use these methods not only have to pick a basis set and functional, but also select the coarseness or fineness of the grid. Standard grids and functionals are available in most computer programs, and still more are developed each year. The form of the functional itself can be very complicated, and they have been derived much like basis sets, by fitting adjustable parameters to yield the exact density in simple model problems where the exact density is known.

For many situations, the use of DFT methods can greatly improve the results of HF calculations. For other cases, it shifts results too much. For instance, a calculation of the atomization energy for triplet oxygen using BLYP (a popular exchange-correlation functional) yields a value of 131.3 kcal/mol. This is much closer to experiment than the HF results (see above), but it is now 10% too high. To remedy this, a third class of models, called hybrid methods, has been developed. These define the energy as a linear combination of HF and DFT theories, and recent studies have shown them to be quite adequate given the level of computation required. If the B3LYP model (a DFT–HF hybrid) is used, the atomization energy of triplet oxygen is calculated to be 118.4 kcal/mol, in excellent agreement with experiment. B3LYP also does well on the ozone problem (see Table 2).

It is dangerous to generalize too much from just a few individual cases. Thus, a test suite[3] of molecules has been developed that attempts to include many of the difficult types of bond-forming and bond-breaking processes. Over 100 thermodynamic quantities are included in this list, which involves elements from the first three rows of the periodic table. The mean absolute deviation from experiment for these using the B3LYP model is 4 kcal/mol, demonstrating that it is a reasonably accurate method. Remember, this evaluation is somewhat of a worst-case scenario. Many of the experimental energies are reproduced to a greater accuracy.

Beyond Hartree–Fock and Density Functional Theory

HF and DFT methods have as their strength the ability to be applied to molecular systems that are quite large. If one desires very accurate results (such as heats of formation to within 0.5 kcal/mol), more specialized ab initio procedures will be needed. These attempt to solve the electron correlation problem more exactly instead of replacing it with a simple functional of the density. However, this adds substantially to the time and resources required. The best way to think about these more expensive approaches is to realize that the HF procedure leads to just one possible way of distributing the electrons of the system among the calculated orbitals. There would be other configurations formed by moving electrons out of the orbitals that they occupy and into virtual orbitals instead. These configurations would have a higher energy, but when mixed with the ground state, they may improve the overall quantum mechanical description of the system.

In the complete active space self-consistent field (CASSCF) formalism, a group of occupied and virtual orbitals is chosen along with a certain number of electrons. The complete wave function is then constructed as a combination of all the possible configurations made by distributing the electrons among the orbitals. Each configuration will lead to a set of HF-like equations. The end result produces a set of MOs for each configuration. This procedure accounts for some of the electron correlation neglected at the simple HF level. It also is a procedure useful for studying systems where there are excited electronic states very near the ground state (as in transition metal systems and molecules distorted in a way that places the ground-state potential energy surface near one of the low-lying excited states). The main problem with these methods is that some expertise is still required to decide which electrons are allowed to float into the virtual orbital space and which virtuals they are allowed to occupy. You cannot let them all move since the calculation would be too time-consuming. The best way to learn this technique is to follow examples given in program manuals and tutorials. A clear understanding of the chemistry of the system being studied is essential before attempting this type of calculation.

Another way to mix in the higher configurations is to construct an exponential operator that acts on the HF ground state to produce configurations from only certain classes of excitations. These are usually limited to single, double, and triple substitutions. Since carrying out the exponential as a sum will generate many new configurations that are sums of products of the configurations, this technique is known as coupled cluster singles, doubles, and triples (CCSDT) theory. Triple excitations can also be included in an approximate but less time-consuming way, leading to the CCSD(T) model. These procedures can lead to very accurate energies and structures for molecules including those with very unusual bonding characteristics (such as O_3, FOOF, and Cr_2) that have plagued theorists for years. See Table 2 for an example of how great this accuracy can be for the ozone molecule.[4] Unfortunately, these calculations are extremely costly in terms of time and resources. For now (1996) and for several years to come, they will be practical only for systems containing fewer than 10 atoms.

Finally, Møller–Plesset perturbation theory can be applied to the Hartree–Fock wave function to obtain the effects of the higher configurations to a certain order (MP2, MP3, MP4, etc.). These are one-time corrections to the HF energy and so are generally less-time consuming than the approaches that iteratively solve for the actual weights of the higher configurations in the total wave function.

Molecules in Solution

In all of what we have discussed here, it has been assumed that the molecular system is in the gas phase. No attempt has been made to account for external interactions (neighboring molecules of the liquid or solid, solvent interactions in the case of solutions). Another recent development that has made ab initio methods more useful to chemists is the ability to study these systems by embedding the molecule in a cavity surrounded by a continuum dielectric representing the electric field produced by the neighboring molecules. All that needs to be specified is the dielectric constant for this field and the cavity shape. Cavity shapes such as spheres and ellipsoids lead to interactions that can be evaluated quickly and easily by analytical techniques, but these are not always appropriate if the molecules are not spheres or ellipsoids. A better solution is to define the cavity as a function of the electron density of the solute (an isodensity surface, for instance) that would have the actual shape of the molecule. This method is referred to as the polarizable continuum model (PCM) and many variants of it exist. It has been shown to model the solvated system extremely well as long as there are no specific interactions present, such as hydrogen bonding (i.e., situations where the continuum model is not a viable representation of the interactions present). For these systems it may be possible to include specific solvent molecules within the cavity and then surround the entire system with a continuum dielectric.

Another category of methods is being developed that will not only allow for the calculation of solvent effects, but also the extension to systems with extremely large numbers of atoms. Several groups[5,6] are working on models that partition the molecular system into a small inner region (solute) and a much larger "inactive" region (solvent). Ab initio techniques can then be applied to the inner region, while molecular mechanics or molecular dynamics (much less computationally demanding approaches) is used to treat the outer region. This mixture of quantum and classical methodology shows much promise in terms of being able to model complex chemical systems in their entirety.

Chemical Properties

The first reason chemists might decide to employ ab initio modeling techniques is because they need accurate molecular energies or properties and are unable or unwilling to perform an experiment that will yield them. Perhaps the experiment is difficult or time-consuming to perform (involving unstable species or extreme conditions). Maybe it is desirable to see what effects a series of structural modifications have on the computed properties without having to synthesize any new compounds. In the case where new experimental data are available, chemists may be interested in having an explanation for observed results and trends in terms of the electronic features of the molecule.

Thermochemical Data

The calculation of gas-phase thermochemical data for small molecules (fewer than 10 atoms) is probably the most successful type of application to be solved by ab initio calculations. Properties such as atomization energies, electron affinities, and ionization energies can now be computed at an accuracy that rivals experimental estimates. This usually involves obtaining the geometry from a moderate level of theory (such as MP2) then performing a coupled-cluster calculation using a large basis set. Even this elaborate process, however, does not account for all of the important electron correlation effects. There is one final correction to the energy needed that is derived by insisting that the model yield the exact energy for a two-electron system. One such procedure of this type is known as Gaussian-2 Theory.[3] Tabulations of data using this model are now extensive and demonstrate well the reproducibility of experimental results.

Being able to accurately calculate atomization energies, electron affinities, or ionization energies has obvious benefits to chemists who are trying to assess the thermodynamic stability of new compounds. This type of calculation is also helpful to students who are studying the principles of bonding and reactivity, since a connection can be made between the ability for a species to accept an electron and the reactivity of that species. These data would be difficult to obtain from an undergraduate chemistry laboratory experiment, but the computer allows them to be evaluated directly.

Conformational Energy Differences

Another area in which ab initio calculations can greatly assist in the interpretation and prediction of chemical behavior is in evaluating the lowest energy conformation of a molecule. Qualitative rules are often used by chemists to predict that one conformer is more favorable than another, but unless the system is dominated by strong steric interactions, there may be subtle electronic effects that need to be considered before deciding which form is more stable. A more satisfying process for chemists and students of chemistry is to calculate the energy difference and examine how it is influenced by changing the structure of the molecule (adding a substituent, etc.). A simple example would be to look at 1,2-difluoroethene and replace the fluorines with other halogens or other organic groups. How is the stability of trans versus cis modified by the nature of the substituent? How do the results change when going from the gas phase to solution? These are probing questions that get at the important aspects governing the chemistry of these molecules and would be easily investigated with ab initio techniques. Since one is interested only in approximate relative energies, HF or DFT methods would be appropriate (where energetic differences are large). This means very large systems could be studied.

Taking this example one step further, suppose we also would like to know how easily the two forms can interconvert. Evaluating the barrier to rotation would involve an extremely difficult experiment. Simple qualitative theories (or even semiempirical quantum mechanical methods) would also have difficulty evaluating this, since they are normally parameterized for equilibrium structures. However, it is quite routine to use an ab initio calculation to evaluate the energy of a structure like the perpendicular form of 1,1-difluoroethene (the transition state for the rotation). Comparing this for a series of substituents in and out of solution can lead to a better understanding of the nature of a double bond.

Spectroscopic Measurements

Infrared (IR) spectroscopy is a common tool for the identification of unknown compounds. Every chemist is familiar with the correspondence between peaks in the IR spectrum and functional groups of the molecule. Being able to calculate these absorptions offers several advantages. First, the calculation leads not only to the position of the peaks, but also to the exact vibrational motion of the molecule to which the peak corresponds. Assignment of peaks can be assisted by comparing the experimental spectrum with the predicted vibrations. Programs that offer graphical display of these motions are widely available and can greatly aid students who are trying to visualize and understand the origin of the effect they so often see in the laboratory.

When just a qualitative understanding of the IR spectrum is needed, HF theory is quite adequate in computing frequencies of vibration, and so very large systems can be examined. One can scale the computed values by 0.89 (a single factor works since the error in computed HF frequencies is a consistent one) to obtain better agreement with experiment, or a higher level of theory can be used.

Although very few laboratories have the capability of doing Raman spectroscopy, the calculation for the Raman spectrum is no more difficult than that for the IR, so ab initio techniques can generate data that are not easily obtainable from experimentation. This would be useful in undergraduate instruction, since students do not otherwise get a chance to examine and interpret the vibrational modes that are only Raman-active.

Nuclear magnetic resonance (NMR) spectra can also be computed using ab initio theories. Here, the calculation is a bit more complicated than vibrational frequencies since an external magnetic field must be applied to the wave function. A given level of theory will lead you to the isotropic magnetic shielding (in ppm) for each nucleus. To compare to an experimental spectrum, one needs to compute the chemical shift by doing an equivalent calculation on the reference used (usually tetramethyl silane) and take the difference between the two calculations. In practice, it is hard to predict the exact shift for a proton NMR spectrum, since the actual magnitude of the effect is small. C-13 spectra are reproduced much better. HF theory can be used, but a large basis set is needed to fully account for the electronic effects. These ab initio calculations can greatly assist in the assignment of resonances for compounds with complicated structures. Such calculations should also play an important role in instrument simulation for teaching purposes, since multinuclear NMR capabilities are often too costly for small institutions.

Electronic spectra (UV-vis) are much more difficult to model. The problem originates from the fact that to obtain the electron density and energy in an ab initio calculation, the energy is minimized with respect to the density. If an excited-state energy and density are sought in this manner, there must be something in the theory to prevent the solution from collapsing to the ground state. CASSCF calculations are useful in this regard, since one can simultaneously solve for several electronic states. Another procedure, configuration interaction with singles (CIS), forms the excited state as a combination of single excitations from the HF level. This has the advantage of being easier to use (no need to specify active space orbitals or electrons), but can only be used to treat excited states that are largely single excitations. Still, the gross features of many electronic spectra can be reproduced well. An essential feature of these calculations is that they report oscillator strengths for the transitions predicted. This is to help in comparing to actual spectra, since there are many excited states that can be calculated for a molecule but that have zero or near-zero intensities (and therefore are not observed in the normal optical experiment).

Chemical Reactivity

Calculation of Reaction Enthalpies

The first way to begin to explore real chemistry with ab initio methods is to simply look at differences in energies between reactants and products. Heats of formation for many substances are known from calorimetric experiments, so one way to validate a particular level of theory is to compare this energy difference directly with experiment. The first problem one will encounter is that the calculation leads to the energy of the system in the gas phase at a temperature of zero Kelvin. The zero-point energy (the energy of the lowest vibrational quantum state) is not even included. A separate calculation must be performed that evaluates the frequencies of vibration and, from them, allows the calculation of zero-point energies, thermal energies, enthalpies, and free energies.

For a certain class of reactions, the computed enthalpies are actually better than one would expect at HF theory. This is because the neglect of electron correlation results in approximately the same error for reactants and products, so by taking differences in energies the error is removed. This is especially true of reactions where products have the same number of like bonds (single, double, triple) as reactants.

Searching for the Transition State

Now that you have a level of theory that reproduces the correct energies for a reaction, you can begin to study the mechanism and kinetics of the reaction. This involves finding the geometry of the transition state. Ab initio calculations can be used to compute the energy of a system for any set of nuclear coordinates. A plot of this energy versus any combination of geometric parameters will generate a two-dimensional potential energy surface (PES) for the system. In general, potential energy surfaces are N-dimensional, where N is the number of degrees of freedom for the molecule. Chemical reactivity can be related to the way in which molecules move along the PES. Molecules need to distort themselves beyond their equilibrium geometries if they are going to participate in a chemical reaction. This is energetically unfavorable and therefore requires energy to be put into the system. The amount of energy needed to produce the exact distortion necessary for the molecule to react is the activation energy. The geometry at that point on the PES is the transition structure. In searching for this point, algorithms have been designed that can minimize the energy with respect to all geometric coordinates of the system except the reaction coordinate, which it maximizes.

Since the resultant geometry is in many cases far from the equilibrium structure, the HF method may not be very accurate. Usually, a method that partially includes electron correlation is needed to compute the transition structures and activation energies. It is an interesting computational experiment to consider how the magnitude of the activation energy changes based on changes in the molecular structure and solvent effects. This information is readily obtainable from ab initio calculations.

The Intrinsic Reaction Coordinate

Once the transition structure has been obtained, it may be useful to consider the exact path that will lead back to the reactants and forward to the products. The intrinsic reaction coordinate (IRC) method does just this. One must be careful not to associate this path with the actual mechanism of the reaction. This is just the path that the reaction is predicted to take at a temperature of zero Kelvin. Real reactions occur at much higher temperatures, and so many energy paths would be sampled by the vibrating molecule. The IRC does, however, provide a nice rubric for thinking about the course of a reaction. It also helps characterize transition structures (it is not always obvious exactly what is downhill from them). Finally, combined with variational state theory, the IRC can be used to calculate reaction rates.

Visualization of Data

Types of Volumetric Data

Programs for the three-dimensional visualization of scientific data are quite commonplace in the chemist's workplace. Many quantum chemistry programs have graphical interfaces that allow the direct examination of molecular properties. What do all of these elaborate and colorful pictures really mean? To gain a better appreciation of these renderings, one must first consider the form of the data that emerges from an ab initio calculation.

Any molecular property is a result of the electron density. This is computed from the density matrix and basis set of the method. Visualize a three-dimensional grid of points surrounding a molecule with the atoms completely buried within the cube of points. Imagine that these points are very close together. Now at any given point, the value of a molecular property can be evaluated. Examples of properties include the electron density (related to the probability that an electron will occupy that precise point in space), electrostatic potential (a scalar quantity related to the value of a point charge at that position in space), and electric field (a vector quantity that indicates how the electrostatic potential would be affected by an electric field). All of these quantities can be used to rationalize the observed properties of the molecule and its reactivity.

To view all of the data for a given property, you would have to somehow be able to see inside the surrounding cube to the points within. To make sense of it all, graphical programs can be great aids. For instance, one could map the range of values for a given property to a color map. The visible spectrum from blue to red could represent low to high values. In order to see within the cube, a transparency value could be set for the points near the outside. Another popular way to view this type of data is to show a three-dimensional object defined by requiring that the value of property be the same everywhere on its surface. This is called an isosurface. For instance, the 0.001 electron density isosurface is often used to show the spatial extent of a molecule. It surrounds that part of space that has the largest probability of containing the electrons. This surface can then be "painted" by using the values of another property mapped to the colors of the visible spectrum.

Visualization in Research Applications

It is often desirable to consider some property of a molecule and attempt to relate it to its reactivity (for example, the identification of lone pairs for nucleophilic addition). A very illustrative way of representing this is by computing the electrostatic potential at points on a density isosurface. Colors can be used to show the value of the potential, and therefore regions of positive and negative charge can be easily identified. Comparing two molecules depicted in this fashion is useful in understanding the differences in reactivity between the two species.

Another technique is to compute changes in the electron density for a molecule undergoing some process. For instance, when studying electronic transitions, it is very insightful to compute the difference between the ground- and excited-state densities for a molecule and paint an isodensity of the ground state with those values. This would clearly identify regions in the molecule from which the electrons are departing, and to which they are going, during an electronic transition. Another useful demonstration is to compute the difference between the electron density computed at a very high level of theory and that at a low level of theory (for example, MP2 versus HF densities). With this information you can begin to explain the physical nature of the deficiencies in the less sophisticated model.

Visualization in Education

In introducing students to electronic structure calculations, it is essential that graphical depictions of molecules and properties be possible so that students can focus on the chemistry of a problem and not be lost in the myriad of numbers that emerges from a given computation. There is a rich level of detail that results from an ab initio calculation and it is probably best left to the student to explore this on his or her own. This is greatly facilitated by having an interface that can display a wide variety of properties in many different ways. Most ab initio programs can easily generate the data (a cube of points with the value of the property at each point) in a format that is accessible by a generic scientific visualization program. By manipulation and comparison of volumetric data, a student can discover important aspects of electronic structure. This type of technology will soon make ball-and-stick model sets obsolete for use in introductory chemistry courses, since a laptop computer can deliver a much more realistic and meaningful rendition of the molecule.

Limitations of Ab Initio Methods

Throughout the discussion above, there are several warnings concerning the application of ab initio techniques to problems that are still out of reach. The size of the system that can be realistically studied depends on the questions being asked in the investigation. It is clear that some of these barriers will be removed by faster computers and more clever algorithms, but even for the near future, the very accurate energetic calculations will be limited to relatively small (fewer than 20 atoms) systems.

The less expensive models such as HF and DFT will be able to be applied to extremely large systems. Predictions at these levels are still quite useful, even if they are less quantitative in nature.

There are many scientists who would prefer that molecular modeling techniques be "black box" in nature. This may be both an unrealistic and unnecessary goal. In order to truly gain any value from these methods, one still needs to have some chemical intuition in the same way that understanding an NMR spectrum requires a knowledge of chemical structure. Beyond this warning, however, chemists who have no experience or background in this type of molecular modeling should be encouraged to experiment on their own to gain a feeling for how the models can assist them in their research or teaching.

In order to get started, novices should challenge the computer to confirm their most basic notions concerning the system of study. Ask questions such as: What is the lowest energy conformer? What are the general features of its IR and C-13 NMR spectra? Where are the electrons located in this molecule (places of greatest electron density)? Are these observations consistent with the experimental data for the compounds? If the model does not agree with your notions, then either the model is useless for the system you are studying or you need to rethink your notions in light of the computational experiment. The interpretation of the results of one ab initio calculation should be given the same (or less) consideration as one would give a single piece of experimental data. Chemists seldom draw conclusions based on one measurement collected from one instrument under one set of conditions. However, when several experiments are conducted with varying techniques and conditions, and correlations can be drawn among the data, then conclusions begin to emerge. Likewise, several theoretical models must be applied and results compared to what is known experimentally before the calculation is of any use. Validation of the model on systems that are similar to the one being investigated, and for which experimental data are well known, is a key first step in any thorough theoretical study.

As ab initio methods become more and more accurate, it is certain that there will be even more cooperation between theory and experiment in the discovery process of chemistry. This can only come about if experimental chemists (in particular nonspecialists) are willing to begin using ab initio methods in conjunction with their wet chemistry work. The best way to facilitate this is by placing more emphasis on the methods in undergraduate courses. There is much activity in this area, and I am quite hopeful that the current generation of chemists being trained will consider ab initio techniques as just one more set of tools they can routinely rely upon to bring them closer to a solution for a particular chemical problem.

References

1. Foresman, J. B.; Frisch, A. E.; *Exploring Chemistry with Electronic Structure Methods;* Gaussian, Inc.: Pittsburgh, 1993.
2. Szabo, A.; Ostlund, N. S.; *Modern Quantum Chemistry;* McGraw-Hill: New York, 1982.
3. Curtiss, L. A.; Raghavachari, K.; Trucks, G. W.; Pople, J. A.; *J. Chem. Phys.* 1991, *94,* 7221.
4. Lee, T. J.; Scuseria, G. E.; *J. Chem. Phys., 93,* 489.
5. Maseras, F.; Morokuma, K.; *J. Comput. Chem.* 1995, *16,* 1170.
6. Gao, J.; Furlani, T. R.; *IEEE Computational Science and Engineering* 1995, *2,* 24.

Computer Chemistry: A Curse or the Way of the Future?

Warren J. Hehre

The more progress physical sciences make, the more they tend to enter the domain of mathematics, which is a kind of centre to which they all converge. We may even judge the degree of perfection to which a science has arrived by the facility with which it may be submitted to calculation.

Adolphe Quetelet 1796–1874

Every attempt to employ mathematical methods in the study of chemical questions must be considered profoundly irrational and contrary to the spirit of chemistry. If mathematical analysis should ever hold a prominent place in chemistry—an aberration which is happily almost impossible—it would occasion a rapid and widespread degeneration of that science.

A. Compte 1798–1857

To see if computation is poised to play an important—even vital—role in the chemistry of the future, one must ask a simple yet rather fundamental question.

What Do Chemists Do?

While no single answer suffices, one response which would receive broad acceptance is "most chemists spend most of their time dreaming up and then making useful materials". How do they do this? While there is no magical recipe, there is a general strategy which many if not most chemists follow. It is to establish connections between structure and the chemical, physical, biological or other property of interest.

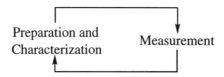

This usually involves some sort of a "loop", in which results from some measurement are fed back to help better formulate a new candidate to be prepared and characterized. The chemist stands right in the middle of the loop. Experience, patience, a keen sense of observation and intuition are necessary ingredients for success.

There is no reason why this kind of design cycle could not be based on computation instead of experiment.

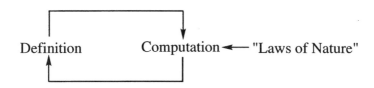

Here preparation and characterization (the "hard parts") would be replaced by simple definition of the system, and the various measurements would be replaced by computation. As before, the chemist is right in the middle of things.

The two approaches are similar in fundamental ways. They both require careful scrutiny of the data, be it measured or computed. They both need to be repeated either until the desired result is achieved or until the chemist runs out of patience or time or both. There are also fundamental differences. Perhaps most obvious, is that computations do not actually lead to "useful materials" (or any materials for that matter). Once an appropriate system has actually been defined it needs to be prepared (and characterized). Another important difference is that the "laws of nature" or some simplified model of natural behavior need to be incorporated in an explicit manner into the computations.

What are the advantages and what are the disadvantages of computation over experiment?

Computations, offer some significant advantages over experiments, among them:

Generality	Choice of systems is without regard to stability or even existence.
Ease	Calculations are easy to perform.
Cost	Calculations are becoming less and less costly while experimental chemistry is more and more costly.
Safety	Experimental chemistry is dangerous.

Of course, there are disadvantages too, and there are limitations. The most obvious disadvantage is that computational models have yet to be formulated for all properties and phenomena, and those that have been formulated may not be applicable to all systems and/or not yield results of acceptable quality.

Computation should not be viewed as a replacement for experiment. Rather, it should be thought of, just as NMR and mass spectrometry, as a tool for chemists to do chemistry. It is a very powerful tool, but like other chemical tools it cannot stand entirely on its own.

Are We Ready for Computer Chemistry?

We are perhaps not far removed from the time when we shall be able to submit the bulk of chemical phenomena to calculation.

Joesph Louis Gay-Lussac 1778–1850

We've certainly jumped the gun before, but probably not this time. Chemists already make use of computers in their daily routines—if only for word processing and chemical structure drawing—and the computers which now sit on their desks are already powerful enough to tackle real problems. We need to be patient, for it is not likely that computation will establish itself "overnight" as an essential component of chemistry. After all, it took NMR and mass spectroscopy a decade or more! As we think about introducing computation to chemists, with the end goal being to establish it as an essential component of the science, we should keep a number of things in mind:

Chemistry Not Computation Chemists want to do chemistry not computations. The emphasis needs to be placed on how computations can be used to explain and rationalize known chemistry better than existing qualitative languages, and how they can be employed to discover new chemistry.

   The improper focus on computation and not on chemistry is, in the opinion of this author, a primary reason why computation has not yet experienced rapid acceptance among the majority of chemists.

Accessibility to Chemists In the sense of "hands on": computational chemistry like experimental chemistry is a "laboratory" science, and must be learned by "doing" and not just reading. In the sense of non-intimidating: the underpinnings of computational chemistry (quantum mechanics) are certainly intimidating to chemists, but calculations are actually rather easy to perform given currently available software.

Standards There is only one experimental "result", but there are many different computational results corresponding to the many different levels of computation. It is necessary to define a small number of standard computation levels and to fully document the performance of each. These standards should span a full range from very simple, which may not be very accurate but are applicable to very large molecular systems, to very sophisticated, which are very accurate but applicable only to very simple molecular systems.

Overall Picture and Not Details As with any emerging and rapidly growing discipline, there has been a tendency to focus on the "latest and greatest" and lose sight of the overall objective. While the existing methodology is not perfect, it is now good enough such that it can be of great value in learning and researching chemistry.

Relationship with Experimental Chemistry Computation should not be viewed as a replacement for experiment, but rather as a companion to experiment.

Because calculations provide an imprecise description of a precisely defined quantity, whereas experiments provide a precise description of something which is as often as not imprecisely defined, the two approaches both complement and supplement one another.

In short, the focus must remain on chemistry. New generations of chemists need to be made to feel comfortable with computation and be provided realistic appraisal of the performance and limits of available techniques. The key is education. Computation must be brought into the undergraduate curriculum, not only as a "textbook science" but also as a laboratory science.

A Bit of Theory

The underlying physical laws necessary for the mathematical theory of a large part of physics and the whole of chemistry are thus completely known, and the difficulty is only that the exact application of these laws leads to equations much too complicated to be soluble.

P.A.M. Dirac 1902–1984

Modern electronic structure methods have their roots in the quantum physics of 1920's and in the deceptively simple looking equation of Schrödinger which now bears his name.

$$\hat{H}\Psi = \varepsilon\Psi$$

In this equation, $\hat{H}$ is termed the Hamiltonian operator; it describes both the kinetic energies of the particles which make up the molecule, i.e., its nuclei and electrons, as well as the electrostatic interactions felt between individual particles. That is, nuclei, which are positively charged, repel other nuclei, and electrons, which are negatively charged, repel other electrons, but nuclei attract electrons. The quantity ε in the Schrödinger equation is the energy of the system and Ψ is a function (a **wave function**) of the positions and momenta of all the particles.

The Schrödinger equation has been solved exactly only for the hydrogen atom and its solutions, termed atomic orbitals, are actually quite familiar to chemists (s-, p-, d- orbitals). The Schrödinger equation may easily be written down for both many-electron atoms and molecules, although its exact solution is beyond reach. Approximations need to be made.

Hartree-Fock Models[1]

Three approximations take us from the full many-electron Schrödinger equation to the simplest practical models, termed Hartree-Fock models:

1. Separation of nuclear and electron motions (the **Born Oppenheimer approximation**). In effect, what this says is that "from the point of view of the electrons, the nuclei are stationary". This eliminates the nuclear kinetic energy term in the Hamiltonian and leads to a constant nuclear-nuclear potential energy term. In

so doing, it eliminates the mass dependence in what is now referred to as the electronic Schrödinger equation. Known mass dependencies on molecular properties and on reactivity and selectivity (isotope effects) need to be accounted for on other grounds.

2. Separation of electron motions (the **Hartree-Fock approximation**). What is actually done is to represent the many-electron wave function as a sum of products (in the form of a determinant) of one-electron wave functions, the spatial parts of which are termed **molecular orbitals**.

3. Representation of the individual molecular orbitals in terms of linear combinations of atom-centered basis functions or **atomic orbitals** (the **LCAO approximation**). This reduces the problem of finding the best functional form to a much simpler problem of finding the best set of linear coefficients.

These three approximations lead directly to the **Roothaan-Hall equations**[2], the solution of which may not actually be accomplished in closed form but instead requires an iterative self-consistent field (SCF) procedure. In effect, what is done is to find the best description of each electron in a static field made up of all the other electrons and all the nuclei, and then to incorporate this description back into the field to be used to find the best description of the next electron and so forth.

What remains to be specified is the nature of the atomic orbitals (the so-called **basis set**), both in terms of the number and kinds of functions employed. All present day molecular orbital methods make use of **Gaussian basis functions**, written in terms of polynomials in x,y,z multiplying an exponential in r^2, i.e.,

$$x^l y^m z^n \exp (\alpha r^2),$$

where α is a constant which dictates the radial extent of the function. These functions are by convention named s, p, d, etc., depending on the order of the polynomial. The sum of the integers l, m, n is 0 for an s function, 1 for a p function, 2 for a d function, etc.

The simplest (smallest) possible atomic orbital representation is termed a **minimal basis set**. This comprises only that number of functions required to accommodate all of the electrons of the atom, while still maintaining overall spherical symmetry. In practice, this requires a single (1s) function for hydrogen, five functions (1s, 2s, $2p_x$, $2p_y$, $2p_z$) for first-row elements, nine functions (1s, 2s, $2p_x$, $2p_y$, $2p_z$, 3s, $3p_x$, $3p_y$, $3p_z$) for second-row elements, and so forth. Among the most widely used minimal basis sets is STO-3G[3]. Here the individual atomic orbitals are represented by sums of three Gaussian functions chosen to most closely fit exponential functions, i.e., hydrogen atom solutions.

Minimal basis sets are not capable of adequately describing non-spherical (anisotropic) electron distributions in molecules. This is because the individual basis functions are either themselves spherical (s functions), or come in sets of equivalent functions which uniformly span a spherical space (p and d functions). The simplest remedy for any problems which such a restriction might cause it to "split" the valence description into "inner" and "outer" components. In a so-called **split valence basis set**, the valence manifolds of first- or second-row elements are represented by two complete sets of s- and p-type functions. You can think of the effect this has by recognizing that the coefficients which multiply the individual basis functions are "variables" in the calculation. For example, to describe a p function which is to be used in construction of a "tight" σ bond, one needs to mix a large amount of a highly contracted basis function with a small amount of a more diffuse function, i.e.,

$$p_\sigma = \text{large} * \bigcirc\!\!\!\blacksquare + \text{small} * \bigcirc\!\!\!\blacksquare \Rightarrow \bigcirc\!\!\!\blacksquare ,$$

while the reverse would be true for construction of a p function to be employed in making a "loser" π bond, i.e.,

$$p_\pi = \text{small} * \,\begin{smallmatrix}\circ\\\bullet\end{smallmatrix} + \text{large} * \,\begin{smallmatrix}\circ\\\bullet\end{smallmatrix} \Rightarrow \begin{smallmatrix}\circ\\\bullet\end{smallmatrix} .$$

In practice, first-row elements are described by nine functions ($1s$, $2s$, $2p_x^i$, $2p_y^i$, 2_z^i, $2s^o$, $2p_x^o$, $2p_y^o$, $2p_z^o$) and second-row elements by thirteen functions ($1s$, $2s$, $2p_x$, $2p_y$, $2p_z$, $3s^i$, $3p_x^i$, $3p_y^i$, $3p_z^i$, $3s^o$, $3p_x^o$, $3p_y^o$, $3p_z^o$). For completeness, split-valence basis sets represent hydrogen in terms of a pair of (contracted and diffuse) s-type basis functions.

Among the most simple split-valence basis sets is 3-21G[3]. This uses three Gaussian functions to represent each non-valence atomic orbital, and two and one Gaussians to represent the "inner" and "outer" components of each valence shell orbital. While 3-21G performs quite well for structure and energetic comparisons involving molecules comprising hydrogen and first-row elements only, it generally performs poorly in the description of molecules incorporating second-row elements. Adequate descriptions here have been found to require incorporation of unoccupied (in the atom) but energetically low-lying d-type functions. A very simple representation of this type is the 3-21G[(*)] basis set, which is used in conjunction with 3-21G for hydrogen and first-row elements.

Practical molecular orbital methods restrict basis functions to be centered on atoms. There is actually no reason for this except for the ambiguity of locating off-atom basis functions. It is possible to overcome any consequences which might arise because of this restriction by incorporating d-functions on main-group elements (where they are not occupied). These allow for small displacements of the center of electronic charge contained in valence p-type functions away from the nuclear positions (on which the basis functions are centered), i.e.,

$$\begin{smallmatrix}\circ\\\bullet\end{smallmatrix} + \lambda\, \text{\Large\bowtie} \Rightarrow \text{\Large\bowtie} .$$

What we are doing here is very closely related to Pauling's hybridization arguments, where the common example relates to the construction of an "sp hybrid" from mixture of a p-type function into a valence s orbital, i.e.,

$$\bullet + \lambda\, \bigcirc\!\!\!\blacksquare \Rightarrow \infty\!\!\!\blacksquare .$$

Functions added to basis sets in order to achieve charge displacement are generally termed polarization functions, and basis sets which incorporate polarization functions are called **polarization basis sets**. The simplest representations of this kind require 15 basis functions for first-row elements (split valence basis + a set of six d-type functions), and nineteen basis functions for second-row elements. Polarization functions (p-functions) may also be added to hydrogen atom descriptions if desired. Among the most popular polarization basis sets in widespread use is the 6-31G* representation.[3] Here, six Gaussian functions make up each non-valence atomic orbital, and the valence description is split into "inner" (three Gaussian) and "outer" (one Gaussian) parts. Polarization functions are included only on heavy atoms.

Beyond Hartree-Fock Models[1]

The key to understanding how to proceed beyond Hartree-Fock models lies in the LCAO approximation. Specifically, it needs to be noted that far more molecular orbitals are constructed than are actually required to hold all the electrons. The full set of molecular orbitals may be subdivided into two subsets: lower energy molecular orbitals which are occupied and higher energy molecular orbitals which are unoccupied, i.e.,

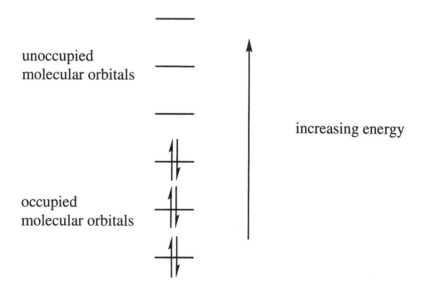

While the number of occupied molecular orbitals for a given molecule does not vary with basis set, the number of unoccupied orbitals increases with increasing size of the underlying atomic basis set.

The Hartree-Fock energy depends only on the occupied molecular orbitals. This energy can be lowered and the Hartree-Fock wave function can be improved by combining it with other wave functions formed from the Hartree-Fock wave function by promotion of one or more electrons from occupied into unoccupied molecular orbitals, i.e.,

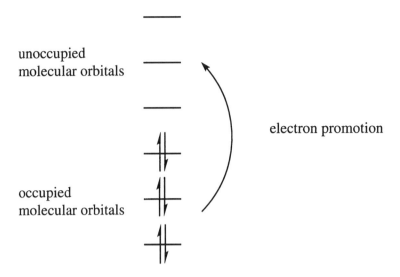

In the limit of all possible single and multiple electron promotions, and assuming a "complete" basis set (where the number of unoccupied molecular orbitals would actually be infinite), we can actually show that the "best" resulting wave function is in fact that corresponding to the exact solution of the electronic Schrödinger equation. While this cannot of course be achieved, practical methods have been devised to yield significant improvements over Hartree-Fock models.

The most commonly employed model beyond Hartree-Fock is the MP2 (second order Møller Plesset) model. Like Hartree-Fock models, this requires an underlying basis set, but unlike Hartree-Fock models this basis set needs to properly represent not only molecular orbitals which are formally occupied but also those which are vacant (corresponding to "excited state" electron configurations). The simplest model for practical applications, referred to as MP2/6-31G*, makes use of the 6-31G* polarization basis set.

There is an important consequence of thinking about solution of the electronic Schrödinger equation in terms of the Hartree-Fock approximation (and its extensions) and the LCAO approximation. To clearly see what it is, let's represent these approximations in terms of progression along the axes of two- dimensional chart, i.e.,

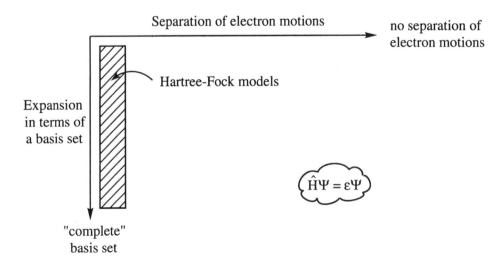

What is important to recognize is that every point on the chart corresponds to some particular model, and that moving down or across the chart should lead to improvement. In the event that movement down or across leads to no further change then we can conclude that for the particular property of the particular system at hand we have in fact solved the electronic Schrödinger equation. That is to say we can use convergence of a property in lieu of agreement with experiment to judge the performance of the model.

Density Functional Models[4]

The left hand column of our two-dimensional chart corresponds to the series of Hartree-Fock models. The limit (**Hartree-Fock limit**) for the hypothetical case of a complete basis set yields a well defined energy (**Hartree-Fock energy**). The fact that the limiting Hartree-Fock energy is not the experimental energy follows from the "separation of electron motions approximation", which does not allow the individual electrons sufficient flexibility "to get out of each other's way". The difference between the Hartree-Fock energy and the experimental energy is termed the **correlation energy**, meaning the energy lowering resulting upon correlation of the motions of the electrons.

The approach outlined in the previous section is not the only way to account for the correlation energy in a many-electron system. An alternative is provided by so-called **density functional models**, which incorporate the correlation explicitly into the Hamiltonian. Perhaps the best way to think about density functional models is to briefly revisit Hartree-Fock theory, and in particular to consider the description of the total energy of a molecular system from the standpoint of Hartree-Fock theory. This involves the kinetic energy of the individual electrons (recall that the nuclear kinetic energy is zero within the Born-Oppenheimer approximation), the repulsive potential energy between nuclei (a constant for a given nuclear configuration), the attractive potential energy between nuclei and electrons, and finally the repulsive potential energy between electrons.

Two terms involving electron-electron interaction actually appear in the Hartree-Fock Hamiltonian. One of these clearly contributes in a repulsive manner, and is labeled the Coulomb term. The other term appears to contribute in an attractive manner, and therefore cannot easily be given a classical physical interpretation, except to say that it somehow compensates for the apparent overestimation of electron-electron repulsion given by the **Coulomb term**. This second term is called the **exchange term**.

Density functional theory takes the electron kinetic term, the nuclear-nuclear and nuclear-electron potential terms and the Coulomb term directly from Hartree-Fock theory. It replaces the exchange term and adds a correlation term. (In the event that the exchange term is set to the Hartree-Fock exchange and the correlation term is set to zero, the density functional energy is the Hartree-Fock energy. In other words, Hartree-Fock theory may be thought of as a special case of density functional theory).

The exchange and correlation terms used in so-called local density functional models originate from the "exact" (numerical) solution of an idealized many-electron problem, mainly an electron gas of constant total electron density. (While this may be a "reasonable" model for highly-extended systems, e.g., metals, its validity for calculations on molecules remains to be established.) In practice, what is first done is to establish functional relationships between the exchange and correlation energies for such an "idealized gas" and the total electron density, i.e., to establish exchange and correlation functionals, and then to incorporate these relationships into the Hamiltonian describing a general many-electron system. Density functional models which allow for non-uniformity of the overall electron distribution (so-called non-local methods) have also been proposed, as have "hybrid schemes", whereby the total exchange energy is arrived at by combining the Hartree-Fock exchange energy and the exchange energy resulting from some functional. One or more parameters adjusted to provide "best fit" with specific experimental observables may be utilized. Unfortunately, in none of the schemes proposed thus far is there a unique "recipe" guaranteeing "improvement" with systematic progression from one functional to another.

Because the form of the correlation and exchange functionals are typically quite complex, they cannot be dealt with analytically, and practical density-functional calculations involve numerical integration steps. While these can lead to significant loss of precision (compared with ab initio methods which are strictly analytical), in principle the computational cost of such numerical integrations and hence of a density functional calculation may be made to scale as the square of the number of basis functions times the number of integration points. The latter is a large number, and density functional calculations on very small systems will be slower than Hartree-Fock calculations (but not conventional correlation energy calculations), although the two methods will eventually cross in performance. For large systems, density functional methods will offer significant cost savings.

Like Hartree-Fock and correlated ab initio calculations (and semi-empirical calculations as well), density functional calculations make use of explicit atomic basis sets. The same considerations apply in basis set selection as for these other methods, although in practice basis sets for density functional calculations are typically generated "on-the-fly" as numerical representations of best atom solutions.

Semi-Empirical Models[5]

Semi-empirical models follow in a straightforward way from Hartree-Fock models. In effect, a single additional approximation, termed the **NDDO** (neglect of diatomic differential overlap) **approximation**, is made. This is very severe, in that it eliminates overlap of atomic basis functions on different atoms (atomic basis functions on the same atom already do not overlap because of orthogonality). However, it leads directly to a reduction in computation effort from the fourth power of the total number of basis functions (in Hartree-Fock models) to the square of the number of basis functions. Other steps in practical semi-empirical models, e.g., matrix diagonalization, scale as the cube of the total number of basis functions and usually dominate the calculation.

Most present generation semi-empirical models are restricted to a minimal valence basis set of atomic functions. Inner-shell functions are not included explicitly, and because of this, the cost of doing a calculation involving a second-row element, e.g., silicon, is no more than that incurred for the corresponding first-row element, e.g., carbon. So-called **Slater-type orbitals** (STO's, which are closely related to the exact solutions for the hydrogen atom) are used in place of Gaussian functions employed for ab initio calculations.

Additional numerical approximations are invoked to simplify the calculations, and more importantly adjustable parameters are introduced in order to reproduce specific experimental data as closely as possible. Choice of parameters is the key to successful semi-empirical methods.

Selection of an Appropriate Model

Computational methods, in particular, methods based on quantum mechanics, are now rapidly moving into the mainstream of chemistry. They are still very new, however, and the collective experience in their application to "real chemical problems" is still very limited. We need to question their reliability in preface to applications to new chemistry, and we need to be concerned with practical issues of whether or not the calculations can be performed sufficiently rapidly to actually be of use. All of this falls under a general heading of model selection.

Most important among many considerations involved in the selection of an appropriate molecular mechanics or electronic structure model are the level of confidence required in the results and the computational resources available. Only rarely will it be possible to utilize the most sophisticated theoretical treatment available, and even then this may not be sufficient to guarantee the level of accuracy desired. Typically, practical concerns will dictate use of lower levels of calculation. Thus, it is important for the user to understand in some depth the capabilities and limitations of available models, from semi-empirical models which can be applied to molecules comprising more than a hundred atoms, to the simplest minimal basis set Hartree-Fock treatments, which may be applied to systems comprising up to one hundred atoms, to correlated models (including density functional models) with large basis sets which are currently practical only for very simple molecules. Indeed, there will be problems where not even the simplest quantum chemical model can be used and only mechanics-based procedures are practical.

In the final analysis, there can be no substitute for experience, and this can be gained only through repeated applications.

What Can We Calculate and How Well Can We Calculate It?

Calculations will not find their rightful place unless and until they are routinely able to provide reliable and accurate information about broad areas of interest and importance to chemists, among them molecular structure and conformation, thermochemical stability, chemical reactivity, and product selectivity. Here we remark briefly on the performance of practical ab initio and semi-empirical molecular orbital models and of local density functional models with regard to each of these areas. Our objective is modest in that we only wish to give an overall impression of the present state of affairs.[6]

Equilibrium Geometries

Sufficient experimental structural data exist to allow thorough assessment of theoretical methods for the calculation of equilibrium geometries of organic molecules.

Ab Initio Models

The following general conclusions may be drawn from the extensive comparisons which have been made:

1. Accurate equilibrium structures may be obtained from ab initio molecular orbital theory. Use of moderately large basis sets (6-31G* or larger) and Møller-Plesset treatment of electron correlation truncated at second-order (MP2) generally guarantees that calculated structural parameters will be close to measured equilibrium values. Smaller basis set MP2 calculations generally lead to bond distances which are longer than experimental values, sometimes significantly so, and cannot be recommended. Limiting MP2 bond lengths are typically, but not always, longer than experimental distances, although errors are quite small.

2. Hartree-Fock models using the 6-31G* polarization basis set generally overestimate bond lengths, typically by 0.01–0.02Å, although errors involving bonds between two highly electronegative elements can be much larger, e.g., ~0.1Å in F_2. These discrepancies reflect the behavior of the Hartree-Fock model rather than limitations in the underlying basis set. Limiting Hartree-Fock bond lengths are nearly always shorter than experimental values. Skeletal bond angles and dihedral angles calculated at the Hartree-Fock level are generally in good accord with experimental data.

3. Hartree-Fock models using small to medium size basis sets are generally successful in accounting for equilibrium structures of organic molecules. The 3-21G split-valence basis set (3-21G$^{(*)}$ for molecules incorporating second-row and heavier elements) appears to be the simplest method of choice for widespread application. Even the STO-3G minimal basis set generally yields equilibrium geometries in good accord with experimental data.

Local Density Functional Models

There is much less experience with local density functional models, although some conclusions may be drawn regarding their performance:

1. Accurate equilibrium structures may be obtained from local density functional models with moderate to large basis sets including one or more sets of polarization functions. In the limit, local density functional models generally give bond lengths which are shorter than experimental values (typically by 0.01–0.02Å), the same result as noted for limiting Hartree-Fock models, but not the same as noted for comparable MP2 models.

2. Local density functional models with smaller (minimal and split-valence) basis sets without polarization functions almost always lead to bond distances which are longer than experimental values, often by as much as 0.1Å. They do not provide a reliable account of equilibrium geometry and cannot be recommended.

Semi-Empirical Models

The following general conclusions may be drawn from the extensive comparisons both to experiment and to high-level ab initio calculations:

1. All three present-generation semi-empirical models (MNDO, AM1 and PM3) are generally suitable for the calculation of equilibrium geometries. All yield similar errors both for bond distances involving heavy atoms and for skeletal bond angles. None of the models is as reliable as even the simplest ab initio schemes, and bond length and angle errors are typically twice as large as those resulting from Hartree-Fock 3-21G calculations.

TABLE I   RMS errors in bond distances connecting heavy atoms (ångstroms) and in skeletal bond angles (degrees).

Model	Bond Distances	Bond Angles
AM1 semi-empirical	0.048	3.3
PM3 semi-empirical	0.037	3.9
3-21G Hartree-Fock	0.028	1.7
6-31G* Hartree-Fock	0.028	1.4
MP2/6-31G* correlated	0.018	1.5
SVWN/6-31G* local density functional	0.018	1.6

2. The PM3 model is generally the best of the three schemes, and MNDO generally the worst, for structure determination. In particular, it is significantly better than MNDO and AM1 for the calculation of the geometries of molecules incorporating second-row and heavier elements, most conspicuously for hypervalent molecules, i.e., molecules which exceed the normal complement of eight valence electrons. Limited data for the newly described MNDO/d model suggests that it too is successful for the geometries of hypervalent molecules.

3. All semi-empirical models perform somewhat (but not markedly) less well for ions and free radicals than they do for "conventional" molecules. This is most likely due to the sparse representation of these types of molecules in the training sets used for parameterization, rather than to any inherent limitations of the models. MNDO and to a lesser extent AM1 models are unreliable in their description of hydrogen-bonded systems, while PM3 appears to perform relatively well in this task.

4. Each of the methods has its own "quirks". For example, the PM3 model is known to improperly account for the planarity of amides and to show weak "attractions" between non-bonded hydrogens. In general, known problems are restricted to specific classes of molecules, and more than anything else, are probably a result of a lack of representation of these compounds in the original parameterizations.

Comparison Errors in Equilibrium Geometries for Different Models

RMS errors in bond lengths and skeletal bond angles for molecules comprising first- or second-row elements, and which are well described in terms of conventional valence structures are provided in Table 1. Included in the comparison are the AM1 and PM3 semi-empirical models, Hartree-Fock models with 3-21G and 6-31G* basis sets, the correlated MP2/6-31G* model and the local density functional SVWN/6-31G* model. Other data sets would yield similar results, except perhaps that the two semi-empirical models may present difficulties for classes of molecules for which it has not been explicitly parameterized, e.g., charged species and radicals. Note, however, that situations where calculated structures at semi-empirical levels are outlandish are actually quite rare.

TABLE 2   Conformational energy differences (kcal/mol).

Molecule	Low Energy/ High Energy Conformer	AM1	PM3	3-21G	6-31G*	MP2/ 6-31G*	SVWN/ 6-31G*	expt.
n-butane	*trans/gauche*	0.7	0.5	0.8	0.9	0.7	0.4	0.77
1-butene	*skew/cis*	0.6	1.0	0.8	0.7	0.5	−0.8	0.2
1,3 butadiene	*trans/gauche*	0.8	0.7	3.5	3.0	2.6	3.6	1.7,>2, 2.5
acrolein	*trans/cis*	−0.2	0.4	0.0	1.7	1.5	1.7	2.0, 2.06
glyoxal	*trans/cis*	0.0	0.1	5.1	5.6	4.3	4.1	3.2
methylcyclohexane	*equatorial/axial*	1.4	1.1	1.9	2.3	1.9	1.6	1.8
tert-butylcyclohexane	*equatorial/axial*	5.1	1.1	6.6	6.3	—	4.6	5.4
2-chloroterahydropyran	*axial/equatorial*	3.6	3.1	3.6	2.5	2.8	4.6	1.8

Equilibrium Conformations

Searching for the lowest-energy conformation of a flexible molecule is even more demanding than obtaining equilibrium geometry (bond lengths and angles) once conformation is known. The performance of molecular mechanics and electronic structure methods in properly identifying equilibrium conformation is also not as easy to judge as the performance in reproducing experimental equilibrium geometries. Only for very simple systems are lowest-energy conformers known with certainty, and even here thermochemical differences from higher-energy conformers may have large uncertainties.

Table 2 compares conformational energy differences for a few very simple systems from AM1 and PM3 semi-empirical models, Hartree-Fock models with 3-21G and 6-31G* basis sets, the MP2/6-31G* correlated model and the local density functional SVWN/6-31G* model.

Both AM1 and PM3 methods properly account for the *axial* preference in 2-chlorotetrahydropyran, as well as for the conformation of saturated and unsaturated hydrocarbons. However, both fail to reproduce the experimental conformations of acrolein and glyoxal. Note the very low energy difference between equatorial and axial conformers of *tert*-butylcyclohexane obtained at the PM3 level. This is a direct consequence of the tendency for non-bonded hydrogens to attract each other.

3-21G and 6-31G* Hartree-Fock models provide a superior account for all systems, the only notable exception being the failure of 3-21G to show a difference between *cis* and *trans* conformations of acrolein.

Both MP2/6-31G* and SVWN/6-31G* models perform favorably in their description of relative conformer energies. The only poor result in the table is the incorrect assignment of the ground state conformer of 1-butene at the SVWN/6-31G* level.

Transition State Geometries

The geometries of transition states on the pathway between reactants and products are not easily anticipated. This is not to say that they do not exhibit systematics as do "normal" molecules, but rather that we do not yet have sufficient experience to identify what systematics do exist, and more importantly how to capitalize on structural similarities. The problem is that transition states cannot even be detected let alone characterized experimentally, at least not directly. While measured activation energies relate to the energies of transition states above reactants, and while activation entropies and activation volumes as well as kinetic isotope effects imply some aspects of transition-state structure, no experiment can actually provide direct information about the detailed geometries of transition states.

Our only guide to the performance of theory in dealing with transition state geometries (aside from "reasonableness") is convergence with increasing level of calculation. This of course applies only to ab initio methods, which then need to serve as "benchmarks" for other calculations.

Only a few general remarks are appropriate at this time:

1. *Ab initio* Hartree-Fock models generally show a smooth progression in transition state geometries with increasing size of basis set. Bond lengths and skeletal bond angles typically show two to three times the sensitivity to changes in basis set than observed for equilibrium species.
2. Correlated (MP2) models generally, but not always, give "looser" transition states than Hartree-Fock schemes with the same basis set. There is insufficient experience to comment on convergence of MP2 models with respect to increasing size of basis set.
3. Semi-empirical models and to some extent minimal basis set ab initio models are unreliable in the description of transition state geometries. While transition states for many reactions, in particular for concerted processes, are similar to those from higher-level calculations, transition state geometries for other reactions can be quite different. Caution needs to be exercised in its use for this purpose.
4. There is insufficient experience to judge the performance of local density functional models for the determination of transition state geometries, although there are some indications that here transition structures may be too tightly bound. Caution needs to be exercised in their use.

Reaction Thermochemistry

Thermochemical comparisons may conveniently be placed into one of several categories depending on the extent to which bonds and non-bonded lone pairs are conserved (Table 3). This distinction is important as electronic structure models differ most in the way that they treat electron correlation, i.e., the coupling of motions of electrons. Correlation effects would be expected to be most important for electrons which are paired. Hartree-Fock models completely ignore correlation, while semi-empirical models which have been parameterized to reproduce experimental data, may be thought of as methods in which electron correlation is taken into account implicitly. The same may be said for local density functional models, although here explicit parameters are not necessarily introduced, but rather the form of the underlying model, i.e., the exchange and correlation functionals, are adjusted to best represent a specific model.

At one extreme are processes in which not even the total number of electron pairs (bonds and non-bonded lone pairs) is conserved. Homolytic bond dissociation processes, e.g.,

$$H\text{–}F \Rightarrow H\cdot + F\cdot \qquad \text{homolytic bond dissociation}$$

TABLE 3 Performance of theoretical models for description of reaction thermochemistry.

Type of Process	Examples	Minimum Level of Calculation
No conservation of total number of electron pairs	Homolytic bond dissociation	Correlated models excluding local density functional models; large basis sets
Conservation of total number of electron pairs, but no conservation of total number of bonds	Heterolytic bond dissociation	Hartree-Fock and local density functional models; moderate to large basis sets with diffuse functions if anions are involved
Conservation of total number of bonds and total number of non-bonded electron pairs, but no conservation of number of each kind of bond, or number of each kind of non-bonded electron pair	Hydrogenation, structural isomerism	Hartree-Fock and local density functional models; moderate to large basis sets
Conservation of number of each kind of bond and number of each type of non-bonded electron pair (*isodesmic* reactions)	Substituent effects, regio- and stereochemical comparisons, conformation changes	Hartree-Fock models; small to moderate basis sets (local density functional models require moderate to large basis sets)

are an example. Comparisons of transition states and reactants (as required for the calculation of absolute activation energies) are also candidates for processes in which the total number of electron pairs is not conserved. This will be discussed later.

Less drastic are reactions in which the total number of electron pairs is maintained, but chemical bonds are converted to non-bonded lone pairs or vice versa. Heterolytic bond dissociation reactions, e.g.,

$$Na - F \Rightarrow Na^+ + \overset{..}{F}{}^-$$

and some structural isomerizations, e.g.,

$$H_2C{=}O \Rightarrow H\overset{..}{C}OH$$

are examples. The most important examples may again involve comparisons between transition states and reactants.

Even more "gentle" are reactions in which both the number of bonds and the total number of non-bonded lone pairs are conserved. This type of reaction is very commonly encountered. Several examples are given below.

$$H_2C=CH_2 + 2H_2 \Rightarrow 2CH_4 \qquad \text{hydrogenation}$$

$$\overline{CH_2CH_2CH_2} \Rightarrow CH_3CH=CH_2 \qquad \text{structural isomerism}$$

$$2CH_2=CH_2 \Rightarrow H_3C\text{-}CH_3 + HC\equiv CH \qquad \text{disproportionation}$$

Most gentle are reactions in which the number of each kind of formal chemical bond (and each kind of non-bonded lone pair) are conserved. These are *isodesmic* ("equal bond") reactions. Examples include the processes below.

$$H_3C\text{-}C\equiv CH + CH_4 \Rightarrow H_3C\text{-}CH_3 + HC\equiv CH$$

$$(CH_3)_3NH^+ + NH_3 \Rightarrow (CH_3)_3N + NH_4^+$$

In addition, all regio- and stereochemical comparisons are *isodesmic* reactions, as are conformation changes. Thus, *isodesmic* processes constitute a large class of reactions of considerable importance.

Many examples exist of all these types of reactions, and it is possible to offer some general remarks regarding the performance of various levels of calculation:

1. Correlated models (excluding local density functional models) are able to provide accurate descriptions of homolytic bond dissociation reactions. Large basis sets, with one or more sets of polarization functions and perhaps as well diffuse functions, are required. Hartree-Fock models and local density functional models give unsatisfactory results. Hartree-Fock models underestimate the magnitudes of homolytic bond dissociation energies, while local density functional models overestimate their magnitudes. Present-generation semi-empirical models lead to unsatisfactory descriptions of the energetics of homolytic bond dissociation.

2. The energetics of reactions in which the total number of electron pairs are conserved, including heterolytic bond dissociation in which a bond is exchanged for a non-bonded lone pair, are generally well described using Hartree-Fock models. Moderate to large basis sets including polarization functions are required. Correlated models (including local density functional models) also perform well, although basis sets which are even larger than those needed for Hartree-Fock models may be required. Semi-empirical models give unsatisfactory results.

3. The energetics of *isodesmic* reactions are generally well described using both Hartree-Fock and correlated models (including local density functional models). Small to moderate basis sets usually give acceptable results for Hartree-Fock models, although larger basis sets may be required for use with correlated models (including local density functional models). Semi-empirical models give unsatisfactory results.

Some additional comments follow from the observation above:

1. The fact that Hartree-Fock models provide reasonable descriptions of the energetics of processes in which the total number of electron pairs is conserved, together with the observation that Hartree-Fock reaction energies in these cases converge toward their limiting values much more rapidly than do correlated

models (including local density functional models) with increasing size of basis set, suggests that they may be more suitable for this purpose than correlated models. That is to say, the description of the energetics of processes for which correlation effects largely cancel may be better accomplished with models which do not explicitly take electron correlation into account.

2. The poor performance of present-generation semi-empirical models for all types of thermochemical comparisons, even *isodesmic* reactions which are reasonably well described using minimal basis set Hartree-Fock models, may be due to the fact that they have been explicitly parameterized to minimize absolute errors in heats of formation rather than to minimize errors in reaction energies. Random errors in individual heats of formation are large enough (>7 kcal/mol in the case of AM1) such that the overall error in a given reaction will be unacceptable. Future generation semi-empirical models might be more successful for energetic comparisons were they to be parameterized to reproduce energies for specific classes of reactions.

The primary recommendation to follow from our comments (aside from needing to exercise caution in the use of semi-empirical models for energetic comparisons of any kind) is to make use of *isodesmic* reactions whenever possible. When this is not possible, try to write reactions in which the total number of chemical bonds is conserved.

Reaction Kinetics

Below we offer a few comments regarding the performance of different levels of calculation for both absolute and relative activation energies. Our remarks for relative activation energies in particular are based on very limited data and must be looked on as tentative.

1. Proper description of absolute activation energies requires correlated models (excluding local density functional models) with large underlying basis sets. Lower-level treatments are unreliable. Hartree-Fock models generally give absolute activation energies which are too large irrespective of basis set, while local density functional models usually underestimate absolute activation energies, sometimes significantly so. Semi-empirical models are unreliable in their account of absolute activation energies.

2. Hartree-Fock models appear to provide a reasonable account of substituent effects on activation energies. Split-valence and polarization basis sets lead to acceptable results, but minimal basis set Hartree-Fock models and semi-empirical models do not. There is insufficient experience to judge the performance of correlated models (including local density functional models) for this purpose, although there is no reason to believe that they would not lead to acceptable results.

3. Hartree-Fock models, even with minimal basis sets, appear to give reasonable descriptions of the relative energies of transition states which differ only in regio- and/or stereochemistry. Semi-empirical models also seems to fare well in this regard. There is too little experience to assess the performance of correlated models (including local density functional models), although there is no reason to believe that they too will not lead to acceptable results.

Graphical Models Making Chemistry Visual

Computer generated graphical displays may serve not only to convey structural information, but also to provide insight into the results of quantum chemical calculations. Quantities which have proven of interest include the molecular orbitals, the total electron density and the molecular electrostatic potential. One way to render these quantities, all of which are functions in three dimensions, is to define a surface of constant value, a so-called **isovalue surface** or more simply **isosurface**. This is done by setting the value of the function to a constant, the value of which is arbitrary, but may be chosen to reflect a particular physical observable of interest, e.g., the "size" of a molecule in the case of display of total electron density.

Molecular Orbitals

Chemists are very familiar with the π orbitals of ethylene, benzene and other planar molecules. They recognize that the nodal structure of the valence molecular orbital manifold provides a key to understanding why certain chemical reactions proceed easily whereas other do not. With some notable and important exceptions, e.g., the Woodward-Hoffmann rules[7], such qualitative considerations have not, however, extended very greatly beyond planar π systems. In great part, this is due to the difficulty of constructing by hand and visualizing molecular orbitals of three-dimensional systems, a situation which modern computer graphics has now completely altered.

Molecular orbitals are written as linear combinations of nuclear-centered atomic orbitals. Even though, they are completely delocalized throughout the molecule, they are often readily interpretable in terms of familiar chemical concepts, such as two-center bonds or non-bonded lone pairs. For example, inspection of the highest-occupied molecular orbital of sulfur tetrafluoride,

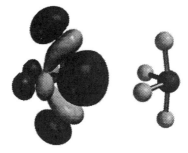

clearly reveals that it incorporates a lone pair on sulfur, in accord with its classical valence structure.[8]

Molecular orbital descriptions are often clearly related to "conventional" chemical indicators, and as such may be useful in interpreting chemical phenomena. For example, comparison of the lowest-unoccupied molecular orbital (LUMO) of planar (left) and perpendicular (right) benzyl cation,

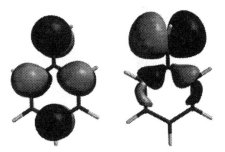

clearly reveals the difference in charge delocalization of the two systems. Recall that it is into the LUMO, the energetically most accessible unfilled molecular orbital, that any further electrons will go. Hence, it may be thought of as demarking the location of positive charge in a molecule. The picture above reveals that the LUMO in planar benzyl cation is delocalized away from the cation center and onto the *ortho* and *para* ring carbons, in accord with classical resonance structures,

while the LUMO in the perpendicular structure is almost entirely localized on the benzylic carbon.

What advantages if any do orbital pictures, such as those above for benzyl cation, offer over familiar resonance structures? There are two. First, they are quantitative; the magnitude of the orbital contribution for a given atomic center can be measured and related to charge. The resonance picture tells us that a particular center is charged but does not tell us anything about its magnitude. Second, orbital pictures are easily applicable to any system, not just simple planar molecules.

Total Electron Densities

The total electron density for a molecule is defined in terms of a sum of the squares of all occupied molecular orbitals. Depending on its value, a density isosurface (isodensity surface) may either serve to locate chemical bonds, or alternatively to indicate overall molecular size and shape. For example, a 0.08 electrons/au^3 surface of total electron density for cyclohexanone conveys very much the same information as a conventional skeletal structure model, that is, it depicts the locations of bonds,

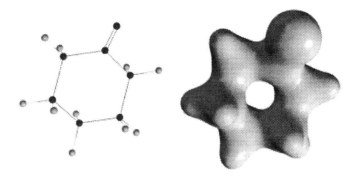

while a 0.002 electrons/au³ surface for the same molecule is much like a space-filling (CPK) model in that it depicts overall size and shape.[9]

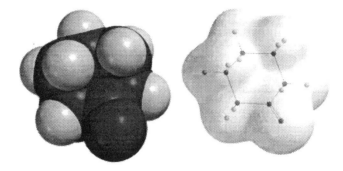

The obvious application of the former type of surface is to delineate bonding in situations where it might not be apparent. For example, the total electron density representation for diborane,

makes clear that the appropriate structural representation is one which lacks a boron-boron bond, rather than one in which the two borons are directly bonded, i.e.,

Perhaps the most interesting application of representations of this kind is to the description of transition states for chemical reactions. Here, one should be able to see clearly which bonds are being cleaved and which are being formed, e.g., in the pyrolysis of ethyl formate leading to ethylene and formic acid.

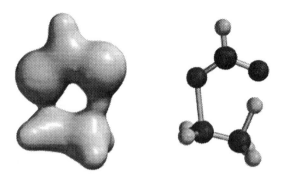

In this case, we see clear evidence of a "late transition state". The CO bond is nearly fully cleaved and the migrating hydrogen seems more tightly bound to oxygen (as in the product) than to carbon (as in the reactant). Further information, in particular about the timing of the overall reaction, may be obtained by replacing the static picture above by a "movie", i.e., animation along the reaction coordinate.[10]

Total electron density surfaces constructed to portray overall molecular size also have interesting applications. One of these might be to indicate relative ionic character. For example, the images below clearly convey the impression that lithium acetylide (left) is less ionic than sodium acetylide (right).

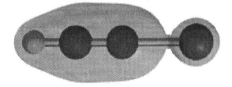
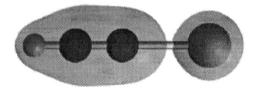

Electrostatic Potentials

The electrostatic potential is defined as the energy of interaction of a point positive charge with the nuclei and electrons of a molecule. Surfaces of constant molecular electrostatic potential map out areas which are both electron poor and electron rich. Positive isopotential surfaces indicate electron-poor regions where nucleophilic attack is likely, while negative isopotential surfaces indicate electron-rich regions where electrophilic attack is likely. For example, surfaces of constant negative electrostatic potential for trimethylamine (left), dimethyl ether (middle) and fluoromethane (right) are an artifact of the non-bonded "lone pairs" of electrons.

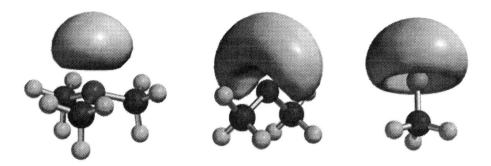

While the first of these results from a single non-bonded valence molecular orbital (the highest-occupied molecular orbital), the electrostatic potential in dimethylether and fluoromethane result from a combination of two and three high-lying molecular orbitals, respectively.

Composite Surfaces

Having defined an isosurface, it is possible to encode the value of some property as a function of location on a particular isosurface onto the isosurface. This leads to a representation which conveys four dimensions of information (three dimensions conveying structure and one dimension conveying the property). For example, the value of the electrostatic potential may be color-encoded onto the total electron density isosurface (corresponding to the van der Waals contact surface), thereby, indicating which of the accessible regions on a molecule are electron rich and which are electron poor. This kind of imagery can also help to simply convey what might otherwise be difficult chemical concepts. For example, such an encoded surface for planar and perpendicular forms of benzyl cation would show extensive charge delocalization in the former but high buildup of positive charge centers in the latter.

Use of Graphical Models to Describe Molecular Structure and Chemical Reactivity and Selectivity

What is the best way to employ graphical models to describe molecular structure or chemical reactivity and selectivity? There are some obvious answers. For example, the total electron density for a molecule provides a quantitative measure of its size and shape. As previously commented, it can be employed in much the same manner as a conventional space-filling (CPK) model, which it closely resembles, e.g.,

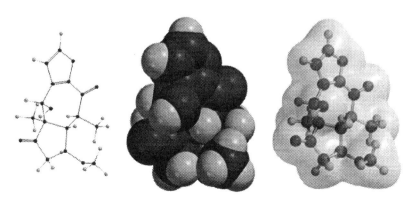

to demark steric requirements, and (like space-filling models) thereby assist assignment of preferred products on the basis of sterics. Chemical reactivity and product selectivity, however, may depend not only on size and shape (whether reagent and substrate can approach each other and, if they can, what approach geometries are permissible), but also on electronic considerations. That is to say, reagent and substrate will be most likely to react when their individual electronic demands are complementary, e.g., a nucleophile with an electrophile. Electronic effects, which may or may not be completely independent of steric factors, also need to be properly taken into account in any realistic treatment of chemical reactivity and selectivity.

It is not yet obvious how to best gauge electronic preferences or, more importantly, to gauge relative preferences from one system to another. Relative sizes (extensions) of key molecular orbitals (the HOMO and LUMO) have been employed with some success, e.g., in the description of regioselectivity in Diels-Alder cycloadditions,[11] although it is now apparent that such treatments are limited in their ability to provide quantitative accounts of product distributions.

It is possible to quantify molecular orbital size and shape, and in particular to examine the asymmetry of key molecular orbitals relative to the total electron density. For example, simultaneous displays of the total electron density surface for an asymmetrical carbonyl compound such as cyclohexanone and the lowest unoccupied molecular orbital have been successfully interpreted as indicating not only where a nucleophile might best approach, i.e., where the LUMO is of greatest magnitude, but also where it would likely avoid, i.e., areas of high electron density.[12]

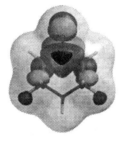

In this case the axial face (left) is the more accessible, in accord with the usual preference for nucleophilic attack.

An alternative indicator of chemical reactivity and product selectivity is the electrostatic potential. We have already suggested that this reflects overall charge distribution, and as such serves to delineate areas which are electron rich and hence subject to electrophilic attack, and those which are electron poor and hence subject to nucleophilic attack. Several examples, in which electrostatic potential is mapped onto an isodensity surface (demarking regions which are accessible to an approaching reagent) have already appeared in the literature.[11,13]

Conclusion

It's been a long time in waiting, but computation is now poised to play an important role in chemistry. No, the computer will not solve all our chemical problems. Indeed, it will probably raise more questions than it will answer. This is no doubt, a symptom of any new technology. Nor will computation cause us to abandon our laboratories. Some experiments will no doubt be superceded by computation (some already have been), but others will appear in their place stimulated by the results of computations.

Computation is very much a part of the natural evolution of chemistry. The initial excitement, and the inevitable apprehension and skepticism will eventually give rise to acceptance. The situation is perhaps very little different from that following the introduction of NMR spectroscopy several decades ago.

References

1. For a review, see: W. J. Hehre, L. Radom, P. v. R. Schleyer and J. A. Pople, **Ab Initio Molecular Orbital Theory**, Wiley, New York, 1986, Chapter 2.
2. C. C. J. Roothaan, Rev. Mod. Phys., **23**, 69 (1951); G. G. Hall, Proc. Roy. Soc. (London), **A205**, 541 (1951).
3. For a review, see ref. 1, Chapter 4.
4. For reviews, see: E. Wimmer in **Density Functional Methods in Chemistry**, J. K. Labanowski and J. W. Andzelm, ed., Springer-Verlag, New York, 1991, Chapter 2; R. G. Parr and W. Yang, **Density Functional Theory of Atoms and Molecules**, Oxford University Press, New York, 1989.
5. For reviews see: T. Clark, **A Handbook of Computational Chemistry**, Wiley, New York, 1986; J. J. R. Stewart, J. Computer Aided Molecular Design, **4**, 1 (1990).
6. Much more extensive surveys can be found in ref.1, chapter 6, and W. J. Hehre, **Critical Assessment of Modern Electronic Structure Methods**, Wavefunction, Irvine, California, 1996.
7. R. B. Woodward and R. Hoffmann, **The Conservation of Orbital Symmetry**, Verlag Chemie GmbH, Weinham, 1970.
8. This is of course why the molecule adopts a trigonal bipyamidal as opposed to a tetrahedral equilibrium geometry.
9. The radii used to define CPK models have been chosen to reflect the space which molecules take up when they pack in solids (or associate in liquids).
10. Examples are provided in a recent CD Rom. T. S. Hehre, L. D. Burke, W. W. Huang and W. J. Hehre, SPARTAN Live!, Wavefunction Inc., Irvine, California, 1995.
11. For a discussion, see: S. D. Kahn, C. F. Pau and W. J. Hehre, J. Am. Chem. Soc., **108**, 7381 (1986), and references therein.
12. S. Fielder, Ph.D. thesis, University of California, Irvine, 1993.
13. S. D. Kahn, C. F. Pau and W. J. Hehre, J. Am. Chem. Soc., **108**, 7396 (1986); S. D. Kahn, C. F. Pau, A. R. Chamberlin and W. J. Hehre, **ibid.**, **109**, 650 (1987); S. D. Kahn and W. J. Hehre, **ibid.**, **109**, 666 (1987); A. R. Chamberlin, R. L. Mulholland, Jr., S. D. Kahn and W. J. Hehre, **ibid.**, **109**, 672 (1987); S. D. Kahn and W. J. Hehre, **ibid.**, **108**, 7399 (1986); S. D. Kahn, K. D. Dobbs and W. J. Hehre, **ibid.**, **110**, 4602 (1988); S. D. Kahn and W. J. Hehre, Tetrahedron Lett., **27**, 6041 (1986); S. D. Kahn and W. J. Hehre, J. Am. Chem., Soc. **109**, 663 (1987).

Molecular Graphics and Simulation of Proteins and Nucleic Acids

Scott H. Northrup

Introduction

A revolution is currently underway in our understanding of biological processes on the atomic and molecular level.[1] This revolution has resulted from the synergy of precise x-ray crystallographic information about the atomic structure of proteins and nucleic acids, coupled with an explosion of computational power and methodology to realistically model biomolecular mechanisms based on atomic coordinates. Bringing together chemistry, physics, biochemistry, and computer science, a new generation of *computational biochemists* is using computers to solve the fundamental equations of physics for whole proteins, nucleic acids, and their various complexes in order to provide an important complement to the powerful experimental techniques that are simultaneously being developed. With an order-of-magnitude increase in computational power every five years, computations now make it possible to study a wide variety of complex biochemical processes with increasing detail and realism. The current most practical ramification of computational biochemistry is in the area of drug design, which is discussed in Chapter 17. Beyond this is the philosophical goal of a deeper and more fundamental understanding of the molecular processes of life, which will give way in the not-too-distant future to scientific and technological developments for the benefit of mankind that are beyond our current ability to imagine.

In this chapter we will discuss the following topics: (1) the appropriation of biomolecular structural information gained from crystallography and NMR to perform computer modeling; (2) the power of computer graphics to enhance visualization of macromolecules; (3) the empirical potential energy functions and their utilization to compute the energetics and forces of molecules; and (4) a variety of standard computational methodologies including energy minimization of structures,

TABLE I   A small segment of the Protein Data Bank file for baker's yeast *cytochrome c peroxidase* enzyme,[5] showing residues 4 and 5, leucine and valine, resp. The first three columns of floating point numbers are the Cartesian coordinates. These are followed by occupation numbers and temperature factors, which measure atomic thermal fluctuations.

ATOM	38	N	LEU	4	11.033	78.632	22.870	1.00	77.58	1CCP	143
ATOM	39	CA	LEU	4	12.270	77.839	22.750	1.00	66.55	1CCP	144
ATOM	40	C	LEU	4	12.477	77.728	21.226	1.00	59.64	1CCP	145
ATOM	41	O	LEU	4	11.491	77.896	20.487	1.00	59.74	1CCP	146
ATOM	42	CB	LEU	4	11.931	76.407	23.203	1.00	68.16	1CCP	147
ATOM	43	CG	LEU	4	11.574	76.128	24.635	1.00	69.45	1CCP	148
ATOM	44	CD1	LEU	4	12.880	76.043	25.442	1.00	71.55	1CCP	149
ATOM	45	CD2	LEU	4	10.680	77.231	25.191	1.00	71.63	1CCP	150
ATOM	46	N	VAL	5	13.656	77.384	20.819	1.00	52.68	1CCP	151
ATOM	47	CA	VAL	5	13.906	77.091	19.383	1.00	47.89	1CCP	152
ATOM	48	C	VAL	5	14.559	75.708	19.482	1.00	46.05	1CCP	153
ATOM	49	O	VAL	5	15.363	75.567	20.418	1.00	47.71	1CCP	154
ATOM	50	CB	VAL	5	14.661	78.179	18.665	1.00	46.89	1CCP	155
ATOM	51	CG1	VAL	5	14.679	78.002	17.149	1.00	44.42	1CCP	156
ATOM	52	CG2	VAL	5	14.024	79.530	19.018	1.00	50.24	1CCP	157

molecular dynamics simulations, numerical electrostatic calculations, protein-folding algorithms, and Brownian dynamics.

Macromolecule Coordinate Files

Computer modeling of proteins and nucleic acids[2,3] is made possible by the atomic coordinates of a macromolecule determined by x-ray crystallography or NMR solution studies deposited in the Protein Data Bank (PDB) at Brookhaven National Laboratory.[4] Coordinates of hundreds of proteins, protein complexes, protein–nucleic acid complexes, and nucleic acid fragments can be conveniently downloaded across the Internet to a local site in a matter of minutes. The PDB file generally contains basic information about the molecule in the header records, including bibliographic information, structure resolution, properties of the unit cell, method of structure determination and refinement, amino acid sequence, regions of secondary structure, and, most importantly, the coordinates of every structurally resolved heavy (nonhydrogen) atom. In Table 1 is an example of a few atom records from a PDB file. In addition the PDB has a library of free software for the manipulation and analysis of atomic coordinate files that can be downloaded and executed on a personal computer or workstation. Such libraries are also becoming available over the Internet from the computational laboratories of many major research institutions doing macromolecular modeling around the world. The National Institutes of Health has a software list that can be accessed via the Internet at URL: http://molbio.info.nih. gov:80/modeling/gateway.html.

Molecular Graphics Representation

Many graphics packages exist for personal computers and high-performance graphics workstations and are available both commercially at a modest price (e.g., MidasPlus[6]) and free of charge over the Internet. Virtually all of these are capable of importing unmodified PDB-formatted files of proteins or nucleic acids and presenting them in color graphics on the workstation monitor. The graphics programs offer various choices of renderings of the model (e.g., wire-frame, ball-and-stick, ribbon, space-filling), with color coding of atoms or groups, and with selective labeling. The user may manipulate the model with a mouse or dial-and-button box, and even see stereo representations through special glasses. A simple wire-frame representation of the structure of cro repressor protein binding to DNA is shown in Figure 1, while a space-filling rendering of a four-helix bundle is shown in Figure 2.

With robust graphics programs it is a simple matter to isolate and recognize the three-dimensional structure of alpha helices, beta sheets, and turns of proteins, and the major and minor binding grooves of DNA double helices. Students can much more readily comprehend secondary structural elements of biomolecules with interactive graphics than with textbook drawings. This ability coupled with the low cost of graphics hardware has essentially transformed the field of molecular biology and biochemistry to the extent that most experimental laboratories doing structural and mechanistic studies of biomolecules routinely employ graphics workstations to perform structural analysis and obtain important insights.

Many graphical display programs also offer the user the ability to construct and modify models interactively. A short polypeptide fragment can easily be constructed on the screen using a data bank of standard residue coordinates, and torsion angles can be rotated to produce, for example, an alpha helix. Existing protein structures can be "mutated" by least-squares fitting of standard side-chain coordinates of a residue onto the backbone scaffolding, all in a few keystrokes. Fairly complicated organic ligand molecules can be interactively drawn on the screen with a mouse, and then be caused to conform with accepted values of bond lengths, angles, and torsion angles to produce a tentative structure.

Potential Energy Functions

For detailed calculations of the flexibility, stability, and motion of a biomolecule, we need to know the potential energy of the system as a function of the atomic coordinates.[2] It is impractical to use quantum mechanics to provide this description for large molecules, so various *empirical potential energy functions* have been developed of the "molecular mechanics" type.[9] These allow the very rapid computation of the energy and its spatial derivative (i.e., atomic forces) for a given configuration. A typical potential function U takes the following form:

$$U = \sum_{bonds} \frac{K_b}{2}\left(b - b_o\right)^2 + \sum_{bonds\_angles} \frac{K_\theta}{2}\left(\theta - \theta_o\right)^2$$

$$+ \sum_{dihedral\_angles} \frac{K_\phi}{2}\left[\left(1 + \cos\left(n\phi - \delta\right)\right)\right] + \sum_{pairs}^{nonbonded} \left[\frac{A}{r^{12}} - \frac{C}{r^6} + \frac{q_1 q_2}{\varepsilon r}\right].$$

(1)

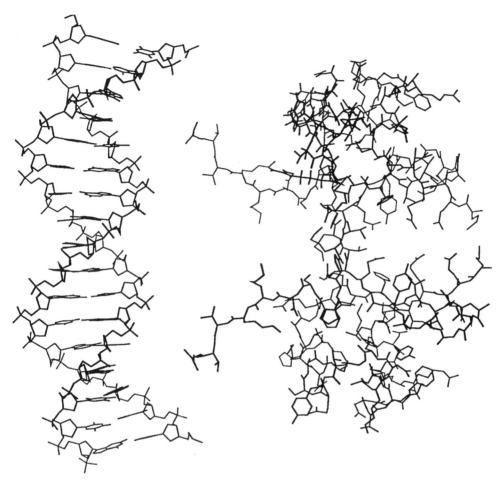

UCSF MidasPlus

FIGURE 1   Simple wire-frame representation of cro repressor-DNA docking generated by MidasPlus[6] using coordinates derived from the PDB.[7] This figure may be viewed using Netscape on the World Wide Web at URL http://pirn.chem.tntech.edu/acsbook/index.html.

The first term is a summation over every covalent bond in the molecule, with each bond being treated as a harmonic spring with force constant K_b and equilibrium length b_0. Since typical magnitudes of bond vibrations at room temperature are small, the harmonic approximation is a good one. The second term expresses the energy of bending covalent bond angles θ, while the third term accounts for the torsion energy of rotating about a bond axis ϕ. The last summation involves nonbonded interactions, which constitute the most computationally demanding portion of the potential energy. This term involves the distance- (r-) dependent Lennard–Jones 6–12 interaction between nonbonded pairs separated by at least two other atoms in a covalent chain, and the Coulombic interaction between charges. Ordinarily, a smaller list of nonbonded interacting pairs (i.e., those within some cutoff distance ~ 6–10 Å) is kept and frequently updated during a simulation to decrease computer time.

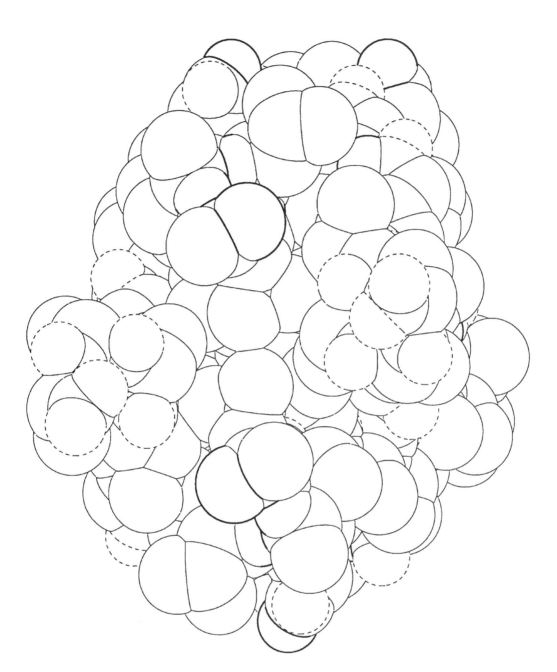

FIGURE 2   Space-filling representation of the atomic packing along the axis of a human-designed four-helix bundle, generated by Artist for Quanta.[8] This figure may be viewed using Netscape on the World Wide Web at URL http://pirn.chem.tntech.edu/acsbook/index.html.

The charge pair interactions are more problematic, since Coulombic interactions are long-ranged, and various truncation/correction schemes are employed.[10]

This potential function is implemented on a computer by combining data from several files. *Residue topology files* (RTF) contain information on the covalent bonding

TABLE 2 A representative listing of just of few of the many software packages that simulate the energy and dynamics of macromolecules.

Software Package	Source	Reference
CHARMM	Molecular Simulations, Inc./Harvard University	11
AMBER	University of California, San Francisco	12
GROMOS	Biomos/University of Groningen	13
DISCOVER	Biosym	14
SYBYL	Tripos	15
MM3	N. Allinger, Dept. of Chemistry/University of Georgia	16
MACROMODEL	Columbia University	17

connectedness within residue types, nucleotides, and other organic prosthetic groups. Along with the amino acid sequence (or nucleotide sequence) and knowledge of how prosthetic groups are ligated to proteins, the RTF data are used to construct a list of covalent bonds, covalent angles, torsion angles, and atom types, a nonbonded interaction exclusion list, and a charge interaction exclusion list for a whole macromolecule, the *molecular topology file* (MTF). This information is combined with *parameter files* containing the various interaction and force constants in eq 1 for the large variety of atom types. Empirical potential functions are constantly being improved and updated, largely by finding more realistic and accurate sets of parameters and defining new atom types. Quite a large number of molecular simulation programs are now in existence that incorporate similar empirical potential functions, including but not limited to the more well-known packages listed in Table 2.

Because of the large number of solvent atoms required to "solvate" a macromolecule, the greatest limitation in calculating an accurate energy of a molecular configuration is the treatment of solvent–solvent and solvent–biomolecule interactions. For instance, to "solvate" the trypsin enzyme (containing 1,620 heavy atoms) with four layers of water requires 4,785 water molecules.[18] For this reason, it has been standard to include the solvent only implicitly by an appropriate adjustment of the parameters of the potential function. A choice of dielectric constant $\varepsilon = 80$ representing water is a reasonable start but clearly ignores the behavior of motionally restricted waters such as in active site channels and surrounding ionic groups, as well as the low dielectric interior ($\varepsilon \sim 4$) of the biomolecule. Screened Coulomb potentials can be incorporated in pairwise charge–charge interactions to treat the presence of salt in the medium. Explicit water can be incorporated in simulations in a limited way.[10] Even when explicit water is being used, the intrinsic polarizability of the water molecule is not adequately treated with current empirical water model potentials[10] (e.g., SPC, TIP3P, F3C), and better modeling of water is currently an intensely active area of study.[19]

The choice of potential energy function and methodology for treatment of solvent interactions should be dictated by which method most faithfully describes the particular process of interest. For instance, the motion of a ligand in an enzyme-active site cleft might require the explicit incorporation of water in the cleft, while events in the interior of a protein may not be so demanding of water inclusion.

Energy Refinement of Structures

The most basic application of energy functions described above is *energy minimization*. Since the potential energy function U is a differential function of the Cartesian coordinates of atoms, the gradient of the potential can be evaluated in terms of simple functions of the coordinates and used to compute the force $\bar{F}$ on every atom.

$$\bar{F}\left(\bar{x}_1 \ldots \bar{x}_N\right) = -\nabla U\left(\bar{x}_1 \ldots \bar{x}_N\right). \tag{2}$$

Atoms may then be moved small distances in the directions dictated by these forces, and a lower energy configuration can be obtained. By repeating this process in an iterative fashion, the structure can be energy minimized; that is, the optimum structure can be obtained corresponding to the empirical potential function being used. A number of algorithms exist to carry out this basic process, the first-order method of steepest descents being the most common, followed by the conjugate gradient method and the second-order adopted basis set Newton–Raphson (ABNR) minimization method.[2] These methods can be useful for resolving steric conflicts and refining the geometry of model compounds or poorly resolved regions of the crystal structure. Proteins that have been "mutated" by the superposition of a different side chain on the backbone can be refined by applying energy minimization. Model compound geometries generated in approximate fashion can be refined into a more physically realistic structure. Many protein structures have been refined by crystallographers using the Konnert–Hendrickson procedure,[20] which optimizes bond lengths and angles as a part of structure determination.

Because finding the absolute minimum of a multivariate function such as the potential energy function is a highly nonlinear problem, there are limitations in the usefulness of energy minimization. For instance, one cannot be sure that a global minimum has been achieved, and in fact, for proteins is it practically impossible. This is the notorious *multiple minimum problem* of macromolecular structure. Only a very small fraction of the conformation space can be explored by these methods. For illustrative purposes, consider the energy contour map for two torsional degrees of freedom (Ramachandran plot) in the simple alanine dipeptide[3] in Figure 3. Even for only two torsional degrees of freedom one notes the presence of multiple minima.

A related application of energy minimization is the *adiabatic mapping* technique. Here, rather than considering a static energy optimum, one wishes to assess the energy cost of a structural transition within a protein or nucleic acid. Examples include a hinge-bending motion in a multidomain protein[21] or nucleic acid,[22] and the insertion of an intercalating agent within DNA.[23] A reaction coordinate is defined that prescribes the transition of interest. The biomolecule is deformed by an external constraint to move along the reaction coordinate, while at every position the balance of the molecule is allowed to relax by energy minimization. The total energy of the molecule is thus determined as a function of the reaction coordinate. This procedure is complicated by the choice of reaction coordinate, which may need to be a collective coordinate rather than a single covalent term in the potential function. An illustrative example is provided in Figure 4, showing the hinge-bending motion in transfer-RNA.[22]

Molecular Dynamics Simulations

Flexibility and motion are clearly important to the biological functioning of proteins and nucleic acids.[23] These molecules are not static structures, but exhibit a

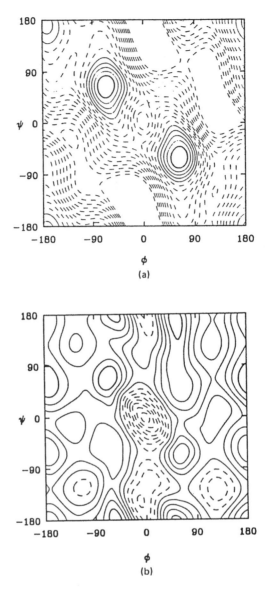

FIGURE 3   Energy contour map (Ramachandran plot) for ϕ versus ψ of the alanine dipeptide. Solid lines mark the five lowest energy contours at 1 kcal/mol, and the bottom contour is marked with a heavy line; dashed lines mark the higher contours at 1-kcal/mol intervals: a, vacuum potential surface; b, solvent-modified potential surface from MD simulation.[3] (Reproduced by permission from John Wiley and Sons, 1988.)

variety of complex motions both in solution and in the crystalline state. The most commonly employed simulation method used to study the motion of proteins and nucleic acids on the atomic level is the *molecular dynamics* (MD) method. Rather than being confined to a single low-energy conformation, MD allows the sampling of a thermally distributed range of intramolecular conformations. In MD, one numerically and iteratively integrates the classical equations of motion for every explicit

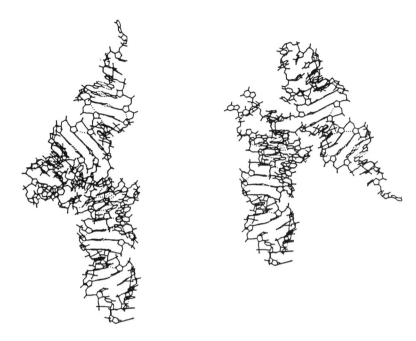

FIGURE 4 Fully extended model and closed model of tRNA[phe] from the adiabatic mapping study of Tung et al.[22] This figure may be viewed using Netscape on the World Wide Web at URL http://pirn.chem.tntech.edu/acsbook/index.html.

atom N in the system by marching forward in time by tiny time increments, Δt. Quite a number of algorithms exist for this purpose,[2,3] but the simplest is shown below:

$$\bar{x}_i\left(t + \Delta t\right) = \bar{x}_i\left(t\right) + \bar{v}_i\left(t\right)\Delta t$$

$$\bar{v}_i\left(t + \Delta t\right) = \bar{v}_i\left(t\right) + \frac{\bar{F}\left(\bar{x}_1 \ldots \bar{x}_N, t\right)}{m}\Delta t \ . \tag{3}$$

The time increment must be sufficiently small that errors in integrating these 6N equations (3N velocities and 3N positions) are kept manageably small, as manifested by conservation of the energy. As a result, Δt must be kept on the order of a femtosecond (10^{-15} s). Furthermore, since the forces must be recalculated for every time step, including nonbonded energy terms on the order of N × N terms, MD is a computation-intensive task. Thus, only a few picoseconds to a few nanoseconds can be conveniently simulated, depending on the number of atoms and the speed and memory of the computer. The method was originally developed for simulation of bulk fluids, taking advantage of periodic boundary conditions. But with the advent of faster and larger computers, it has become feasible to simulate the dynamics of whole proteins and nucleic acids. The first protein simulation was on bovine pancreatic trypsin inhibitor (458 heavy atoms) in 1976.[24] Since then, MD simulations have become commonplace to such an extent that much larger proteins are now being simulated in an explicit water environment.[10] For example, one of the largest simulations is HIV protease (over 1,500 atoms) simulated for 96 ps in 6,990 explicit waters.[25,26] In another MD simulation the binding of an antiviral agent to a human cold virus was performed for 1,000 ps.[27]

We will now describe the typical process of performing an MD simulation on a protein.[2,3] Coordinates are downloaded from the Protein Data Bank or other source. These are imported into the MD program where the sequence information is combined with RTFs to produce an MTF. Special prosthetic groups such as hemes or other organic functional entities must be provided for by defining special residues within the RTF set, and ordinarily one needs to provide the program with special instructions on covalently connecting nonpolypeptide groups or ions into the structural model. Typically, 100 or more cycles of steepest-descents energy minimization are performed on the starting structure to relax any steric conflicts and allow the molecule to adjust to the potential function. This causes a stabilization of the structure by perhaps thousands of kJ/mol, resulting in a molecule that is effectively at a very low "temperature". The model is then subjected to a heating and equilibration period of MD simulation. Velocities are assigned to all the atoms from a Maxwellian distribution at some low temperature (e.g., 50 K), and MD is allowed to proceed for a short period (e.g., 1–2 ps). The velocities are then reassigned or rescaled up to some higher temperature incrementally until eventually the final simulation temperature is achieved. A gentle warm-up process is necessary to allow proteins to maintain tertiary structures very close to their crystal structures, generally with a 1–2 Å rms deviation. The protein is then subjected to tens of picoseconds of equilibration with constant temperature constraints applied until an equilibrium partitioning of the energy exists between kinetic and potential components. At the end of the equilibration period, there should be no systematic drifts of potential energy nor serious changes in the global structure of the model. At this point a long period of simulation can be observed, usually still with constant-temperature or constant-pressure constraints, during which thermal averaging over desired quantities is collected. The trajectory is stored on magnetic media for subsequent analysis.

Applications and Limitations of Molecular Dynamics

Figure 5 shows the reasonable correspondence of rms fluctuations in atomic positions in tRNA[Phe] comparing a 96-ps MD simulation[28] with fluctuations derived from Debye–Waller factors in the crystallographic study.[29] Thus, MD has demonstrated itself to be extremely powerful and useful for the study of thermal motions in proteins and nucleic acids. However, the small time increment and many force terms required for MD simulations remain one of the most fundamental limitations to the usefulness of the method as applied to biomolecules. Table 3 shows the characteristic time scales of internal motions of proteins and nucleic acids relative to the MD time scale.[2] One can see that only small-amplitude local motions of proteins are sampled on the MD time scale, such as the rotation of small side chains at the surface. Much more biologically relevant processes, such as allosteric transitions, diffusion of substrate into and out of enzyme active sites, docking interactions between antibodies and antigens, and the like, are on much longer time scales and are currently inaccessible to *direct* simulation by MD methods. Because of the stiffness of DNA, one may observe global deformations of nucleic acids such as stretching and twisting in MD, but not global bending, which actually may be more important to function.[30] Still, MD affords one a much more robust sampling of conformational

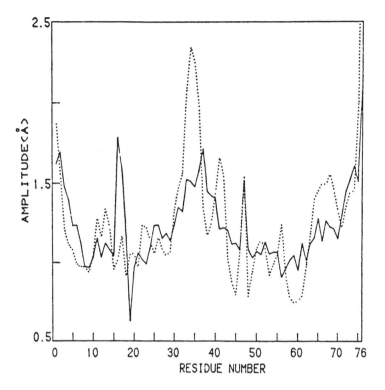

FIGURE 5 Average rms amplitudes of atomic motions in tRNA[phe] on a residue-by-residue basis. The solid line shows the values from the thermal factor analysis of the crystallographic data,[29] while the broken line gives the results of MD simulation.[28] (Reproduced by permission.)

space than Monte Carlo (MC) sampling methods, since MD motions are naturally directed by the forces, while MC steps are more randomly generated.

By clever modification and application of constraints in MD simulations, useful and biologically relevant information has been obtained about events in biomolecules that are beyond the time scale of direct simulations. *Umbrella sampling* techniques, in which an external potential is applied to a salient reaction coordinate, have allowed the simulation of rare events such as the crossing of high *activation barriers* within the molecules.[2,3] By forcing the molecule to remain constrained in small sampling windows, which are slowly varying to drive the reaction coordinate to cross a barrier, potentials of mean-force (i.e., free-energy) barrier profiles have been computed in which the MD provides efficient sampling over the thermal or bath motions. This method was applied to compute the activation barrier to ring flipping of tyrosine in BPTI[31] (see Figure 6). By starting the MD trajectory on the barrier top and running it forward and backward in time, the *activated dynamics* of barrier crossing can be observed, and transmission coefficients entering transition-state rate constants can be determined directly.[32]

Perhaps the most powerful application of MD simulation methods to date is the *thermodynamic cycle perturbation method*.[33] This and related methods are becoming useful tools in the design of drugs and enzymes with tailored specificities. Let's suppose that one is interested in the equilibrium constant K_b of binding of severally

TABLE 3   Typical features of some internal motions of proteins and nucleic acids.

Motion	Spatial Extent (Å)	Amplitude (Å)	Characteristic Time
Relative vibration of bonded atoms	2 to 5	0.01 to 0.1	0.01 to 0.1 ps
Longitudinal motions of bases in double helices (nucleic acids)	5	0.1	0.01 to 0.1 ps
Lateral motions of bases in double helices (nucleic acids)	5	1	0.1 to 1 ps
Global stretching (nucleic acids)	10 to 300	0.3 to 3	0.1 to 10 ps
Global twisting (nucleic acids)	10 to 300	1 to 10	0.1 to 10 ps
Elastic vibration of globular region	10 to 20	0.05 to 0.5	1 to 10 ps
Sugar repuckering (nucleic acids)	5	2	1 to 10 ps
Rotation of side chains at surface (protein)	5 to 10	5 to 10	10 to 100 ps
Torsional libration of buried groups	5 to 10	0.5	10 ps to 1 ns
Relative motion of different globular regions (hinge bending)	10 to 20	1 to 5	10 ps to 100 ns
Global bending (nucleic acids)	100 to 1000	50 to 200	100 ps to 100 ns
Rotation of medium-sized side chains in interior (protein)	5	5	0.1 ms to 1 s
Allosteric transitions	5 to 40	1 to 5	0.01 ms to 1 s
Local denaturation	5 to 10	5 to 10	0.01 ms to 0.1 s

Adapted from Reference 2.

and structurally homologous drug ligands L, L', L'' to a protein P in some well-characterized fashion.

$$P + L \xleftrightarrow{\;K_b\;} PL \tag{4}$$

The direct calculation of a binding constant is quite difficult in practice. However, it is a relatively simple matter in an MD simulation to calculate the free-energy change of perturbing or transmuting ligand L into a related ligand L'. This can be accomplished both in the enzyme active site and in the free state in solution.

$$L \xrightarrow{\;\Delta G_{free}\;} L'$$
$$PL \xrightarrow{\;\Delta G_{bound}\;} PL' \tag{5}$$

By forming a thermodynamic cycle, one can then calculate the change in free energy of binding $\Delta\Delta G_b$ comparing one ligand to the other, as follows:

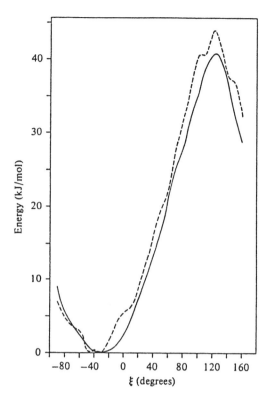

FIGURE 6   The potential mean force A(ξ) (——) and mean potential energy <V(ξ)> (-----) as functions of the tyrosine 35-ring rotational isomerization reaction coordinate ξ from an MD study by Northrup et al.[31] The energies are given in units of kcal/mol (1 kcal = 4.184 kJ).

$$P + L \xrightarrow{\Delta G_b} PL$$

$$\downarrow \Delta G_{free} \quad \downarrow \Delta G_{bound} \tag{6}$$

$$P + L' \xrightarrow{\Delta G'_b} PL'$$

Then

$$\Delta \Delta G_b = \Delta G_b - \Delta G'_b = \Delta G_{free} - \Delta G_{bound} \tag{7}$$

Finally, one can determine the ratio of binding constants between two ligands binding to the same protein:

$$\frac{K_b}{K'_b} = e^{-\Delta \Delta G_b / RT} \tag{8}$$

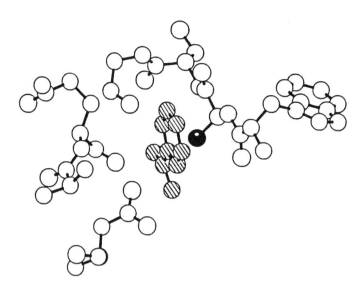

FIGURE 7   Single-site mutation in the trypsin specificity pocket, viewed from the surface of the enzyme. The methyl group (solid sphere) added in the Gly 216 → Ala 216 mutation projects toward the benzamidine inhibitor (hatched spheres), showing a steric conflict predicted to reduce the affinity of the enzyme for the inhibitor. The coordinates are from one instantaneous configuration generated in an MD study of free energy of inhibitor binding. (Reproduced by permission from reference 18.)

A similar calculation can be performed in which an enzyme is mutated to an altered form at the active site, and the binding of the altered form is compared to the native form. An example of such a calculation is shown in Figure 7.

Another very powerful application of MD simulation is the *distance geometry* procedure, which is used as a refinement tool in NMR determinations of *solution* structures of proteins and nucleic acids and their complexes.[34] The experimental NOE measurements yield a partial set of distance constraints between atoms. Ordinarily, some structural information is already available for the macromolecule, such as existence of certain regions of secondary structure. This information is incorporated into an initial model structure, which is then refined by MD simulation subject to these distance constraints by extra terms in the potential function. MD provides a much more efficient means of sampling configuration space than energy minimization or Monte Carlo choices for atomic motions. This approach has been used successfully in numerous studies, such as those of the solution structure of DNA[35,36] and lac repressor operator complex in solution.[37]

Electrostatic Calculations

Virtually every property of a protein or nucleic acid varies with the pH and salt concentration of the medium, indicating the tantamount importance of electrostatic effects in these biomolecules.[38] The overall stability of proteins and nucleic acids, as well as their interactions with other molecules, is a balance between a variety of forces in which electrostatic interactions is among the most important factors. Unfortunately, the greatest errors in empirical potential energy functions are in the electrostatic terms. This is because the electric fields are propagating over long distances through an inhomogeneous medium rather than bulk fluid. Numerical methods are being used to successfully deal with the problem.

There are two aspects of protein electrostatics that play a role in pH and ionic strength dependent properties. First, the titratable sites (e.g., Lys, Arg, Glu, Asp, His) of a protein have pK values that depend on their local environments, including their solvent exposure or burial factor and their interactions with the electrostatic field generated by other charged titratable sites.[39] Thus, to predict whether a particular acidic or basic group will be in its dissociated or protonated form, respectively, at a particular pH is a difficult problem involving the coupling of a large number of spatially distributed titrating sites. Second, given that one knows which sites are charged and which are not, it is still a challenging problem to compute the interaction between a protein and another reaction partner in solution owing to the inhomogeneity of the medium. In both cases one cannot simply evaluate the interaction between charge pairs via Coulomb's law, since the dielectric of the medium is varying over space, and mobile salt ions may be present. However, reasonably accurate results can be obtained by computer calculations based on theories that treat the system as a dielectric continuum with a space-varying dielectric constant and an electrolyte screening term. These calculations involve the solution of the Poisson–Boltzmann (PB) equation of electrostatics:

$$\nabla \cdot \varepsilon(x)\nabla\phi - \varepsilon(x)\kappa^2(x)\sinh(\phi) = -4\pi\rho(\bar{x}) \qquad (9)$$

or its oft-used linearized form in which the sinh function is approximated by its argument. Here, ϕ is the electrostatic potential around a set of charges distributed spatially according to $\rho(\bar{x})$, where $\varepsilon(\bar{x})$ is the space-varying dielectric coefficient and $\kappa(\bar{x})$ is the space-varying Debye screening coefficient. For simple geometries such as uniformly charged cylinders (to represent DNA) and charged spheres (to represent proteins), the PB equation can be solved in terms of known functions. For more complicated and realistic topologies, one resorts to numerical simulations, the most typical of which is the finite difference solution on a cubic grid.

Tanford and Kirkwood tackled the computation of pKs of titratable sites on proteins by solving the PB equation for a spherical low-dielectric domain with imbedded charges surrounded by a high-dielectric medium.[40] In a fraction of a second, the PDB coordinate file of a protein can be processed to determine effective spherical size and distribution of titratable sites, and a "titration" can be performed to determine the net charge of the molecule and mean charge of each site at any specified pH, ionic strength, and temperature. Titration curves can be generated rapidly by this method, and with modest empirical improvements,[41] it quite accurately represents the electrostatics of a protein.

More modern computational techniques exist not only to more accurately compute pKs of titratable residues but to calculate electrostatic fields surrounding proteins and nucleic acids and, hence, intermolecular electrostatic interactions. These are based on numerical solution of the finite difference PB equation on a grid originally developed by Warwicker and Watson,[42] or by use of alternative finite element methods.[43] These programs are readily available at low cost from a variety of sources and

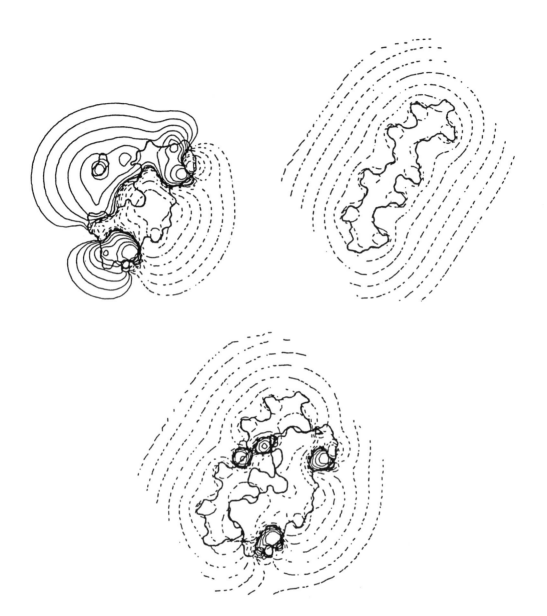

FIGURE 8   Electrostatic potential and ion distributions around λ*bd*/DNA. A slice was taken through λ*bd* (top left), DNA (top right), and the λ*bd*/DNA complex (bottom) in the plane of the approximate twofold symmetry axis of the protein. The bold solid line indicates the molecular surfaces of the protein and DNA. Isopotential contours ranging from 1/8 to 8 kT/e in factors of two are depicted. Positive contours are solid; negative contours are dashed. (Adapted by permission from reference 44.)

are easy to use, and include the following: DELPHI, MEAD, HYBRID, UHBD, POLARIS, and others. An excellent review of electrostatic effects in macromolecular modeling is provided by Harvey,[38] who references these various programs. Beautiful and instructive color representations of electrostatic field surfaces of constant potential are easily produced by existing software. A black-and-white contour map of the electrostatic field around λ*bd*/DNA complex[44] is shown in Figure 8.

Protein Folding and Structure Prediction

One of the most significant outstanding problems in 20th-century molecular biology is the prediction of protein structure from amino acid sequence.[45,46] The intractability of protein structure prediction stems from the multiple minimum problem of biomolecular conformation, combined with the problem of "frustration" induced by incorrect folds early in the folding process. Progress is being made simultaneously along two fundamentally different computational tracks, the physics-based track and the information theoretical track. In the former paradigm, physical factors such as hydrogen bonding, hydrophobicity, and packing interactions are incorporated into simple models, where a whole residue is represented by a single interaction bead on diamond or cubic lattices, and penalties to various primitive folding moves on the lattice are assigned from these physical factors. Quite a number of algorithms now exist in this vein, even employing neural network processing.[47] Figure 9 illustrates the unfolding of bovine pancreatic trypsin inhibitor from a dynamic simulation using a novel approach.[48] In the information theoretical approach, large databases of protein sequence and structure are statistically analyzed to determine the correlation between pairs of amino acids. These pair correlations, which are based purely on local regions of structure, are again incorporated into bead movement algorithms. The accuracy of predicting even local secondary structure from statistical approaches has reached only about 65%,[50] mainly because no torsion angle restrictions (a physical factor) are included in an informational approach, nor is the database on known high-resolution structures large enough to encompass a statistically significant number of sequences (an information factor). It can be optimistically stated that the reliable protein-folding algorithm is a reasonable hope for the future. Computational biochemists will need to continue to rely on the a priori existence of crystallographic and NMR experimental structural data on a protein or nucleic acid of interest as a necessary prerequisite to most modeling studies.

Brownian Dynamics

An alternative method to molecular dynamics that is able to simulate much longer and more biologically relevant time scales is the method called *Brownian dynamics* (BD). The BD method[51] coarse-grains the molecular structure (i.e., abandons a purely atomistic treatment) and treats the diffusive motion of whole interacting subunits in solution by generating trajectories composed of a series of small displacements obeying the diffusion equation with forces. These subunits could be spheres that represent whole residues linked together to form a polypeptide chain, as in a study of protein-folding dynamics.[52] The subunits could be whole proteins with irregular shapes dictated by the atomic structure, as in simulations of protein–protein interactions.[53] A wide variety of scenarios is possible. The algorithm for free displacements Δr in the subunit centers of mass in time step Δt is given by the Ermak and McCammon algorithm:[54]

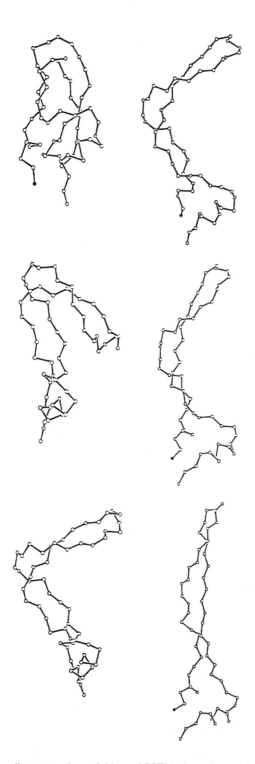

FIGURE 9   Diagrams illustrating the unfolding of BPTI lacking the residue 30–51 disulfide bond from a dynamic simulation. The conformations of BPTI are represented by the C^α trajectory of the protein. The conformations are obtained at time frames of 0, 5, 10, 16, 22, and 33 ps in the simulation. (Adapted by permission from reference 48.)

$$\Delta r = DF\left(k_B T\right)^{-1} \Delta t + S .$$
(10)

Here, D is the subunit translational diffusion coefficient, F is the force, $k_B T$ is the Boltzmann constant times absolute temperature, and S is the stochastic component of the displacement arising from collisions with solvent molecules. This is generated by taking normally distributed random numbers obeying the average relationship $<S^2> = 2D\Delta t$. A similar equation governs the independent rotational Brownian motion of each subunit, where force is replaced by torque, and D is replaced by an isotropic rotational diffusion coefficient D_{ri} for each particle i.

A very important application of the BD method is to obtain rate constants for bimolecular reactions in solution.[55] The two reactant species could be an enzyme and small diffusing ligand, such as the simulations of superoxide anion reacting with the enzyme superoxide dismutase,[56] or two whole proteins, such as in electron-transfer reactions between proteins cytochrome c and cytochrome c peroxidase.[57] In either case the general procedure[55] is to divide the space around the central or target molecule into two concentric regions. The inner region is inside a radius b in which the interactions between the two molecules are complicated by electrostatic forces and steric interactions. Outside this region is an outer region in which the diffusional behavior can be described by simple analytical formulas for centrosymmetric diffusion. The bimolecular rate of association is then given by[55]

$$k = k_D\left(b\right)\beta$$
(11)

Here, $k_D(b)$ is the steady-state rate constant at which species II reacts with a perfectly absorbing sphere of radius b from outside in an infinite medium. The term β is the fraction of trajectories started at distance b that "react" with the target molecule rather than escaping to infinity. Reaction criteria are chosen usually on geometric grounds for the specific case of interest. Trajectories are actually truncated at some large outer distance, and an exact truncation correction to β is applied.

BD simulations provide rate constants that can be compared directly to experiment as a function of pH, ionic strength, viscosity, and temperature. Computer "mutations" can be performed with simplicity and compared with analogous site-directed mutagenic experiments in order to explore the relationship between structure and biological function.[58] Not only are rate constants obtainable, but trajectories provide a large number of protein–protein and protein–DNA docking conformations weighted according to their importance to the kinetics of the associated process. Figure 10 shows the docking profile of incoming cytochrome c as it docks with its partner cytochrome c peroxidase.[57] Two or three regions are indicated rather than one specific docking domain. BD rigid docking provides starting structures for subsequent MD simulations to relax the interfaces and allow observation of intimate protein–protein binding details beyond what BD can show.[59] Figure 11 shows representative docked complexes of two distinct types between cytochrome c and cytochrome b5.

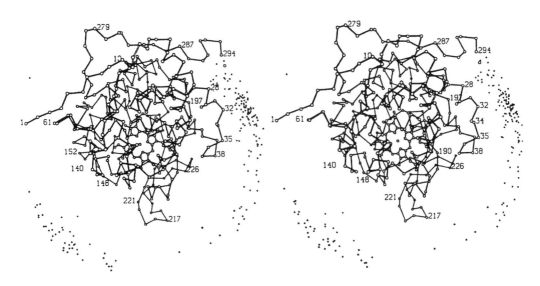

FIGURE 10   Stereoscopic projection of the C^α skeleton and heme atoms of cytochrome c peroxidase, surrounded by points representing the center of mass of incoming electron transfer partner cytochrome c at a point in space where docking criteria for electron transfer are met. Accordingly, electron transfer may occur from a wide range of encounter complexes rather than from a single association complex. This figure may be viewed using Netscape on the World Wide Web at URL http://pirn.chem.tntech.edu/acs-book/index.html. (Reproduced by permission from reference 57.)

Teaching Macromolecular Modeling

The contents of this chapter have arguably demonstrated the need for formal course work to introduce macromolecular modeling concepts and methodologies to novices. The interested reader is strongly encouraged to read a recent article by Harvey and Tan, which has aptly addressed this need.[61] In this article they (1) outline a proposed one-semester introductory course on modeling at the advanced graduate level to students with largely a biology background, and (2) suggest a brief (1–2 lectures) introduction of modeling in undergraduate courses such as biochemistry or physical chemistry. They advise an emphasis less on methodology than on the legitimacy of molecular modeling as a scientific discipline that serves as an important complement to experimentation and analytical theory. In Table 4 one possible syllabus for the graduate-level macromolecular modeling course is displayed, with footnotes indicating material that could be incorporated into other courses to provide a brief, scaled-down, 1–2 lecture introduction.

There are a number of short courses and workshops that teach introductory and/or advanced courses in macromolecular modeling. The ACS has a short course available by satellite, and the Biosym/MSI company offers regular workshops on modeling, to mention just a couple. The NIH Molecular Modeling Home Page lists a variety of services available on the Web at URL http://cmm.info.nih.gov/model-ing/net_services.html. There are numerous hypermedia tutorials available on the Web. For an index of these, one may point to the University of Sheffield's ChemDex Web site; the URL is http://www.shef.ac.uk/~chem/chemdex/chem-educa-tion.html#Computers.

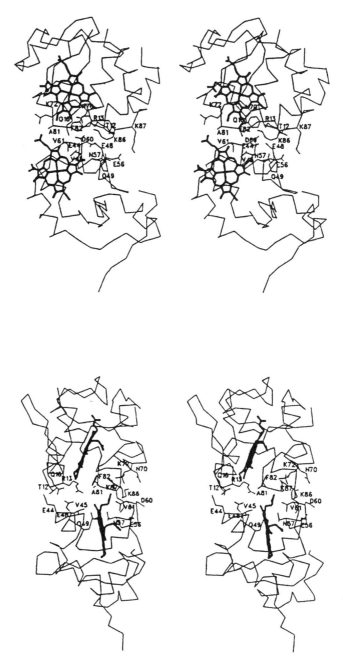

FIGURE 11   Stereodiagrams of the energy-minimized, putative complex formed by yeast iso-1-cytochrome c and trypsin-solubilized bovine liver microsomal cytochrome b₅ based on the electrostatically "dominant" complex predicted by BD simulations. This figure may be viewed using Netscape on the World Wide Web at URL http://pirn.chem.tntech.edu/acsbook/index.html. (Reproduced by permission from reference 60.)

TABLE 4   A one-semester introductory graduate course in macromolecular modeling.

Course Goals:
To introduce the basic concepts of molecular modeling and to provide students an understanding of
  the kinds of problems that can be approached through molecular modeling methods, and the
  limitations of the same. Students would have the opportunity to work weekly with a commercial
  modeling package to demonstrate the basic concepts of the course.

Course Outline:
  Introduction†
    Modeling role as an adjunct to experiment
    Basic requirements for successful modeling†
  The Coordinate Database and Its Organization
  Representation of Molecules on Computer†
    Interactive graphical representations†
    Topology files
    Covalent and noncovalent interactions
  Potential Functions and Energy Parameters†
    Simple organic molecule example (relationship between potential energy and conformation)
  Energy Minimization†
    Method of steepest descents
    Example with hindered rotation in ethane (trans-gauche isomerism)
    Simple dipeptide energy landscape (ϕ,ψ dihedrals)
    Adiabatic mapping of energy surfaces
  Molecular Dynamics†
    Algorithm†
    Energy conservation, energy components, and time step
    Boundary conditions
    Thermalization techniques
    Application to a small polypeptide
    Types of motions in experiments and MD†
  Electrostatic Interactions
    Pairwise interaction schemes
    Continuum methods by numerical techniques
  Treatment of Solvent Effects in Aqueous Solutions
    Continuum models
    Hydrophobic effect
    Explicit solvent treatments
  Free Energy Calculations
    Free energy versus potential function energy
    Thermodynamic cycles and perturbation techniques
  Specialty Algorithms
    Monte Carlo
    Brownian Dynamics
    Distance geometry method
  Protein Structure Prediction
    The multiple minima problem
    Protein folding algorithms overview
  Rational Drug Discovery Methods
  Conclusion†
    Strengths and limitations of molecular modeling†
    Important unsolved problems†
    Future directions for research

---

† A scaled-down 2-lecture overview of macromolecular modeling would introduce these items.

TABLE 4 (continued)   A one-semester introductory graduate course in macromolecular modeling.

Course Prerequisites:
One semester undergraduate biochemistry, two semesters of physical chemistry, or a one-semester survey
    course in physical chemistry for life sciences, one year of undergraduate organic chemistry. The student
    should have some familiarity with the use of computers, though a formal programming course is not
    necessary.

Materials Required:
Students work on a sign-up basis for approximately 2 hours per week on a high-performance workstation
    with at least the capability of a Silicon Graphics Personal Iris or Indy. A good commercial software
    package is required such as MSI Quanta/Charmm or Biosym Insight/Discover having an easy-to-
    use graphics interface, good molecular graphics rendering capabilities, the ability to perform energy
    minimizations and molecular dynamics, and a solid tutorial for the modeling package. Short weekly
    student exercises should be prepared with carefully detailed instructions aimed at giving students a
    hands-on demonstration of modeling concepts.

Summary

There is no doubt that computers are revolutionizing the way chemists, biochemists, and molecular biologists are coming to understand biological molecules
and their structure, dynamics, and reactivity on the atomic level. The emergence of
the new discipline of macromolecular modeling has created the need for formal course
work in this area for graduate students.[61] Future developments are expected to take
place in several important areas, including: the development and refinement of more
accurate covalent and noncovalent *force fields* for simulation of structure and motion
in physiological environments; the development of more accurate and efficient algorithms for calculating *electrostatic fields* in inhomogeneous aqueous media; the continual progress in the area of *code development,* which makes optimum use of massively
parallel supercomputers; improvement of *interactive visualization* techniques; and
the invention of comprehensive *protein-folding algorithms* that really work (i.e., have
the ability to predict a unique tertiary structure from sequence information). Even
without further methodological developments and without an increase in computational power, the application of existing methodologies on a grander scale could
make tremendous strides in our understanding of molecular mechanisms of biological
processes. Consider the imminent arrival of teraflop machines (10^{12} floating point
operations per second) and the possibilities become staggering.

References

1. Briggs, J. M.; McCammon, J. A.; "Computation Unravels Mysteries of Molecular
    Biophysics"; *Comp. in Phys.* 1992, *6,* 238.
2. McCammon, J. A.; Harvey, S. C.; *Dynamics of Proteins and Nucleic Acids;* Cambridge
    University Press, New York, 1987.
3. Brooks, C. L. III; Karplus, M.; Pettitt, B. M.; *Proteins: A Theoretical Perspective of
    Dynamics, Structure, and Thermodynamics;* Advances in Chemical Physics; John Wiley
    & Sons, New York, 1988; Vol. 71.

4. Bernstein, F. C.; Koetzle, T. F.; Williams, G. J. B.; Meyer, E. F.; Brice, M. D.; Rodgers, J. R.; Kennard, O.; Shimanouchi, T.; Tasumi, M.; "The Protein Data Bank: A Computer-Based Archival File for Macromolecular Structures"; *J. Mol. Biol.* 1977, *112,* 535. Protein Data Bank URL: http://www.pdb.bnl.gov/.

5. Wang, J.; Mauro, J. M.; Edwards, S. L.; Oatley, S. J.; Fishel, L. A.; Ashford, V. A.; Xuong, N.-H.; Kraut, J.; "X-Ray Structures of Recombinant Yeast Cytochrome c Peroxidase and Three Heme-Cleft Mutants Prepared by Site-Directed Mutagenesis"; *Biochemistry* 1990, *29,* 7160.

6. Ferrin, T. E.; Conrad, C. H.; Jarvis, L. E.; Langridge, R.; "The MIDAS Display System"; *J. Mol. Graph.* 1988, *6,* 13. World Wide Web URL: http://cgl.ucsf.edu/midasplus.html.

7. Brennan, R. G.; Roderick, S. L.; Takeda, Y.; Matthew, B. W.; "Protein-DNA Conformational Changes in the Crystal Structure of a Cro-Operator Complex"; *Proc. Nat. Acad. Sci.* 1990, *87,* 8165.

8. ARTIST/QUANTA 4.1, Molecular Simulations, Inc., 16 New England Executive Park, Burlington, MA, 01803-5297; http://www.msi.com/.

9. Burkert, U.; Allinger, N. L.; *Molecular Mechanics;* American Chemical Society: Washington, DC, 1982.

10. Daggett, V.; Levitt, M.; "Realistic Simulations of Native-Protein Dynamics in Solution and Beyond"; *Ann. Rev. Biophys. Biomol. Struct.* 1993, *22,* 353.

11. Brooks, B. R.; Bruccoleri, R. E.; Olafson, B. D.; States, D. J.; Swaminathan, S.; Karplus, M.; "CHARMM: A Program for Macromolecular Energy, Minimization, and Dynamics Calculations"; *J. Comp. Chem.* 1983, *4,* 187. MSI/Biosym URL: http://www.msi.com/.

12. Weiner, P. K.; Kollmann, P. A.; "AMBER: Assisted Model Building with Energy Refinement. A General Program for Modeling Molecules and Their Interactions"; *J. Comp. Chem.* 1981, *2,* 287. AMBER URL: http://www.amber.ucsf.edu/amber/amber/html.

13. Herman, J.; Berendsen, H. J. C.; van Gunsteren, W.; Postma, J. P. M.; "A Consistent Empirical Potential for Water-Protein Interactions"; *Biopolymers* 1984, *23,* 1513.

14. BIOSYM Technologies; Discover, User Manual, Version 2.5, BIOSYM Technologies, San Diego, CA, 1988.

15. SYBYL, Tripos, Inc., 1699 S. Hanley Road, St. Louis, MO 63144; http://www.webcom.com/~tripos2/.

16. Allinger, N. L.; Yuh, Y. H.; Lii, J.-H.; "Molecular Mechanics. The MM3 Force Field for Hydrocarbons I, II, and III"; *J. Am. Chem. Soc.* 1989, *111,* 8551.

17. MACROMODEL v.4.5, W. Clark Still, Department of Chemistry, Columbia University, New York; http://www.cc.columbia.edu/cu/chemistry/mmod/mmod.html/

18. Wong, C. F.; McCammon, J. A.; "Computer Simulation and the Design of New Biological Molecules"; *Israel J. Chem.* 1986, *27,* 211.

19. New, M. H.; Berne, B. J.; "Molecular Dynamics Calculation of the Effect of Solvent Polarizability on the Hydrophobic Interaction"; *J. Am. Chem. Soc.* 1995, *117,* 7172.

20. Konnert, J. H.; Hendrickson, W. A.; "A Restrained-Parameter Thermal-Factor Refinement Procedure"; *Acta Crystall.* 1980, *A36,* 344.

21. McCammon, J. A.; Gelin, B. R.; Karplus, M.; Wolynes, P. G.; "The Hinge-Bending Mode in Lysozyme"; *Nature* 1976, *262,* 325.

22. Tung, C. S.; Harvey, S. C.; McCammon, J. A.; "Large-Amplitude Bending Motions in Phenylalanine Transfer RNA"; *Biopolymers* 1984, *23,* 2173.

23. Lybrand, T.; Kollman, P.; "Molecular Mechanical Calculations on the Interaction of Ethidium Cation with Double-Helical DNA"; *Biopolymers* 1985, *24,* 1863.

24. McCammon, J. A.; Gelin, B. R.; Karplus, M.; "Dynamics of Folded Proteins"; *Nature* 1977, *267,* 585.

25. Harte, W. E., Jr.; Swaminathan, S.; Mansuri, M. M.; Martin, J. C.; Rosenberg, I. E.; Beveridge, D. L.; "Domain Communication in the Dynamical Structure of Human Immunodeficiency Virus 1 Protease"; *Proc. Nat. Acad. Sci.* 1990, *87,* 8864.

26. Swaminathan, S.; Harte, W. E., Jr.; Beveridge, D. L.; "Investigation of Domain Structure in Proteins via Molecular Dynamics Simulation: Application to HIV-1 Protease Dimer"; *J. Am. Chem. Soc.* 1991, *113,* 2717.

27. Wade, R. C.; McCammon, J. A.; "Binding of an Antiviral Agent to a Sensitive and a Resistant Human Rhinovirus. Computer Simulation Studies with Sampling of Amino Acid Side-Chain Conformations. II. Calculation of Free-Energy Differences by Thermodynamic Integration"; *J. Mol. Biol.* 1992, *225,* 697.

28. Prabhakaran, M.; Harvey, S. C.; McCammon, J. A.; "Molecular-Dynamics Simulation of Phenylalanine Transfer RNA. II. Amplitudes, Anisotropies, and Anharmonicities of Atomic Motions"; *Biopolymers* 1985, *24,* 1189.

29. Hingerty, B.; Brown, R. S.; Jack, A.; "Further Refinement of the Structure of Yeast tRNA"; *J. Mol. Biol.* 1978, *124,* 523.

30. Erie, D. A.; Yang, G.; Schultz, H. C.; Bustamante, C.; "DNA Bending by Cro Protein in Specific and Nonspecific Complexes: Implications for Protein Site Recognition and Specificity"; *Science* 1994, *266,* 1562.

31. Northrup, S. H.; Pear, M. R.; Lee, C. Y.; McCammon, J. A.; Karplus, M.; "Dynamical Theory of Activated Processes in Globular Proteins"; *Proc. Nat. Acad. Sci.* 1982, *79,* 4035.

32. Ghosh, I.; McCammon, J. A.; "Sidechain Rotational Isomerization in Proteins: Dynamic Simulation with Solvent Surroundings"; *Biophys. J.* 1987, *51,* 637.

33. Straatsma, T. P.; McCammon, J. A.; "Computational Alchemy"; *Ann. Rev. Phys. Chem.* 1992, *43,* 407.

34. Kaptein, R.; Zuiderweg, E. R. P.; Scheek, R. M.; Boelens, R.; van Gunsteren, W. F.; "A Protein Structure from Nuclear Magnetic Resonance Data; Lac Repressor Headpiece"; *J. Mol. Biol.* 1985, *182,* 179.

35. Banks, K. M.; Hare, D. R.; Reid, B. R.; "Three-Dimensional Solution Structure of a DNA Duplex Containing the BclI Restriction Sequence: Two-Dimensional NMR Studies, Distance Geometry Calculations, and Refinement by Back-Calculation of the NOESY Spectrum"; *Biochemistry* 1989, *28,* 6996.

36. Withka, J. M.; Swaminathan, S.; Srinivasan, J.; Beveridge, D. L.; Bolton, P. H.; "Toward a Dynamical Structure of DNA: Comparison of Theoretical and Experimental NOE Intensities"; *Science* 1992, *255,* 597.

37. Lamerichs, R. M. J. N.; Boelens, R.; van der Marel, G. A.; van Boom, J. H.; Kaptein, R.; "Assignment of the N-NMR Spectrum of a Lac Repressor Headpiece-Operator Complex in H_2O and Identification of NOEs"; *Eur. J. Biochem.* 1990, *194,* 629.

38. Harvey, S. C.; "Treatment of Electrostatic Effects in Macromolecular Modeling"; *Proteins* 1989, *5,* 78.

39. Matthew, J. B.; "Electrostatic Effects in Proteins"; *Ann. Rev. Biophys. Biophys. Chem.* 1985, *14,* 387.

40. Tanford, C.; Kirkwood, J. G.; "Theory of Protein Titration Curves. I. General Equations for Impenetrable Spheres"; *J. Am. Chem. Soc.* 1957, *79,* 5333.

41. Tanford, C.; Roxby, R.; "Interpretation of Protein Titration Curves. Application to Lysozyme"; *Biochemistry* 1972, *11,* 2192.

42. Warwicker, J.; Watson, H. C.; "Calculation of the Electric Potential in the Active Site Cleft Due to α-helix Dipoles"; *J. Mol. Biol.* 1982, *157,* 671.

43. You, T. J.; Harvey, S. C.; "Finite Element Approach to the Electrostatics of Macromolecules with Arbitrary Geometries"; *J. Comp. Chem.* 1993, *14,* 484.

44. Sharp, K. A.; Friedman, R. A.; Misra, V.; Hecht, J.; Honig, B.; "Salt Effects on Polyelectrolyte-Ligand Binding: Comparison of Poisson-Boltzmann, and Limiting Law/Counterion Binding Models"; *Biopolymers* 1995, *36,* 245.

45. Fasman, G. D.; "The Development of the Prediction of Protein Structure"; in *Prediction of Protein Structure and the Principles of Protein Conformation;* Fasman, G. D., Ed.; Plenum: New York, 1989, 193–316.

46. Garnier, J.; "Protein Structure Prediction"; *Biochimie* 1990, *72,* 513.

47. Qian, N.; Sejnowski, T. J.; "Predicting the Secondary Structure of Globular Proteins Using Neural Network Models"; *J. Mol. Biol.* 1988, *202*, 865.

48. Hao, M.-H.; Pincus, M. R.; Rackovsky, S.; Scheraga, H. A.; "Unfolding and Refolding of the Native Structure of Bovine Pancreatic Trypsin Inhibitor Studied by Computer Simulation"; *Biochemistry* 1993, *32*, 9614.

49. Gibrat, J. F.; Robson, B.; Garnier, J. "Influence of the Local Amino Acid Sequence upon the Zones of the Torsional Angles-ϕ and Angle-ψ Adopted by Residues in Proteins"; *Biochemistry* 1991, *30*, 1578.

50. Rooman, M J.; Wodak, S. J.; "Identification of Predictive Sequence Motifs Limited by Protein Structure Data Base Size"; *Nature* 1988, *335*, 45.

51. Harvey, S. C.; Cheung, H. C.; "Computer Simulation of Fluorescence Depolarization Due to Brownian Motion"; *Proc. Nat. Acad. Sci.* 1972, *69*, 3670.

52. McCammon, J. A.; Northrup, S. H.; Allison, S. A.; "Diffusional Dynamics of Ligand-Receptor Association"; *J. Phys. Chem.* 1986, *90*, 3901.

53. Northrup, S. H.; Boles, J. O.; Reynolds, J. C. L.; "Electrostatic Effects in the Brownian Dynamics of Association and Orientation of Heme Proteins"; *J. Phys. Chem.* 1987, *91*, 5991.

54. Ermak, D. L.; McCammon, J. A.; "Brownian Dynamics with Hydrodynamic Interactions"; *J. Chem. Phys.* 1978, *69*, 1352.

55. Northrup, S. H.; Allison, S. A.; McCammon, J. A.; "Brownian Dynamics Simulation of Diffusion-Influenced Biomolecular Reactions"; *J. Chem. Phys.* 1984, *80*, 1517.

56. Luty, B. A.; Amrani, S. E.; McCammon, J. A.; "Simulation of the Bimolecular Reaction Between Superoxide and Superoxide Dismutase: Synthesis of the Encounter and Reaction Steps"; *J. Am. Chem. Soc.* 1993, *115*, 11874.

57. Northrup, S. H.; Boles, J. O.; Reynolds, J. C. L.; "Brownian Dynamics of Cytochrome c and Cytochrome c Peroxidase Association"; *Science* 1988, *241*, 67.

58. Northrup, S. H.; Thomasson, K. A.; Miller, C. M.; Barker, P. D.; Eltis, L. D.; Guillemette, J. G.; Mauk, A. G.; "Effects of Charged Amino Acid Mutations on the Bimolecular Kinetics of Reduction of Yeast Iso-1-ferricytochrome c by Bovine Ferrocytochrome b"; *Biochemistry* 1993, *32*, 6613.

59. Guillemette, J. G.; Barker, P. D.; Eltis, L. D.; Lo, T. P.; Smith, M.; Brayer, G. D.; Mauk, A. G.; "Analysis of the Bimolecular Reduction of Ferricytochrome c by Ferrocytochrome b5 Through Mutagenesis and Molecular Modeling"; *Biochimie* 1994, *76*, 592.

60. Mauk, A. G.; Mauk, M. R.; Moore, G. R.; Northrup, S. H.; "Experimental and Theoretical Analysis of the Interaction Between Cytochrome c and Cytochrome b5"; *J. Bioenerg. Biomem.* 1995, *27*, 311.

61. Harvey, S. C.; Tan, R. K.-Z.; "Teaching Macromolecular Modeling"; *Biophys. J.* 1992, *63*, 1683.

Molecular Modeling and Drug Design

Alexander Tropsha
J. Phillip Bowen

Molecular modeling, the combination of computational chemistry and computer graphics, has been a part of some of the most recent and exciting applications of computers in chemistry. Contemporary molecular modeling can be defined as the generation, manipulation, calculation, and prediction of realistic chemical structures and associated physicochemical and biological properties.

The advent of high-speed computers and state-of-the-art computer graphics has made plausible the use of computationally intensive methods such as quantum mechanics, molecular mechanics, and molecular dynamics simulations to determine those physical and structural properties most commonly involved in molecular recognition. The power of molecular modeling rests solidly on a variety of well-established scientific disciplines including computer science, theoretical chemistry, biochemistry, biophysics, and pharmacology. Molecular modeling has become an indispensable complementary tool for most experimental chemistry research.

This chapter will briefly review some of the major computational methods routinely used in molecular modeling and will discuss the advances, applicability to various problems, and limitations of molecular modeling. There will be an emphasis on recent developments and successes of ligand-based and receptor-structure–based computational methods in designing new pharmaceutical agents. In addition, aspects of teaching molecular modeling as an integrated part of the undergraduate and graduate curriculum in chemical and biological sciences will be discussed.

Introduction

Researchers in chemistry, molecular biology, and biotechnology are moving toward a greater understanding of relationships between structure and function of physical and living matter. Computational chemistry and computer-assisted molecular modeling have rapidly become a vital component of such research. Mechanisms of drug–receptor and enzyme–substrate interactions, protein folding, protein–protein and protein–nucleic acid recognition, and de novo protein engineering are but a few

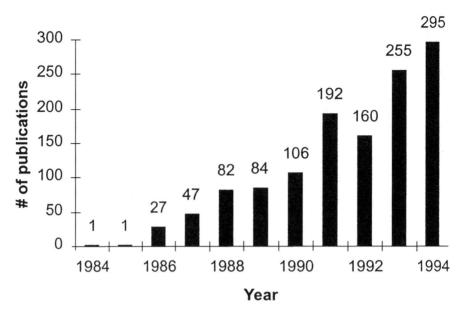

FIGURE I   Rapid growth of publications on molecular modeling.

examples of problems that may be addressed and facilitated by this technology. Recent advances in this field that have direct impact on many experimental scientific disciplines include the use of molecular dynamics for refinement of three-dimensional (3D) protein structures determined by x-ray crystallography and/or NMR, development of approaches based on free-energy perturbation methods to evaluate protein stability change due to point mutations, development of knowledge-based methods for 3D protein modeling and structure prediction, novel methods for structure-based drug design, etc. These novel applications greatly extend the power of molecular modeling beyond the more traditional uses of these techniques for the prediction and optimization of chemical structures of small molecules, drug design, and 3D structure manipulation. The advent of high-speed computers and state-of-the-art computer graphics makes plausible the use of computationally intensive methods such as quantum mechanics, molecular mechanics, and molecular dynamics simulations to determine those physical and structural properties most often involved in molecular recognition. The rapid growth of publications in the field of molecular modeling is reflected in Figure 1. Nowadays, it is not the lack of the ideas but of the availability of adequate computing power and appropriate visualization tools that precludes broader application of molecular modeling as a standard component of experimental chemical and biological research.

Molecular modeling can be broadly described as the generation, visualization, manipulation, and prediction of realistic molecular structure and associated physicochemical and/or biological properties.[1] Contemporary molecular modeling has become possible due to a "happy marriage" between computational chemistry and molecular graphics. The latter provides an exciting opportunity to augment the traditional description of chemical structures via two-dimensional chemical connectivity graphs by allowing the manipulation and observation, in real time and in three dimensions, of both molecular structures and many of their calculable properties.

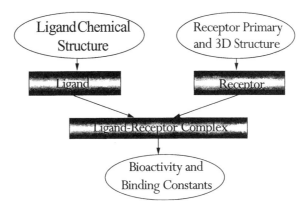

FIGURE 2 The flow chart of experimental information about drug–receptor interaction. The objects of experimental and theoretical investigation are in the rectangles, and the experimental information is in the ovals.

Recent advances in this area allow visualization of even intimate mechanisms of chemical reactions by graphical representation of the distribution and redistribution of electron density in atoms and molecules along the reaction pathway.

It is practically impossible, in a small chapter, to review all major developments, applications, approaches, and caveats of these methods. Several specialized reviews addressing various aspects of molecular modeling have been published in recent years in the ongoing series *Reviews in Computational Chemistry.*[2] Therefore, we have decided to concentrate in this review on one outstanding problem of modern chemistry and theoretical biology, namely, ligand–receptor recognition. By considering this problem as a paradigm, we will introduce major molecular modeling concepts and tools that one can apply to address the problem and discuss the need of integration of these tools in the course of such a project. In order to provide a practical example of a molecular modeling project, we will consider the dopaminergic ligand–receptor system that has been in the focus of our research in recent years.[3,4] We hope that by considering this important practical problem of molecular modeling we can better illustrate the power and limitations of this technique and also provide a practical introduction to major concepts of molecular modeling.

The Relationships Between Experimental and Theoretical Approaches to Studying Drug–Receptor Interactions

Molecular modeling research in drug design starts from the analysis of experimental observables of ligand–receptor interaction. This interaction leads to the formation of ligand–receptor complex, followed by conformational change of the receptor, which constitutes the mechanism of signal transduction. One can obtain either ligand–receptor binding constants from direct measurements in vitro or the biological activity data measured on isolated tissues or on the whole organism. Thus, experimental observables of the ligand–receptor system include chemical structure of ligands, their biological activity or binding constants, and, often, receptor primary sequence and, in some cases, receptor 3D structure (Figure 2).

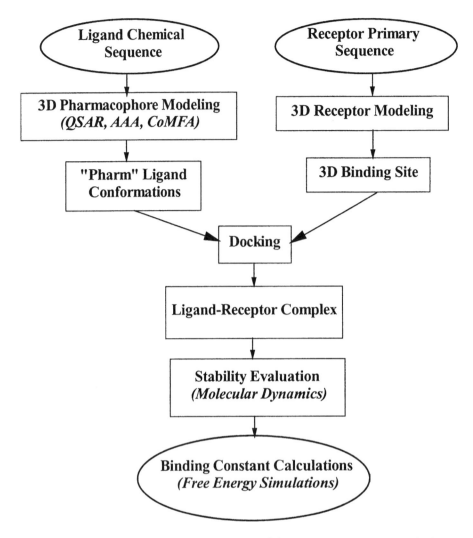

FIGURE 3 The flow chart of theoretical modeling of drug–receptor interaction and relation to experiment. The objects of theoretical investigation are in the rectangles, and the experimental information is in the ovals.

Molecular modeling methods use the experimental observations as input data to programs (Figure 3). The ultimate goal of molecular modeling as a pharmacological and medicinal chemistry tool is to predict, in advance of any laboratory testing, the biological activity of novel compounds. In general, there are two broad approaches toward this goal. First, if the receptor structure has been characterized by either x-ray crystallography or NMR, computer graphics methods can be utilized to design novel compounds based on complementarity between the surfaces of the ligand and of the receptor binding site. Energy calculations may be applied afterwards to quantify and refine the results of molecular graphics experiments. Second, if the receptor structure is unknown, one is left to infer the shape of the receptor binding site. This is, in fact, a more general case. Several negative imaging receptor mapping techniques have been developed, such as conformational search, active analog approach (AAA),[5] and, more recently, comparative molecular field analysis (CoMFA).[6] The idea of these approaches is to construct a pharmacophore that integrates all important structural and physicochemical features of the active compounds required for their successful binding to the receptor. Thus, these methods afford prediction of active (pharmacophoric) conformations for each of the biologically active ligands. However, although the pharmacophore itself has a significant predictive power as far as rational drug design is concerned, the design can be greatly facilitated if the receptor model can be predicted or derived independently.

3D macromolecular modeling is perhaps the most challenging example of molecular modeling. Despite enormous efforts to decipher the stereochemical code that determines 3D folding of protein structure from the primary structure, the problem is still unresolved. Only in a few cases were the folded protein structures successfully designed de novo; these usually include aggregates of well-defined secondary structures such as, for example, the four-helical bundle protein, Felix.[7] In the lack of the computerized "protein folding" algorithms, several knowledge-based 3D macromolecular modeling techniques have been developed. All of them require information about the primary sequence of the modeled protein and also utilize the structural information on related proteins whose structures are obtained from the Brookhaven protein database.[8] The most functionally and structurally related protein serves as a template that is gradually modified to develop the new model. Recent successes in deciphering primary sequences of many receptors and the development of several computerized protein modeling tools enabled researchers to devise putative 3D models of the receptors.

The ligand-structure–based methods and 3D protein structure modeling methods considered above afford independent generation of 3D models of ligands (in their pharmacophoric, i.e., receptor-bound, conformations) and their receptors. At this point, the ligand should be brought into the receptor binding to generate a model of ligand–receptor complex. This is achieved via a special computer modeling routine called "docking", whereby the ligand is literally "docked" into the binding site either manually or in an automated way. Thus, docking completes the circle of computer representation of the natural process of ligand–receptor interaction.

One can see that, at this point, only two types of experimental data are used for receptor modeling: the ligand chemical structure (from which one generates the 3D pharmacophore conformation) and the primary receptor sequence (which is used to generate the 3D receptor model). The interaction between ligands and their receptors is clearly a dynamic process. Once the static model of ligand–receptor interaction has been obtained, the stability of ligand–receptor complexes should be evaluated by the means of molecular dynamics simulations.[9] The most important property of the receptors, however, is their ability to discriminate ligands on the basis of their chemical structure; we quantify this as ligand binding constants. The ability to reproduce binding constants, or at least their relative order, is the most sensitive test of any putative receptor model. The technique that, in principle, can provide such data is the free-energy simulations/thermodynamic cycle approach, a method that relies on full atomistic representation of both ligand and receptor in the molecular model.[10] This approach is very robust but requires substantial computational resources that limit its practical applicability for drug design. In several reported cases, however, it was able to reproduce accurately the experimental binding constants.[11,12] The whole modeling process, including tools used for evaluation and refinement of the putative ligand–receptor model, is summarized in Figure 3.

The successful, realistic modeling of ligand–receptor interactions can only be based on a combined approach incorporating several computer-assisted modeling tools. As outlined previously, these tools must include ligand-based "negative image" receptor modeling, primary structure-based "real" receptor modeling, docking, structure refinement (molecular dynamics), and relative binding constant calculations. We will discuss these techniques in the order in which one may proceed while working on the problem of ligand–receptor interaction.

Computational Approaches to Modeling Ligand–Receptor Interactions

Ligand-Based Receptor Modeling

The application of computers and computerized tools in drug design started from the development and active use of the quantitative structure–activity relationships (QSAR) approach. This approach correlates biological activity (or binding constants if available) of known compounds with their measurable or calculable physicochemical and structural parameters. The typical QSAR equation looks as follows:

$$\log(1/C) = b_0 + \sum_i b_i D_i \,, \tag{1}$$

where C is drug concentration, b_0 is constant, b_i is regression coefficient, and D_i is a molecular descriptor such as Hansch π parameter (hydrophobicity), Hammett σ parameter (electron donating or accepting properties), molar refractivity (which describes the volume and electronic polarizability), etc. This approach has a long and, to some extent, successful history of applications[13] and will not be considered in more details here. Although not directly involved with receptor modeling per se, this approach has generated many useful and, in several cases, predictive QSAR equations and led to several documented drug discoveries.[13]

The advent of faster computers and development of new algorithms led in the late 1970s and early 1980s to the development and active use of 3D negative receptor image modeling tools such as the AAA method briefly mentioned earlier. In this approach, first proposed by G. Marshall,[5] the researcher infers the size, the shape, and some physicochemical parameters of the receptor binding site by modeling receptor ligands. The key assumption of this approach is that all active receptor ligands adopt, in the vicinity of the binding site, conformations presenting common pharmacophoric functional groups to the receptor. Thus, the receptor is thought of as a negative image of the generalized pharmacophore that incorporates all structural and physicochemical features of all active analogs superimposed in their pharmacophoric conformations. The latter are found in the course of the conformational search, starting from the most rigid active analog(s) and proceeding with more flexible compounds. Conformational restraints imposed by more rigid compound(s) with respect to the internal geometry of common functional groups are used to facilitate the search for more flexible compounds. The AAA, or similar approaches, has been successfully applied to negative image modeling of many receptors, yielding important leads for rational drug design. Most importantly, these models can be now incorporated into a common database providing the source of ligand structures for ligand–receptor docking studies (see below). Furthermore, the same database can be used to search for potential specific activity or cross-reactivity of independently designed or synthesized ligands by comparing them with known 3D pharmacophores.

In order to understand the details of calculations necessary for practical implementation of AAA, two methods should be described in more detail: molecular mechanics calculations and conformational searching.

Molecular Mechanics

Computational chemistry may be divided into two broad categories: quantum mechanics[14] and molecular mechanics.[15] The basis for this division depends on the incorporation of the Schrödinger equation or its matrix equivalent. Both quantum and molecular mechanics have inherent strengths and weaknesses, which are briefly reviewed. Arguments have been made that one computational approach is better than the other, but in the final analysis there are many examples one could cite where each method has seemed superior to the other. The competition that seemed to exist in earlier days between the advocates of quantum mechanics and molecular mechanics has subsided somewhat to the point that it is now widely recognized that both methods reinforce one another in the attempt to understand chemical and biological behavior at the molecular level. From a purely practical standpoint, the complexity of the problem, time constraints, computer size, and other limiting factors typically determine which method is feasible. It should be pointed out that neither quantum mechanics nor molecular mechanics can strictly be derived, and that there are numerous approximations underlying each method.

Quantum mechanics can be further divided into ab initio, semi-empirical, and density functional theory (DFT). While there may, in fact, be overlap between these three divisions, each of these terms has become associated with a specific computational paradigm. For example, ab initio is generally understood to refer to the Hartree–Fock (HF), Møller–Plesset (MP) perturbation, and/or configuration interaction (CI) paradigms. The ab initio approach uses the Schrödinger equation as the starting point, and in order to include electron correlation, post-HF calculations must be carried out. Various approximations are made in the construction of basis sets or in the optimization procedures used to derive their solution. On the other hand, the term semi-empirical is usually reserved for those calculations where families of difficult-to-solve integrals are replaced by equations and parameters, and then fit to experimental data. DFT may be classified as a separate method altogether, even though one could arguably classify it as a semi-empirical method. There has been a resurgence of interest in DFT calculations because unlike HF theory, electron correlation is taken into account.

Molecular mechanics is a mathematical procedure based on the Born–Oppenheimer approximation. The Born–Oppenheimer approximation states that electronic and nuclear motions can be considered separately, and for most chemical problems, this theorem works very well. In molecular mechanics, the focus is on the equilibrium position of the nuclei within an isolated molecule and the forces that are acting among the nuclei. As a first approximation, the nuclei may be viewed as if they were connected together by harmonic forces, much like the classic ball-and-spring model. This view of a molecule corresponds to diagonal terms in a force-constant matrix. It is a fundamental assumption that the electrons will find their optimum distribution within the molecular framework. Since electrons are not explicitly involved, the repulsive and attractive forces within a molecular environment are represented by potential energy terms.

Another fundamental assumption in molecular mechanics is that the total energy, E_{total}, of an isolated molecule is the summation of individual energy terms (see eq 2). Each of these specific potential energy terms represents some kind of molecular interaction. Stretching energy, E_{st}, bending energy, E_{bnd}, van der Waals, E_{vdw}, torsional energy, E_{tor}, and electrostatic energy, E_{es}, are examples of specific interactions. Other terms may also be added to account for specific interactions. Usually, these other energy terms connect some combination of interactions, such as the coupling of stretching and bending motions.

$$E_{total} = E_{st} + E_{bnd} + E_{vdw} + E_{tor} + E_{es} + E_{CT} \tag{2}$$

Initially, in the early days of molecular mechanics force-field development, the reproduction of molecular geometry and conformational energy differences or vibrational spectra could not be achieved with any reasonable level of accuracy. Through the addition of essential cross terms, nondiagonal terms in a force-constant matrix, it is now possible to calculate simultaneously and accurately molecular structure, potential energy, and vibrational spectra. A new generation of force fields is dramatically increasing the reliability of such calculations. The series of molecular mechanics programs developed by Norman L. Allinger and coworkers, including MM3,[16] and now MM4,[17] can reproduce molecular structure, energy (heats of formation), and vibrational spectra usually to within experimental error. MM3 and MM4 are among the most accurate molecular mechanics programs.

Typically, in these potential energy equations there are many constants or parameters that have been determined laboriously to fit a wide variety of experimental data or high-level quantum mechanical calculations. The combination of the parameters and equations for a specific class of compounds is referred to as the force field. For example, one would speak of a ketone, ester, or phosphate force field. The potential energy terms range from very simple harmonic expressions such as

$$E_{st} = \sum k \left(l_i - l_0 \right)^2 \tag{3}$$

to more complex terms like those found in MM3 (eq 4) or the new MM4 (eq 5), which include higher order terms. The more complex the mathematical formulation, the more flexibility the force field has in reproducing anharmonicity. In equations 3, 4, and 5, k represents the force-constant parameter, l is the actual bond length, and l_0 is the natural bond length. Any deviation in l_0, i.e., Δl, either from bond stretching or bond compression corresponds to an energy increase.

$$E_{st}^{MM3} = \sum 71.94 k \left(l_i - l_0 \right)^2 \left[1 - a \left(l_i - l_0 \right) + \frac{7}{12} a^2 \left(l_i - l_0 \right)^2 \right] \tag{4}$$

$$E_{st}^{MM4} = \sum 71.94 k \left(l_i - l_0 \right)^2 \left[1 - a \left(l_i - l_0 \right) + \frac{7}{12} a^2 \left(l_i - l_0 \right)^2 - \frac{1}{4} a^3 \left(l_i - l_0 \right)^3 \right.$$
$$\left. + \frac{31}{360} a^4 \left(l_i - l_0 \right)^4 \right] \tag{5}$$

It is important to note that MM3 and MM4 are fit to electron diffraction (r_g) bond lengths and microwave moments of inertia. In general, there are many different ways in which a bond length can be defined based on statistical averaging and the experimental method. Low-temperature x-ray values (r_a) for carbon–carbon bonds are approximately 0.002 Å shorter than the r_g distance measured by electron diffraction. Microwave (r_s, r_z, r_o) bond lengths are shorter than those obtained with MM3 and MM4. Moreover, ab initio bond lengths (r_e) are generally shorter than MM3 and MM4 because the bond length is calculated to correspond with the bottom of the potential energy well, not thermally averaged bond lengths. Many examples can be found in the scientific literature where theoretically derived bond lengths are compared to experimental data, with the erroneous conclusion that one computational method is superior to another, when in reality one is comparing differently defined bond lengths—analogous to comparing apples to oranges.

Unlike quantum mechanics, which will optimize the electron distribution within a molecular skeleton and define molecular bonds, molecular mechanics must be specifically "told" at the beginning of a calculation what atoms are bonded to what other atoms. For the most part, in dealing with organic structures this does not present a problem. Additionally, in molecular mechanics each hybridization state of an atom must be specifically defined, and these are referred to as atom types. Carbon atoms, for example, may exist as C_{sp}, $C_{sp}2$, or $C_{sp}3$, and they each may be positive, neutral, or negative. Thus, there are many different chemical environments in which we may say carbon exists, and each one of these states is given a specific atom type with its own set of parameters. Unfortunately, no two molecular mechanics programs use the exact same atom-typing scheme. Initially in the development of molecular mechanics, there was concern that each unique bonding combination would require a different bond length and force-constant parameters. For example, the carbon–carbon bond in ethane might be different from that in butane, cyclohexane, or cholesterol. As a first approximation, however, all carbon–carbon bonds in similar chemical environments can use the same force-constant parameters. Without this simplifying feature, molecular mechanics calculations would require new parameters for each new molecule.

Without explicit torsional energy terms taken into account, the energy differences found in *staggered* and *eclipsed* ethane could not be duplicated in early molecular mechanics calculations. The torsional terms in MM3 take the form of a truncated three-term Fourier series (eq 6). V_1, V_2, and V_3 are adjustable parameters that may be assigned either positive or negative values. Each unique dihedral angle has different torsional parameters. Like ethane, the reproduction of the $C_{sp}3$–$C_{sp}3$–$C_{sp}3$–$C_{sp}3$ dihedral angle of butane could not be accomplished without the inclusion of torsional energy terms. Knowing the $C_{sp}3$–$C_{sp}3$–$C_{sp}3$–$C_{sp}3$ dihedral angle is important because it plays a critical role in the energy differences in the axial-equatorial preferences for methylcyclohexane, which is the cornerstone of conformational analysis.

$$E_{tor}^{MM3} = \frac{V_1}{2}\left(1 + \cos\theta\right) + \frac{V_2}{2}\left(1 - \cos 2\theta\right) + \frac{V_3}{2}\left(1 + \cos 3\theta\right) \qquad (6)$$

Two of the three terms have intuitive physical meaning. The third term is a mathematical relationship indicative of ethane-like barriers that have threefold periodicity and may be attributed to what one would call steric interactions. Obviously, this term makes a major contribution to saturated hydrocarbons and their derivatives. The second term is useful for ethylene-like structures. The use of this term constrains unsaturated systems of the type $X–C_{sp}2–C_{sp}2–Y$ to have a minima with dihedral angles of 0° and 180°. The first-term is more difficult to grasp. The first-order cosine term may be attributed to dipole–dipole interactions not accounted for in the electrostatic terms. In MM4, two additional terms have been added but are presently only used for $C_{sp}2–C_{sp}2–C_{sp}2–C_{sp}2$ and $H–C_{sp}3–C_{sp}3–H$ (eq 7).

$$E_{tor}^{MM4} = E_{tor}^{MM3} + \frac{V_4}{2}\left(1 - \cos 4\theta\right) + \frac{V_6}{2}\left(1 - \cos 4\theta\right) \tag{7}$$

The assignment of the torsional parameters is the most time-consuming process in molecular mechanics. This can be readily understood considering a generalized torsion A–B–C–D. If there are 200 atom types defined in a molecular mechanics program, then there will be approximately $(200)^4$ dihedral angle combinations that must be determined.

Molecular mechanics parameterization is a major drawback to this method. In order to calculate the molecular features of any new molecule, someone had to develop the parameters for each unique combination of atoms. For most of the commonly encountered organic functional groups, MM3 has been parameterized, but there are still many complex heterocycles useful to pharmaceutical research. What happens when MM3 encounters a unique combination of atoms? A new routine has been programmed into MM3 that will allow it to pick parameters with built-in artificial intelligence. While these new parameters have not been specifically defined, they are a reasonable first approximation. Care should be exercised in using them. Typically, molecular mechanics parameterization is done in a trial-and-error manner, examining the effects on the structure and its energy and vibrational spectra, one parameter at a time. A new approach[18] using a program called PARTS allows the computer to simulate the human trial-and-error inspection method with greater accuracy and in significantly less time.

Molecular mechanics calculations are much faster—by orders of magnitude—than quantum mechanics calculations. Typically, low-level ab initio calculations are proportional to n^4, where n represents the number of orbitals. As the number of atoms increases, the orbitals dramatically increase. With molecular mechanics, however, the time is proportional to N^2, where N represents the number of atoms. Presently, for many molecules of biomedical interest ab initio calculations are just not as practical as molecular mechanics.

Conformational Search.   Most of the energy-minimization algorithms incorporated into molecular mechanics programs, including MM3 and MM4, were designed to seek the molecular configuration that corresponds to the *nearest* low-energy state. This means that if you have introduced into a molecular mechanics program the molecular coordinates that correspond to a *gauche*-like butane conformation, the final geometry will correspond to *gauche* butane, not *anti* butane. The minimizer will not jump across barriers to find the absolute lowest energy structure, known as the global minimum conformation. Thus, in order to calculate *anti* butane, one would need to begin with a structure that resembles the *anti* conformation. For structures with one or two flexible dihedral angles, it is possible using molecular mechanics and quantum mechanical programs to search a part of conformational space by forcing the dihedral angle to move from 0° to 360° in some defined increment. This is referred to as a dihedral angle driver search.

TABLE I Pharmacological analysis of DHX and related compounds.

Drug	D_1 affinity $K_{0.5}$ (nM)	D_2 affinity $K_{0.5}$ (nM)	Adenylate Cyclase EC50 (nM)	Max. Stimulation of Adenylate Cyclase (% vs. DA)
Dopamine	267	36	5000	100
(+)DHX	2.3	43.8	30	120
SKF89626	61	142	700	120
SKF82958	4	73	491	94
A70108 [A]	0.9	41	1.95	96
A77636 [B]	31.7	1290	5.1	92
cis-DHX	$>10^3$	$>10^3$	$>10^4$	17
N-Propyl-DHX	326	27	$>10^4$	36
N-Allyl-DHX	328	182	$>10^4$	32
Ro 21-7767	477	61	$>10^4$	22
N-benz-5, 6-ADTN	$>10^3$	$>10^3$	$>10^3$	38
N-benz-6, 7-ADTN	$>10^3$	335	$>10^4$	25

In the case of butane, and with modern computational chemistry software, this is relatively easy to carry out. We know that *anti* butane is more stable than *gauche* butane, but we may rapidly discover that our chemical intuition cannot keep pace with the myriad conformational possibilities in molecules of biomedical interest that typically have much more than one torsion angle of interest. Since many physical and biological properties may be attributed to a statistical average of all the major conformations, there must be some means available to sample or search all of the conformational space. Programs have been created that will allow users to search conformational space. The underlying methods may be divided into systematic searching and random searching. Like all methods, there are strengths and weakness. The AAA used in pharmacophore design is based on systematically searching conformational space at specific increments. Most conformational programs require significant computer resources.

Practical Application of Active Analog Approach

For illustration, let us consider the application of AAA to an important pharmacological class of dopaminergic ligands. The analysis of pharmacological activity of ligands leads to their classification as active or inactive (Table 1 and Figure 4).[3] Six compounds were chosen as active based on three criteria: they had high affinity for the D_1 receptor ($K_{0.5} < 300$ nM); they could increase cAMP synthesis in rat striatal membranes to the same degree as dopamine; and this increase could be blocked completely by the D_1 antagonist SCH23390. The compounds that met these criteria were dopamine, DHX, SKF89626, SKF82958, A70168, and A77636. As is shown in Table 1, all of these compounds caused similar (complete) activation of dopamine-sensitive adenylate cyclase in this preparation and had $K_{0.5}$ values ranging from 267 nM (dopamine) to 0.9 nM (A70108).

The molecular modeling studies with this set of dopaminergic ligands involved the following analytical steps: (1) The tentative pharmacophoric elements of the D_1 receptor were determined based on known structure–activity relationships. (2) A rigorous conformational search on the active compounds was performed to determine their lowest energy conformation(s). (3) Conformationally flexible superimposition of these compounds was done to determine their common (pharmacophoric) conformation. (4) Similar conformational analyses were performed for inactive compounds, and the superimposition of inactive compounds in pharmacophoric conformations with the active compounds was made to determine steric limitations in the active site. Where appropriate, the geometry of each inactive molecule was obtained by modifying the chemical structure of the relevant active analogs followed by energy minimization of the resulting structure. (5) Finally, an evaluation was made of excluded receptor volume and shape as spatial equivalents of the volume and shape of the pharmacophore. All molecular modeling studies were performed with the multifaceted molecular modeling software package SYBYL (Tripos Associates Inc., St. Louis, Mo.; version 5.5).

Based on the experimental structure–activity relationship data, the following functional groups of agonists were defined as key elements of the D_1 agonist pharmacophore (Figure 4): the two hydroxyl groups of the catechol ring, the nitrogen, and (except for dopamine) an accessory hydrophobic group (e.g., the aromatic ring in dihydrexidine or SKF82958). Thus, the task of molecular modeling analysis was to identify a pharmacophoric conformation for each compound where these key pharmacophoric elements were spatially arranged in a way similar for all active compounds.

Lowest Energy Conformations of Active Compounds: Construction of the Pharmacophore

The evaluation of the D_1 agonist pharmacophore was based on the following three-step routine. Step I: Perform conformational search on each of the agonists to identify their lowest energy conformation(s). Step II: Find common low-energy conformations for all of the compounds; the commonality was assessed by comparing the distances between each of the hydroxyl oxygens and the nitrogen, and the angle between the planes of the catechol ring and the accessory ring. Step III: Superimpose all of the agonists in their most common conformations using dihydrexidine as a template compound by superimposing equivalent pharmacophoric atoms of all the agonists and those of DHX.

D_1 Receptor Mapping and Agonist Pharmacophore Development

The "pharm" configurations of the active molecules also were used to map the volume of the receptor site available for ligand binding. The steric mapping of the D_1 receptor site, using the MVolume routine in SYBYL, involved the construction of a pseudoelectron density map for each of the active analogs superimposed in its pharmacophore conformation. A union of the van der Waals density maps of the active compounds defines the receptor excluded volume.[5]

The essential feature of AAA is a comparison of active and inactive molecules. A commonly accepted hypothesis to explain the lack of activity of inactive molecules

Active Compounds

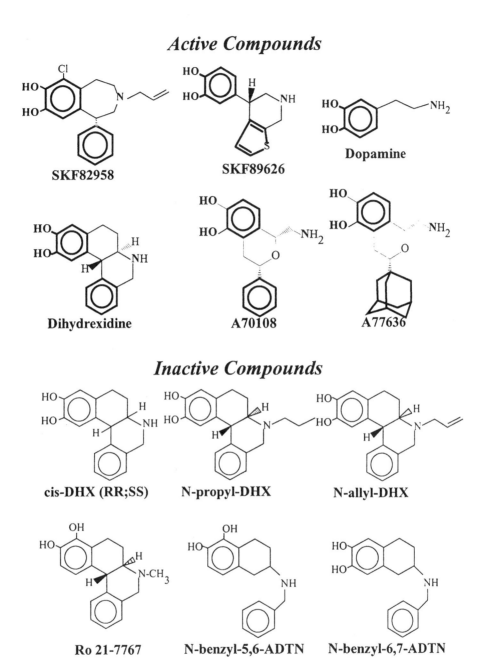

SKF82958

SKF89626

Dopamine

Dihydrexidine

A70108

A77636

Inactive Compounds

cis-DHX (RR;SS)

N-propyl-DHX

N-allyl-DHX

Ro 21-7767

N-benzyl-5,6-ADTN

N-benzyl-6,7-ADTN

FIGURE 4 The chemical structures of the ligands used in the molecular modeling study of the D_1 dopamine receptor. The ligands were divided into two groups (active and inactive) based on their pharmacological properties. The hypothesized pharmacophoric elements are shown in bold.

that possess the pharmacophoric conformation is that their molecular volumes, when presenting the pharmacophore, exceed the receptor excluded volume. This additional volume apparently is filled by the receptor and is unavailable for ligand binding; this volume is termed the receptor essential volume.[5] Following this approach, the density

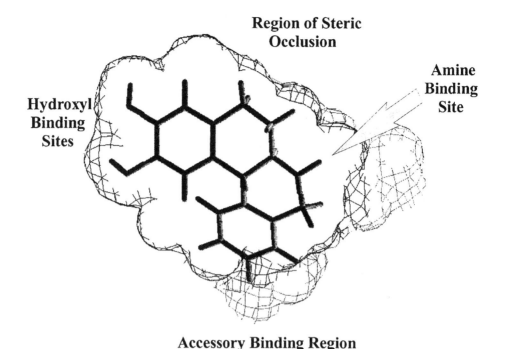

Region of Steric
Occlusion

Amine
Binding
Site

Hydroxyl
Binding
Sites

Accessory Binding Region

FIGURE 5 Excluded volume for the D_1 agonist pharmacophore. The mesh volume shown by the black lines is a cross section of the excluded volume representing the receptor binding pocket. Dihydrexidine (see text) is shown in the receptor pocket. The gray mesh represents the receptor essential volume of inactive analogs. The hydroxyl binding, amine binding, and accessory regions are labeled, as is the steric region.

maps for each of the inactive compounds (in their "pharm" conformations superimposed with those of active compounds) were constructed; the difference between the combined inactive compounds density map and the receptor excluded volume represents the receptor essential volume. These receptor-mapping techniques supplied detailed topographical data that allowed a steric model of the D_1 receptor site to be proposed.

These modeling efforts relied upon dihydrexidine as a structural template for determining molecular geometry because not only was it a high-affinity full agonist, but it had limited conformational flexibility relative to other more flexible, biologically active agonists. For all full agonists studied (dihydrexidine, SKF89626, SKF82958, A70108, A77636, and dopamine) the energy difference between the lowest energy conformer and those that displayed a common pharmacophore geometry was relatively small (<5 kcal/mol). The pharmacophoric conformations of the full agonists were also used to infer the shape of the receptor binding site. Based on the union of the van der Waals density maps of the active analogs, the excluded receptor volume was calculated. Various inactive analogs (partial agonists with D_1 $K_{0.5} > 300$ nM) subsequently were used to define the receptor essential volume (i.e., sterically intolerable receptor regions). These volumes, together with the pharmacophore results, were integrated into a 3D model estimating the D_1 receptor active site topography (see Figure 5).

Figure 5 represents the typical result of the application of the active analog approach. Based on the steric description of essential receptor volume, new ligands can be designed that fit a geometrical description of the pharmacophore. Indeed, recent research in our group based on AAA led to the design of a novel, highly selective dopamine agonist.

Comparative Molecular Field Analysis (CoMFA)

This method is one of the most recent developments in the area of ligand-based receptor modeling. This approach combines traditional QSAR analysis and three-dimensional ligand alignment central to AAA into a powerful 3D QSAR tool. CoMFA correlates 3D electrostatic and van der Waals fields around sample ligands typically overlapped in their pharmacophoric conformations with their biological activity. This approach, first proposed only several years ago,[6] has already been successfully applied to many classes of ligands.

CoMFA methodology is based on the assumption that, since in most cases the drug–receptor interactions are noncovalent, the changes in biological activity or binding constants of sample compounds correlate with changes in electrostatic and van der Waals fields of these molecules. In order to initiate the CoMFA process, the test molecules should be structurally aligned in their pharmacophoric conformations; the latter are obtained using, for instance, AAA described earlier. After the alignment is fixed, the steric and electrostatic fields of all molecules are sampled with a probe atom, usually carbon sp3 bearing +1 charge, on a rectangular grid that encompasses structurally aligned molecules. The values of both van der Waals and electrostatic interaction between the probe atom and all atoms of each molecule are calculated in every lattice point, using the force-field equation described earlier, and entered into a CoMFA QSAR table. Thus, this table contains thousands of columns, which makes it difficult to analyze the results; however, application of special multivariate statistical analyses routines, such as partial least-squares analysis, cross-validation, and boot-strapping, ensures the statistical significance of the final CoMFA equation. A cross-validated R^2 (q^2) that is obtained as a result of this analysis serves as a quantitative measure of the quality of the final CoMFA model. The statistical meaning of q^2 is different from that of the conventional R^2; a q^2 value of greater than 0.3 is considered significant.

Our experience with application of this important 3D QSAR method provides an important example that illustrates how the integration of teaching and research in the educational environment may lead to improvements in scientific methodology. Incidentally, we have found recently that in several cases we could not reproduce the q^2 values reported in the literature or even obtained in our own laboratory. We first observed this problem during the laboratory sessions of the introductory molecular modeling class taught by the first author of this paper at the University of North Carolina. All students were given the same set of compounds, ten D_1 dopamine receptor ligands. However, the final q^2 values differed by up to 0.5 q^2 units even when all students were finally given the same molecular database containing D_1 dopamine receptor ligands. Puzzled by this result, we have examined closely each student's report and found that the only difference between the analyses was the orientation of rigidly aligned molecules on the student's terminal. Thus, we have discovered a major flaw in the CoMFA methodology: The fact that the key result of the method (q^2) depends on the user-controlled orientation of the molecular aggregate on a computer terminal makes the whole technique user-biased, which is generally not acceptable for a computer algorithm.

Our investigation of this phenomenon in greater detail using several sets of model compounds has confirmed this initial observation. We found that for all data sets the q^2 value was sensitive to the orientation of rigidly aligned molecules on the computer terminal and could vary with the orientation by as much as 0.5 q^2 units. In order to circumvent this problem, we have developed a new CoMFA region optimization routine, which we call cross-validated q^2 guided region selection (q^2GRS).[19] This new method eliminates irrelevant variables in partial least-squares analysis and concentrates probe atoms in the regions where changes in steric and electrostatic field values can be best correlated with changes in biological activity data. We have shown that the application of this new routine to several sets of model compounds leads to reproducible, high q^2 values that did not depend on the orientation of molecular aggregates on the terminal. Thus, the problem that was first discovered in the teaching laboratory led to new method development and significant improvements in the CoMFA methodology.

Structure-Based Drug Design

Ligand-based methods of drug design, reviewed above, provide only limited information about the stereochemical organization of the receptor binding site. The detailed knowledge of the active-site geometry can be obtained experimentally by means of x-ray crystallography or NMR; unfortunately, despite rapid progress in practical applications of these techniques, the structures of very few receptors and ligand–receptor complexes are available experimentally. In the absence of any experimental information on the 3D structure of many receptors, predictions from primary sequence remain the only source of generating the receptor structure. Obviously, the knowledge of the active-site geometry significantly facilitates the design of new drugs by providing real spatial and chemical limitations on the structures of newly designed molecules. We will briefly review first the methods of molecular simulations applicable to the analysis of known ligand–receptor complexes and then the current state of the art in the area of modeling G-protein–coupled receptors, which include dopamine receptors. Excellent reviews of methods and successes in the area of structure-based drug design were published recently.[20,21]

Ligand-Receptor Docking

In this approach, the ligand is brought (manually or in some automated way) in the vicinity of the binding site and oriented so that electrostatic and van der Waals interactions between ligand and receptor (that correspond to Coulomb and dispersion terms, respectively, in molecular mechanics energy expressions) are optimized. Pharmacophoric conformation of ligands is usually used in this study. To the best of our knowledge, all commercially and noncommercially available docking programs currently allow docking of rigid ligand into rigid receptor; i.e., no conformational change of either receptor or ligand is permitted as the latter approaches the former. Clearly, this is not the way the interaction occurs in nature since both ligand and receptor are relatively flexible molecules that adjust their conformation to each other, in the process of binding, to maximize steric and chemical complementarity. However, current programs may not afford flexible docking calculations since these require simultaneous fulfillment of three main criteria: fast graphics, fast graphics–CPU communications, and an extremely fast CPU. Further description of one of the popular docking algorithms can be found in reference 22.

Molecular Dynamics

Once the model of ligand–receptor complex is built, its stability should be evaluated. Simple molecular mechanics optimization of the putative ligand–receptor complex leads only to the identification of the closest local minimum. However, molecular mechanics treatment of molecules lacks two crucial properties of real molecular systems: temperature and, consequently, motion. Molecular dynamics studies the time-dependent evolution of coordinates of complex multimolecular systems as a function of intermolecular and intramolecular interactions. Because simulations are usually performed at normal temperature (~300 K), relatively low energy barriers, on the order of kT (0.6 kcal), can be easily overcome. Thus, if the starting configuration of the whole system (i.e., drug–receptor complex) resulting from docking is separated from a more stable configuration by such a low barrier, molecular dynamics will take the system over the barrier. Molecular dynamics may identify more stable, therefore more realistic, conformational states of ligand–receptor complex. Furthermore, it may provide unique information about conformational changes of the receptor due to ligand binding. It may shed light on the intimate mechanisms of receptor activation that currently may not be studied by any other technique. Unfortunately, due to the inherently very short elementary simulation step size, ~2 femtoseconds, this technique is presently limited to relatively short total simulation times, on the order of hundreds of picoseconds to nanoseconds. These limitations are mainly due to computer power still being inadequate for significantly longer simulation times.

Binding Constant Calculation

The combination of free energy simulations (FES) and the thermodynamic cycle (TC) approach is the most promising modern technique that allows one to calculate ligand–receptor binding constants by simulating ligand–receptor interaction.[10] Experimentally, free energies from the binding of two ligands to the same receptor are evaluated in two independent experiments, according to the following scheme:

$$\textit{Ligand 1} + \text{Receptor} \Leftarrow \Delta G^{\circ}_{1} \Rightarrow \textit{Ligand 1}/\text{Receptor}$$

$$\textit{Ligand 2} + \text{Receptor} \Leftarrow \Delta G^{\circ}_{2} \Rightarrow \textit{Ligand 2}/\text{Receptor},$$

In order to calculate relative binding constants of two ligands, *Ligand 1* and *Ligand 2*, we construct the following cyclic scheme:

$$\textit{Ligand 1} + \text{Receptor} \Rightarrow \Delta G^{\circ}_{1} \Rightarrow \textit{Ligand 1}/\text{Receptor}$$
$$\Downarrow \qquad\qquad\qquad\qquad \Downarrow$$
$$\Delta G^{\circ}_{3} \qquad\qquad\qquad \Delta G^{\circ}_{4}$$
$$\Downarrow \qquad\qquad\qquad\qquad \Downarrow$$
$$\textit{Ligand 2} + \text{Receptor} \Rightarrow \Delta G^{\circ}_{2} \Rightarrow \textit{Ligand 2}/\text{Receptor},$$

where ΔG°_1 and ΔG°_2 correspond to binding free energies of *Ligand1* and *Ligand2*, respectively, and ΔG°_3 and ΔG°_4 correspond to free energy of formal transformation of the chemical structure of *Ligand 1* into chemical structure of *Ligand 2* in solution and in the binding site, respectively. From a thermodynamics viewpoint, the preceding scheme represents a closed thermodynamic cycle that consists of four transformations: binding of *Ligand 1* in solution to a receptor (ΔG°_1); chemical transformation of the bound *Ligand 1* to the bound *Ligand 2* (ΔG°_4); dissociation of *Ligand 2* from the receptor ($-\Delta G^\circ_2$); and chemical transformation of *Ligand 2* to *Ligand 1* in solution ($-\Delta G^\circ_3$). Using the TC relationship, $\Delta\Delta G^\circ$cycle = 0, one obtains the difference in the free energy of binding of *Ligand 1* and *Ligand 2*, i.e., $\Delta G^\circ_1 - \Delta G^\circ_2$). Thus,

$$\Delta\Delta G^\circ \text{ cycle} = \Delta G^\circ_1 + \Delta G^\circ_4 - \Delta G^\circ_3 - \Delta G^\circ_2 = 0, \text{ so}$$

$$\Delta\Delta G^\circ \text{ binding} = \Delta G^\circ_1 - \Delta G^\circ_2 = \Delta G^\circ_3 - \Delta G^\circ_4$$

The latter two free energies of chemical transformation are computed in the course of FES. The advantage of this approach is that we avoid calculations of ligand–receptor binding free energy per se, i.e., ΔG°_1 and ΔG°_2, which would be extremely computationally intensive, without sacrificing the theoretical rigor of calculation of binding constants from molecular simulations of ligand-receptor interaction.

Homology Model Building of G-Protein–Coupled Receptors

The preceding approach is applicable only if the receptor structure is known. In many cases it may be obtained by homology model building (reviewed recently in reference 23). We will briefly describe the application of this method to the important pharmacological class of G-protein–coupled receptors (GPCRs), which includes D_1 receptors.

Earlier attempts to predict the structure of GPCRs were limited to very general seven-transmembrane-cylinder models based on the easily identifiable seven stretches of largely hydrophobic amino acid residues in the primary sequence of these receptors. Recent determination of a middle-resolution crystallographic structure of bacteriorhodopsin,[24] a seven-helix transmembrane protein with very limited sequence homology to GPCRs, generated numerous attempts to model the GPCRs at atomic resolution. The theoretical basis for these attempts was the proposed structural similarity between bacteriorhodopsin and the GPCRs, despite the extremely low sequence homology between the receptors and bacteriorhodopsin (6–11%);[25] it is important to note that in the general case of protein homology modeling, these considerations alone would exclude bacteriorhodopsin as a possible candidate template structure.[26] Yet, the compelling interests and needs of investigators have led to several such models during the last few years.[27] Most of the studies used the helix arrangement of bacteriorhodopsin as a template for model building. In the majority of these studies, only the transmembrane portion of the receptor was modeled; this is justified by the fact that only the transmembrane fragment of bacteriorhodopsin was determined experimentally,[24] and that the binding site is presumed to be buried in the interior of the receptor in a location approximately analogous to the retinal moiety in bacteriorhodopsin. Recently, a low-resolution projection map of bovine rhodopsin was published. In contrast to bacteriorhodopsin, bovine rhodopsin shares modest sequence homology with most of the GPCRs and is a GPCR itself.[28] This map shows basically the same arrangement of seven transmembrane helices as in bacteriorhodopsin, except that the orientation and mutual tilt of the helices is somewhat different. These data provide additional evidence that GPCRs share a seven-transmembrane-helix motif and similar arrangements of the helices.

Basic Principles of Protein Homology Model Building as Applied to GPCRs

The template-based protein homology model building process involves three major steps: (1) sequence alignment of the target (modeled) protein with the template, (2) target structure generation based on template modification, and (3) final model analysis and refinement. In general, proteins with sequence homology to the target protein of less than 30% are not selected as templates,[26] whereas, as mentioned, the sequence homology between bacteriorhodopsin and any of the GPCRs is in the range of 6–11%.[25] As a result, traditional sequence alignment methods are not capable of generating any unambiguous alignment. For the alignment of GPCRs and bacteriorhodopsin, a different approach is used. The analysis of the properties of the side chains reveals that in both bacteriorhodopsin and GPCRs, seven stretches of mostly hydrophobic amino acids can be easily identified in their primary sequences. It was shown for bacteriorhodopsin[24] and postulated for GPCRs[25] that the residues in these stretches constitute seven transmembrane α-helical regions. Thus, the sequence alignment of the target (GPCR) to the template (bacteriorhodopsin) is usually deduced from the alignment of these hydrophobic stretches.

Once the alignment of a GPCR to the bacteriorhodopsin template has been decided, the next step is to assign the GPCR residues to the bacteriorhodopsin structure template. This means that a residue that is aligned against a particular residue of bacteriorhodopsin must assume the position of that residue in the 3D structure of bacteriorhodopsin. This assignment is usually done in one of three ways. In the first, the most simple approach, the side chains of bacteriorhodopsin residues in the 3D bacteriorhodopsin structure are mutated into corresponding residues of a GPCR.[29] Thus, a chimerical protein is generated that has the backbone of bacteriorhodopsin and the side chains of the target protein. This protein is then subjected to structure refinement using molecular mechanics and/or molecular dynamics simulations to relieve bad steric contacts. This approach generates a GPCR structure in a straightforward manner that can be easily automated for practically any GPCR model building.

A different approach was employed in several recent publications.[30,31] Here, the model generation starts from multiple-sequence alignment of homologous GPCRs such as, for example, sequences of cationic amine receptors. Particular care is taken to align residues conserved across all the receptors; it is suggested that the conserved residues play a structural role and are responsible for maintaining similar 3D contacts in all the receptors. Based on this alignment, the boundaries of each helix are determined. The helices are then built individually, similar to the second structure-generation method described previously. However, the structure of bacteriorhodopsin and, more recently, a low-resolution projection map of bovine rhodopsin are used as general structural motifs with respect to helix arrangement. The actual helix orientation and register is selected based on several considerations, such as: the conserved residues in the spatially adjacent helices must make contact with each other; no polar residues found in postulated transmembrane domains should be exposed to the lipid phase but, rather, should face the interior of the receptor and make contact either with each other or with the backbone.

Active-Site Modeling

Early molecular modeling studies of GPCRs[32] have been primarily concerned with finding reasonable orientations, registers, and amino acid contacts between seven-transmembrane helices. However, the major feature of each particular receptor that distinguishes it from other receptors is the ability to bind selectively the ligands specific to that receptor. In early studies a single ligand, most often a native neuro-mediator (e.g., dopamine), was docked into the postulated active site, mainly to demonstrate that it could be reasonably well accommodated within the active site. However, it has been recognized[33,34] that accurate receptor modeling can only be realized if the homology model building is combined with the results of ligand-based pharmacophore receptor modeling (discussed earlier for dopaminergic ligands). Thus, the active site in the proposed model must be able to accommodate all known active receptor ligands in their pharmacophoric conformations.

It is important to realize that the accurate prediction of the 3D structure of GPCRs is an extremely difficult task, and with the present relatively meager knowl-edge of how to predict protein structure from sequence in general, it is almost impossible to claim that anyone has devised an accurate model of any GPCR. The most advanced models obtained so far have been able to combine and explain various pieces of information about these receptors coming from homology model building, site-directed mutagenesis, and pharmacophore modeling. The real utility of these models for drug design, however, can only be assessed if they possess the power to predict structures of novel receptor ligands of high affinity and predictable functional characteristics. Nevertheless, modeling receptor structure is still a very useful excur-sion that, in general, allows the researcher to attempt to think about the receptor protein in three dimensions and relate available experimental information from pro-tein mutagenesis and ligand-binding studies to the receptor 3D organization. The growing body of experimental information on receptors and their ligands, combined with rapid developments in the area of database searching and receptor-based drug design, promises that the quality of the models ultimately will be improved to afford generation of new lead compounds.

Conclusions

In this chapter we have reviewed several methods of computer-aided drug design. We have shown that successful molecular modeling and the design of new drugs can be achieved by integration of different computational chemistry and molecular mod-eling techniques including quantum mechanics, molecular mechanics, macromolec-ular modeling, and specialized drug design approaches such as the active analog approach. Due to the rapidly growing computational power of specialized hardware and software, molecular modeling has become an important integral part of multi-disciplinary efforts to design and synthesize new potent pharmaceuticals. Practical knowledge of these techniques and their limitations is a necessary component of formal training of modern chemists.

References

1. Adopted with modifications from: D. B. Boyd, in Lipkowitz, K. B., Boyd, D. B., Eds.; *Reviews in Computational Chemistry;* VCH Publishers: New York, 1991.

2. Lipkowitz, K. B.; Boyd, D. B., Eds.; *Reviews in Computational Chemistry;* VCH Publishers; New York, V. 1–7, 1991–1996.

3. Mottola, D. M.; Laiter, S.; Watts, V. J.; Tropsha, A.; Wyrick, S. W.; Nichols, D. E.; Mailman, R.; "Conformational analysis of D1 dopamine receptor agonists: pharmacophore assessment and receptor mapping", *J. Med. Chem.* 1996, *39*, 285–296.

4. Mailman, R. B.; Nichols, D. E.; Tropsha, A.; "Molecular Drug Design and Dopamine Receptors"; in K. Neve and R. Neve, Eds., *Dopamine Receptors* (in press).

5. Marshall, G. R.; Barry, C. D.; Bosshard, H. E.; Dammkoehler, R. A.; Dunn, D. A.; "The conformational parameter in drug design: the active analog approach"; *ACS Symposium Series* 1979, *112*, 205–226.

6. Cramer, R. D., III; Patterson, D. E.; Bunce, J. D.; "Comparative Molecular Field Analysis (CoMFA). 1. Effect of Shape on Binding of Steroids to Carrier Proteins"; *J. Am. Chem. Soc.* 1988, *110*, 5959–5967.

7. Hecht, M. H.; Richardson, J. S.; Richardson, D. C.; Ogden, R. C.; "De Novo Design, Expression, and Characterization of Felix: A Four-Helix Bundle Protein of Native-Like Sequence"; *Science* 1990, *249*, 884–891.

8. Bernstein, F. C.; Koetzle, T. F.; Williams, G. J. B.; Meyer, E. F.; Brice, M. D.; Rodgers, J. R.; Kennard, O.; Shimanouchi, T.; Tasumi, M.; "Protein Data Bank—Computer Based Archival File for Macromolecular Structure"; *J. Mol. Biol.* 1977, *112*, 535–542.

9. Karplus, M.; Petsko, G. A.; Molecular dynamics simulations in biology. *Nature* 1990, *347*, 631–637.

10. Tembe, T. L.; McCammon, J. A.; "Ligand-Receptor Interactions"; *Comput. Chem.* 1984, *8*, 281–283.

11. Hirono S.; Kollman P. A.; "Calculation of the relative binding free energy of 2'-GMP and 2'-AMP to ribonuclease T1 using molecular dynamics/free energy perturbation approaches"; *J. Mol. Biol.* 1990, *212*, 197–209.

12. Tropsha, A.; Hermans, J.; "Application of free energy simulations to the binding of a transition-state-analogue inhibitor to HIV protease"; *Protein Engng.* 1992, *51*, 29–34.

13. Boyd, D. B.; "Successes of Computer-Assisted Molecular Design"; in Lipkowitz, K. B., Boyd, D. B., Eds.; *Reviews in Computational Chemistry;* VCH Publishers; New York, V. 1–7, 1991–1995.

14. Hehre, J. W.; Radom, L; Schleyer, P.; Pople, J.; *Ab Initio Molecular Orbital Theory;* John Wiley and Sons, Inc.: New York, 1986.

15. Burkert, U.; Allinger, N. L.; *Molecular Mechanics;* American Chemical Society: Washington, DC, 1982.

16. Bowen, J. P.; Allinger, N. L.; in Molecular Mechanics: The Art and Science of Parameterization; *Reviews In Computational Chemistry;* Lipkowitz and Boyd, Eds., VCH, New York, 1991, vol. 2, pp 81—97.

17. Allinger, L.; Chen, K.; Lii, J. H.; "An Improved Force Field (MM4) for Saturated Hydrocarbons", *J. Comput. Chem.*, in press, 1996.

18. Liang; Fox, P. C.; Bowen, J. P.; "The Parameter Analysis and Refinement Toolkit System (PARTS) and its Application in MM3 Parameterization for Phosphine and its Derivatives;" *J. Comput. Chem.*, in press, 1996.

19. Cho, S. J.; Tropsha, A.; "Cross-Validated R2 Guided Region Selection for Comparative Molecular Field Analysis (CoMFA): A Simple Method to Achieve Consistent Results"; *J. Med. Chem.* 1995, *38*, 1060–1066.

20. Bugg, C. E.; Carson, W. M.; Montgomery, J. A.; "Drugs by Design"; *Scientific American*, Dec. 1993, 92–98.

21. Whittle, P. J.; Blundell, T. L.; "Protein Structure-Based Drug Design"; *Annu. Rev. Biophys. Biomol. Struct.* 1994, *23*, 349–375.

22. Meng, E. C.; Shoichet, B. K.; Kuntz, I. D.; "Automated Docking with Grid-Based Energy Evaluation;" *J. Comp. Chem.* 1992, *13*, 505–524.

23. Johnson, M. S.; Srinivasan, N.; Sowdhamini, R.; Blundell, T. L.; "Knowledge-based protein modeling"; *Crit. Rev. in Biochem. and Mol. Biol.* 1994, *29*, 1–68.

24. Henderson, R.; Baldwin, J. M.; Ceska, T. A.; Zemlin, F.; Beckman, E.; Downing, K. H.; "Model for the structure of bacteriorhodopsin based on high-resolution electron cryo-microscopy"; *J. Mol. Biol.*, 1990, *213*, 899–929.

25. Hibert, M. F.; Trumpp-Kallmeyer, S.; Bruinvels, A.; Hoflack, J.; "Three-dimensional models of neurotransmitter G-binding protein-coupled receptors"; *Mol. Pharmacol.* 1991, *40*, 8–15.

26. Chothia, C.; Lesk, A. M.; "The relation between the divergence of sequence and structure in proteins"; *EMBO J.* 1986, *5*, 823–826.

27. Donnelly, D.; Findlay, J. B. C.; "Seven-helix receptors: structure and modeling"; *Curr. Opinion in Struct. Biol.* 1994, *4*, 582–589.

28. Schertler, G. F. X.; Villa, C.; Henderson, R.; "Projection structure of rhodopsin"; *Nature*, 1993, *362*, 770–772.

29. Teeter, M. M.; Froimowitz, M.; Stec, B.; DuRand, C. J.; Homology modeling of the dopamine D2 receptor and its testing by docking of agonists and tricyclic antagonists. *J. Med. Chem.* 1994, *37*, 2874–2888.

30. Baldwin, J. M.; "The probable arrangement of the helices in G protein-coupled receptors"; *EMBO J.* 1993, *12*, 1693–1703

31. Jones, D. T.; Taylor, W. R.; Thornton, J. M.; "A model recognition approach to the prediction of al-helical membrane protein structure and topology"; *Biochem.* 1994, *33*, 3038–3049.

32. Trumpp-Kallmeyer, S.; Hoflack, J.; Bruinvels, A.; Hibert, M.; "Modeling of G-protein-coupled receptors application to dopamine, adrenaline, serotonin, acetylcholine, and mammalian opsin receptors"; *J. Med. Chem.* 1992, *35*, 3448–3462.

33. Nordvall, G.; Hacksell, U.; "Binding site modeling of the muscarinic m1 receptor: A combination of homology based and indirect approaches"; *J. Med. Chem.* 1993, *36*, 967–976.

34. Teeter, M. M.; Froimowitz, M.; Stec, B.; DuRand, C. J.; "Homology modeling of the dopamine D2 receptor and its testing by docking of agonists and tricyclic antagonists"; *J. Med. Chem.* 1994, *37*, 2874–2888.

Part V
Teaching Chemistry with Computers

18

Teaching Organic Chemistry On-Line— The Promise and the Reality

Carolyn Sweeney Judd
Robert G. Ford

Conducting an English lesson in the computer writing laboratory to improve sophomore organic chemistry laboratory reports led to an expanded project for also using writing for clarity as a learning tool in the organic chemistry lecture. By substituting conferences on networked computers for chalkboard lecture time after students have developed sufficient vocabulary, the class engages in common written discussion about problems, visible to all participants, with ample time for student responses and questions, and instructor follow-up. Anonymous passwords for conferencing and group work with computer tutorials additionally encourage student participation. The result is elevation of the plane of organic instruction from memorization to comprehension; the simplicity and cost efficiency of the method are ideal for community colleges or small colleges.

Introduction and Rationale

Many of us teach chemistry in small colleges or in small campus situations, where the sophisticated computer facilities of the large research institutions may be unavailable. Our departments may also be small, with only one or two full-time chemistry faculty members. As a result, many of us may feel that our entry into a world of innovative instruction, using technology, is limited. However, after almost three years of experience at an urban community college, one with a small chemistry department and limited resources, we know that just the opposite is true. Innovation with technology is made possible by locating the facilities that do exist and by creating partnerships with other disciplines on campus to share resources. We have developed a strategy to teach organic chemistry in a shared, interdisciplinary facility; our results suggest that by looking for the technology available, being willing to experiment, and then proceeding carefully, much can be done to enhance chemical education by using technology, even with limited resources.

Our method is not only feasible for small colleges, but it also leads such departments into, we feel, an advanced method of teaching—one our students labeled "a method for the 21st century". In the last three years, computer tutorial and simulation along with electronic conferencing among students and teachers have become part of the lecture component of a traditional organic chemistry course. The overall goal of this change has been to help students understand the content of the course more thoroughly, as they practice problems and examine reactions using tutorials and simulations and then explain what they've observed and understood to their teacher and their peers in a public electronic conferencing forum.

Electronic conferencing allows students to meet and learn from each other. Much of the class time is spent on electronic conferencing. In this forum, students respond to an instructor's prompt, writing in complete sentences for clarity, offering explanations of reactions being studied, and then examining interpretations from other students. Students' messages are often based on computer tutorials and simulations, along with their textbooks and notes. Because the classes are small (due to a limitation in size of the computer laboratory), students quickly meet each other, and electronic discussions spill over into oral discussions about course content.

The lecture, textbook, computer tutorial, and computer conferencing components of the class allow the course to become integrated, as a flowing system (an idea expressed by others such as Wilson, Redish, and McDaniel[1] and Reif and Morse[2]). However, it is the focus on electronic communication that is the heart of the method we present: The overall course content of the class is not produced for the students (by the instructor's knowledge, the textbook's information, and the tutorials' multiple, colorful screens). Instead, the students' electronic conferencing messages and the oral discussion allow members of the class to practice learning in the classroom, as they write and talk about the concepts of organic chemistry they might previously have only memorized. The emphasis is on the students and their learning, as they are encouraged to bring their doubts or lack of knowledge to the forefront without feeling conspicuous or "wrong". We believe our method reinforces critical thought about organic chemistry. We hope it supports similar thinking about other subjects, for that is what college is all about.

By integrating technology as a tool to facilitate learning in basic academic curricula, our program also enhances our students' computer skills, addressing a need that our urban community college students often face. We also address our own needs as faculty for new technology skills, for the integration of lecture, computer tutorial and simulation, and electronic conferencing forces the teacher to develop strategies for teaching never imagined previously. These needs relate to major general education goals; by focusing on shared, interdisciplinary facilities, they may be addressed even without the latest and greatest computer hardware and software.

Camaraderie has built up around the teachers who use the interdisciplinary laboratory. For the chemistry faculty member attempting to go it alone, this cadre of teachers is a literal lifeline when trying to teach using new technology. A network of interdisciplinary teachers has formed; our project is one example, but others are beginning as faculty become more skilled with the technical tools available. Such beginnings mirror the kinds of collaboration the laboratory supports for the students in the classes: Both groups—students and faculty—are committed to helping each other over the hard places in the techno-world, while using computers to advance their chosen subject matter. (See Johnson, Johnson, and Smith's arguments for visible cooperation among faculty.[3])

Procedure

This project developed from an intersection of unrelated but connectable events: a desire to teach organic chemistry more thoroughly by using technology, desires in the English department to use technology to restructure basic freshman composition, experiences with distance education bulletin boards for asynchronous student discussion, and school administrators strategizing about how to provide technology resources with limited funds. Although the particulars of the chemistry course and the interdisciplinary computer laboratory are tied to one school's experience, together they present an example of how limited resources can lead to collaboration and innovation.

Structural Issues: Creating and Using an Interdisciplinary Laboratory

These limited resources have been supported through an interdisciplinary facility initially designed by one department, but funded by many, and responding to the needs and interests of a variety of disciplines and faculty. This laboratory consists of two rooms, each with 20 Macintosh computers connected on a local area network with a file server. The laboratory contains a range of computer types, from lower to higher performance models. Laser printers, an LCD panel and projector, a laser disc player, CD players, and a scanner are also available. The computers have common word processing software, particular discipline-based software, and electronic conferencing software. The facility serves interdisciplinary clients in the academic and technical departments, with priority use for classes (about 70% of the time) and the remaining time for open individual student use.

The laboratory was created from several forces. A room of underused computers was available as a starting point. Also, particular faculty members in the English department showed interest in offering computer-based freshman courses. English faculty had been influenced by their experience with writing courses taught over the modem in the college's distance education program. Students own and maintain their own equipment to dial into the college's BBS. Teachers found that students benefited not only from exchanging essays with their teacher by modem, but also from exchanging messages through bulletin board software asynchronously. Such software helped create a sense of "class" for a course in which students participate at any time, day or evening. Discovering that synchronous software would allow students enrolled in a class at a particular time and place to have a similar experience, the department requested a laboratory to make such classes available to on-campus students who did not have the funds or knowledge of technology to participate in the distance education program. (See Twigg for a discussion of how changes in instructional technology are enhancing education in general.[4])

Crucially, the dean of instruction and school administration were interested in supporting new programs using computer technology for academic and technical instruction. Because the original computers "belonged" to a particular department, because the English budget could not support a laboratory by itself, and because of a college-wide interest in interdisciplinary projects (and the need to piece together funds from several departments to purchase an additional room of computers), the dean of instruction referred to the new laboratory as an "interdisciplinary" laboratory.

Although initially labeled "interdisciplinary", the laboratory has developed a more assertively interdisciplinary focus gradually and incrementally, with the individual departments, faculty, and laboratory staff creating what it now means: that faculty using the laboratory can experiment with using the computers and that the laboratory staff is willing to work on short- or long-term projects. Because the laboratory is funded from a central college source, using the laboratory does not cost department or faculty members anything except their time and interest. The English faculty member serving as laboratory director and adviser works with faculty interested in using the facility, offering them training, suggesting possible uses for the facility, and creating documentation for support.

Overall, the interdisciplinary focus means that the laboratory operates with a strong spirit of experimentation. Originally structured to serve the text-based interests of the English program, the laboratory began with limited software: word processing and file-sharing networking software, along with the first of two electronic conferencing programs. To attract other departments, the laboratory has also supported a range of additional applications: tutorials, simulations, purchased packages, textbook vendor's "freebies", and instructor-created applications. The goal of the laboratory is to suggest possibilities with the available software and to support instructor requests for additional ideas.

Regularly scheduled classes include English and chemistry, along with government, sociology, marketing, history, psychology, and English as a second language. Each teacher uses the laboratory and its network individually, requiring a laboratory staff versed in the idiosyncrasies of each teacher's uses and class needs. The laboratory has worked to provide sufficient orientation and printed materials and directions, without "wasting" valuable class time. (For examples of excellent documentation, see Oakley.[5]) Often, laboratory staff, faculty technology advisers, and course instructors try to integrate instructions in using the computers with content-based information to make every moment of orientation serve some instructional goal.

Along with this experimentation come the restraints under which the laboratory operates. After its creation from underused computers and pieced-together funds for additional workstations, the laboratory still does not operate under a lavish budget. It is by no means a "state-of-the art" facility, though improvements are made each semester and examples of more powerful equipment are available. However, these fiscal constraints have not been that constraining overall; they've meant that equipment must be used creatively and thoroughly. The laboratory's local area network is an example of this kind of use. Our college system is not yet totally connected to the Internet, e-mail, or the World Wide Web; the interdisciplinary laboratory will require additional funds to make such a connection totally functional. However, the local area network allows faculty and students to begin to experiment with how they might use file sharing, e-mail, and electronic conferencing within the laboratory to prepare for wider uses in the future.

Pedagogical Concerns: Teaching Organic Chemistry

The two-semester sophomore organic chemistry class meets twice a week, for one 3-hour lecture and one 3-hour chemistry wet laboratory. The normal class size is 15 to 25 men and women students, from many different backgrounds, cultures, and nationalities. This course is difficult for our students, as evidenced by the presence of many students who have tried the course before but have had to enroll again.[6]

Because of scheduling issues, the lecture portion of the class occurs in one 3-hour block. Even with a very active teacher at the chalkboard, the logistics of the class lead to very passive students sitting at their desks. Something is needed to get the students active also, to engage their minds during the lecture period. Oral discussions or recitations provide one solution, but they are often too embarrassing for many of our students, including women, minority students, non-native English-speaking students, as well as the shy students. These students face personal, cultural, and oral language-based obstacles before they will freely speak aloud. They are most likely silent in every class they take, especially in a frustrating course such as organic chemistry.

From the perspective of organic chemistry, this cooperative teaching venture began because of two particular problems, with a third following shortly thereafter. One problem concerned students' difficulties understanding hard-to-visualize reactions and molecules. The chemistry teacher wished to address this problem by using computer simulations and tutorials collected even before a facility for using them had been identified. Some programs had assisted the instructor in printing quizzes and tests. Still others begged for a facility for student access. With such a facility, students could practice what they had learned in lecture or from reading. Originally, the instructor's goal was to use a few basic programs to help students see aspects of organic chemistry not easily available from the textbook. To support this initial goal, the college purchased network licenses for three inexpensive chemistry programs: a drawing/nomenclature program, a spectra-simulator program, and a quiz program (see Appendix I for the software list). The chemistry instructor assumed that students would use these programs during open laboratory hours, outside of the official lecture/laboratory hours for the course. The creation of the new interdisciplinary laboratory allowed the chemistry department to use these licenses and programs more quickly than it would have otherwise (because it did not have to fund an entire laboratory).

Gradually, the chemistry department has acquired other software for the laboratory, available on only a limited number of computers: several databases of spectra, an organic tutorial, and some rotating 3D images of molecules. One of the best programs depicts mechanisms in motion— a vast improvement over the printed page, and much better than watching an instructor with waving arms attempting to show an SN2 mechanism. This software represents an excellent use of the computers: It provides explanations not possible on the printed page, and it doesn't grow tired, as a teacher might making something move. None of the software used is very expensive; some has come from the Internet freely dispensed.

The second problem concerned the increasingly dismal organic chemistry laboratory reports. The reports were unorganized, too long, lacked observations, and were almost devoid of any logical conclusions. As the chemistry instructor learned about the new laboratory and negotiated to have software installed and times reserved for student orientations, she learned that the director of the laboratory was also a member of the English faculty. Soon after the first orientation for the chemistry students, the two faculty members realized that addressing the problem of chemistry laboratory reports—though not an immediate part of the laboratory's responsibilities—would help create an interesting interdisciplinary project for the new facility.

The two developed a module on chemistry laboratory reports, with the English faculty member serving as writing consultant, examining samples of strong and weak laboratory reports, developing observations, looking at current student examples, and presenting the findings. A presentation covered appropriate guidelines for technical or scientific writing and argued for the importance of clarity, precision, accuracy, and completion. Examples, coming from the students' initial attempts at laboratory reports, focused on the reader's (here the chemistry teacher's, elsewhere potential clients') need for specific illustrations and directions. The writing lecture ended with the students writing introductions to the next week's chemistry laboratory experiment, with both the chemistry instructor and the writing consultant offering suggestions for improvement (each semester has used a slightly different, but related, procedure).

The result, even the first time, was a great success: The reports were shorter, containing observations (such as color, shape, and smell) and exhibiting conclusions based on sturdier logical thought. The chemistry instructor graded the papers and assigned the grade. To reinforce the writing lesson, 10% of the grade of the laboratory reports was based on writing clarity.[7] The writing consultant informally checked the papers, adding comments for the students. The initial conclusion was that the time spent on the writing lesson was really time well spent. From the examples of these initial and subsequent laboratory reports, we concluded that clarity of writing can lead to clarity of thought. (See Beall's "Report on WPI-NEACT [Worcester Polytechnic Institute—New England Association of Chemistry Teachers] Conference..." for F. Peter Boer's emphasis on the importance of communication skills in the chemistry curriculum.[8])

Originally, the use of the laboratory was to be limited to the out-of-class tutorials and simulations and for the semestral project with the writing consultant (a project that was really not connected to the computer laboratory at all). However, this initial foray, suggesting the importance of writing in teaching organic chemistry, combined with the proximity of other software in the laboratory, proved just a start.

The third problem for the chemistry instructor is that the students are literally overwhelmed with the material covered in an organic chemistry lecture. These are among the brightest of students, yet the seemingly unrelated details bury them. They become more disheartened instead of enlightened as the course enters the second semester. All of the memorization of arrows does not lead to real comprehension of the underlying principles for the majority of the students—and they certainly cannot explain those principles or summarize examples very clearly. The reputation of organic chemistry at our school and elsewhere is that it is an impossible task to master, a make-or-break course. Only blind luck about what to study will guarantee a satisfactory grade in the course.

So, the chemistry instructor and the laboratory director/writing consultant had an idea. Could writing help in lecture? If writing could improve the thinking in the laboratory reports, maybe it would also work in the lecture. Software for electronic conferencing was already available (deriving from the basic planning of the English department). The students and the instructor had already gotten their feet wet and were familiar with the computer laboratory, so learning new software did not seem too much of a burden. The computer laboratory was available, and the laboratory director was soliciting departments to use it to fulfill the "interdisciplinary" directives. Laboratory aides would be there to assist with questions about the computers, unsnarl the cables, and deal with the network. As the next semester schedule for the Macintosh laboratory was set, a section of organic chemistry appeared right after a freshman composition course on Tuesday afternoons: an unusual but appropriate juxtaposition for the new laboratory; to provide a compromise between the "old" and the "new", eventually half of the traditional lecture time was scheduled in the laboratory, with half in the traditional classroom.

Electronic Conferencing as a Teaching Strategy

A typical organic chemistry class in the laboratory begins as follows. Students open a conference for the day and see the first question posed by the instructor for that day's discussion. Often, the question involves mechanisms or spectral interpretation. Students are instructed to make their descriptions and analyses clear by using complete sentences. They send these responses to the instructor through a public electronic conferencing system, which allows students to share their efforts. During the course of the period, the instructor responds, and then uploads additional questions and comments (see Appendix II for a sample course syllabus).

For those teachers who have not used electronic conferencing in the classroom, such a system for "lecture" or even class discussion may seem strange—or even useless. After all, since all of the students are in the room together, why not have them speak to each other, as they do in any class discussion? However, it is the use of written communication from the students that is key: not oral discussion or passive listening to the instructor.

The writing faculty's interest in on-line conferencing derived from wanting students to communicate to each other in writing. There are two key elements here: the writing part and the "to each other" part. Students need experience writing; unfortunately, in traditional classes, most student writing is for grading reasons: major essays and written exams. Ungraded, more spontaneous writing, for prewriting or brainstorming purposes, is crucial to help students learn how writing may be a tool for thinking. Often, though, it is much easier for teachers to focus on the graded writing than on the ungraded writing, due to the structure and the time pressures of their courses.

However, ungraded writing helps students process the ideas of the course as they practice writing skills. Electronic conferencing allows the very structure of the class—discussions about concepts of the course and its reading assignments, practice for future assignments, review for essays and exams—to occur in a written format, in effect satisfying more than one goal at the same time. Students practice expressing themselves in writing as they cover the material the teacher feels needs to be covered.

That students are writing to each other and not just to their instructor is also crucial. Students tend to want to write to their instructor only; years of grading etiquette have trained them well, even when it has not always improved their skills at the main task. Instead, to help students become effective writers they should write to a specific audience, learning how to frame their ideas for different purposes and situations. Electronic conferencing is an excellent tool to emphasize audience, for once a student sends a message, it is subject to review from the rest of the class. Class members may ask questions or raise observations about previous comments.

Electronic conferencing class discussions feel different from oral class discussions. First, they are quieter; students are writing and reading with each other. However, throughout the classes, students will speak to make supportive or astonished comments or interjections about what they have just read. There is a spirit of informality in the classroom; usually the teacher is participating—even if just reading the comments flowing along the computer screen. Often, no one is standing at the front of the room in a teacherly stance; everyone looks the same sitting behind the terminals. (For other ideas of the rhythm of a class using electronic conferencing, see Hughes[9] and Condon.[10])

As mentioned above, but in need of continual emphasis, is an additional benefit of conferencing: that students who are shy or are afraid of speaking aloud may participate more easily. Students who can't type quickly can work at their own speed as they compose their messages. Students who think of something to say may contribute their ideas after the group has moved on to another topic. The new message does not disrupt the flow of the discussion and is not ridiculed; instead, it merely takes its place in the transcript of the conference, and others may respond as they wish. This kind of discussion is especially useful for non-native English language-speaking students; often, they are more able to compose a written message to express their ideas, even if the typing is slow and the grammar skills need work. Orally, such students may have good comments to contribute, but they lack the verbal quickness necessary to make their ideas heard at the precise moment in class.

Overall, an electronic conference just changes the dynamics of a class discussion: reflection and reading take the place of jumping quickly from point to point. Perhaps because conferencing is still new for our students (and teachers), participating in such an experience still has the stamp of the "new" and "special" on it; students take the experience seriously. To get some idea of the kind of attention electronic conferencing has received from English faculty, one might examine Cooper and Selfe. They suggest how conferencing allows students to become more aware of the multiple kinds of discourse present in a class discussion. Part of the conclusion of their argument concerns how we can use "computer technology to create non-traditional forums that allow students the opportunity to re-examine the authoritarian values of the classroom, to resist their socialization into a narrowly conceived form of academic discourse, to learn from the clash of discourses, to learn through engaging in discourse."[11] Such ideas relate to two of the needs faced by the organic chemistry class: the need to develop writing skills about chemistry and the need for students to be active participants in their learning. Electronic conferencing thus seems to offer promise for organic chemistry.

One crucial feature of the particular software used in the laboratory (PacerForum) is its graphics capability: A user may upload a picture to the bulletin board. (See Appendix I for the software list.) Although perhaps not as required for a more fully text-based class, such as English, this feature is crucial for chemistry to post structures and spectra, making questions clearer. Ironically, as part of the interdisciplinary pollination in the laboratory, although the chemistry class came to use electronic conferencing after the English department had moved towards it, English classes began experimenting with the graphics capability (to send images as the basis for writing assignments) only after the chemistry classes had demonstrated that the computer network would support such uses.

Another particular feature of the software, and how it has been installed over the network, is that it allows students to correspond anonymously. Students choose pseudonyms (known only to the instructor) to protect their dignity and still allow for honest appraisal of their work by the teacher and other students. This feature seems to encourage participation in all the classes. Students are not hesitant to send messages; they try to participate instead of merely sitting on the sidelines while others speak up. (See Madden[12] on using pseudonyms. For additional information on electronic conferencing, see "Computer Conferences: Alternative to Lectures"[13] and two bibliographies on electronic conferencing.[14,15])

As mentioned previously, the laboratory also now has a collection of chemistry-specific software. However useful this may be to illuminate organic chemistry for the students, the heart of the use of the laboratory is the electronic conferencing. Students practice developing their knowledge as they phrase their answers—this is higher ordered thinking, not just simple choices made with content-specific software. They are working to create understanding for themselves and for others. The students are more responsible for their own teaching/learning. The notes that they take home are the ones that they created themselves—at the end of class, they may print the conference as a study aid. The written discussion becomes a file, available to all.

Results: Examples from a Class Using Electronic Conferencing

Does this work? Do students do better in the course? Do they learn? The best reply comes from the keyboards of the students themselves. We are very pleased by the quality of the responses. We see high-quality work—work that we are proud of—and we perceive good thinking processes. The method allows the teacher to encourage the student, and, by building on the demonstrated knowledge of the individual student, help direct thoughts to more complex issues. As the course progresses, students draw connections to prior topics, making the various parts of the course seem more connected (see Appendix II for a sample course syllabus). We also see independence building, for we see students who can answer assertively. These skills are especially important considering the numbers of non-native students who are not used to responding assertively in any class.

Interactions from a 1995 class session on ethers will demonstrate more clearly how the method works, from instructor role to student interactions. The session began with a question posed by the teacher: "What are the products of acidic cleavage of methyl isopropyl ether? What acids are best to be used? What is the mechanism for the cleavage?" All students would see this message, as well as all the answers posted by the students, and the comments by the instructor. In these examples, students use their pseudonyms, so their identities are veiled to their peers but known to their instructor.

REX wrote "One would get isopropyl alcohol and Iodomethane: when using HI: this is an sn2 mechanism: 1. protonation of oxygen, 2. sn2 with incoming I- as nucleophile."

Another student, Intrepid, wrote "The products will be an alkyl halide and an alcohol. If HI (acid) is used then an iodomethane and isopropyl alcohol will result. This is an sn1 rxn with a carbocation intermediate."

To this, the instructor replied "Intrepid, Good analysis, however, please help me understand why you decided that the mechanism was SN1 instead of SN2."

Early on in this project, we found that some commendation was important to encourage the students to continue sending messages. Therefore, the first part of the instructor's message is always positive. Then, the second part of the message either asks for clarification about a point or asks the student to consider a slightly more complex modification, to help make those connections that seem to be missing. Thus, the student receives individually tailored responses and follow-up questions.

In the meantime, another student, Ace, joined the conference by posting this message: "The acids to be used will be HBr and HI. HCl will not be used because Br⁻ and I⁻ are more nucleophilic than Cl⁻. The product using HBr would be CH_3Br + $(CH_3)_2CHOH$. The mechanisms include the oxygen being protonated and the Br attacking the least substituted site."

This is a great description—one a teacher should be proud of. However, to reinforce the type of mechanism by name, to help make those connections stronger, the instructor replied to the student "Ace, Fine analysis. Can you go a step further and give the type of mechanism?"

Ace quickly replied " The type of mechanisms will be SN2 because primary and secondary ethers tend to cleave by SN2."

Meanwhile, Intrepid was still involved in the discussion. The student posted a follow-up: " I misread my notes and I now say that the rxn would be an sn2 because of the 2nd degree ether." We are delighted that the student did this self-analysis. We think this is real growth. Perhaps the student's thinking was assisted by the clear presentation of Ace, for which we rejoice, for we want the students to teach each other (see also Wilson.[16])

Another session covered aromatics and NMR spectral interpretation. The graphics option was used to display a spectrum obtained from a software program. The students were asked not to identify the compound but, given the name of the compound, to write an analysis explaining the peak positions and the splitting patterns. Complete sentences were requested.

For the spectrum of ethyl benzene, one student wrote "For ethyl benzene, there are three sets of peaks. The triplet closest to zero corresponds to the three hydrogens of the carbon that is farthest from the benzene ring. The quadruplet in the middle corresponds to the carbon of the ethyl group that is directly attached to the benzene ring. Finally, the triplet that is farthest down field corresponds to the hydrogens of the benzene ring itself."

This is excellent writing, and surely points to the fact that the student has caught on to the interpretation. The glory is that the students are learning by their own actions—and the other students can learn from each other. (see also Black[17]).

Strengths and Weaknesses of the Method

As the class has used the interdisciplinary laboratory for its sessions, certain strengths and weaknesses have appeared.

A key issue is the students' readiness for electronic conferencing. As demonstrated by the preceding examples, students must have a sufficient chemical vocabulary to be able to participate in electronic conferences. The result is that the format for first-semester students is talking in small groups clustered around a computer using the chemistry-specific software. Therefore, commercial software is more appropriate as a focus at the first of the course. Instead of sending messages to the instructor, students work to examine a problem together, using the data available in several pieces of software. This method is best for learning, because the students actually discuss the reactions seen on the computer screen. Working alone at a computer is not a good idea; that seems closer to study hall than classroom. As a side benefit for our low-cost laboratory, by having a few copies of software for the students to share, costs are reduced. Right now, five computers have the full range of tutorial and simulation software for the course. Students share these workstations to help each other understand.

Thus, only toward the end of the first semester, about the time they begin spectral interpretation, are the students ready to participate in conferences. The format for the second semester is writing in a conference, with one student per computer and with small groups orally discussing issues raised by the software when necessary.

Further, for both semesters, holding the entire lecture in the computer laboratory left out the auditory learners. Initially, excited and made bold by the availability of the laboratory and the first chemistry laboratory report project, we moved the class there for the entire three-hour lecture block. However, students who do learn well through lecture were not necessarily helped. As a result, the computer laboratory time was shortened to one and a half hours, following introductory lecture time. Students are even afforded a break as they walk from the lecture room to the laboratory room.

As the class shifts from lecture to laboratory time, during this break, the students' behavior changes. They talk among themselves, but they do not copy from one another. The laboratory is often noisy. They do not fall asleep (even though the class is long and follows lunch). They stay late—and they leave tired, for they have been working. They tend to ask "Why?"—while when they are in the lecture part of the course, they are much more likely to ask "Is it going to be on the test?" Active learning occurs, with the spotlight inconspicuously on the students. The anonymous nature of the conferencing allows students to send responses without fear of ridicule from others.[18] Because everyone is working together, the laboratory seems informal; students confer with each other orally as they send messages in the conference. At times, the laboratory is even a noisy place, with students conferring with each other avidly. They talk with each other about mechanisms—hurrah! They create an NMR spectrum as they look for more data and check their answers; students may use these spectra to determine the unknown one in question. When the student finally posts an answer, he or she does so with confidence.

There are some unusual complaints about the laboratory section of the course. Some students do not believe that this is real teaching. Others profess to be even more confused. Still others are concerned by the lack of notes; for them, the printout of the conference sessions does not fit their idea of "notes".

At the same time, other students respond favorably, as in the following comments: "[using the laboratory]...allows the students to as well, teach each other."; "was beneficial in the sense that we could often see reactions in action."; and " I see this style of teaching becoming a very essential part of learning in the future in Institutions across the World." These commendations and observations show, we feel, that the students value some of the same things that we do. The various departments using electronic conferencing, for writing, literature, government, and sociology classes, report similar findings, as do other sources (see also Oakley for examples of student comments[19]).

Additional issues do create concerns about using the laboratory. Electronic conferencing is a less efficient system than lecturing, and therefore some material must be deleted. Other content-based courses such as history or government have discovered this problem also in the laboratory. It's really much more efficient for an instructor to lecture, without interruptions and without worrying about how to frame a question to get a proper response.[20] Further, having to delete information from what one would previously choose requires one to determine carefully what to leave out. However, others have found that their students can understand significant concepts with a mixture of less lecture and more activity.[17,21-23] The availability of new teaching tools helps encourage this process of selection to allow more activity in the laboratory.

Some students believe this method of teaching is not for them, for they need the standard format. Therefore, this course is identified in the class schedule as "computer assisted", as are all the courses scheduled in the laboratory. Also, in this course, as in the others scheduled in the laboratory, students who wish to drop tend to do so earlier than in non-computer-based classes. Those who remain, though, generally stay until the end of the semester.

Overall, we believe the method we use for teaching organic chemistry in the computer laboratory is an excellent equalizer of the student body, allowing the same quality of access for all. The instructor knows his or her students from their intellectual thoughts—which is a proper way for students to display their plumage. We know for sure that students who will barely bring themselves to address the teacher will write beautifully crafted miniature essays as responses—and we are delighted. We also know that we strengthen our students when we require that they write for us. We have compared the writing abilities as we progress through the semester, and without qualification, the writing improves as the students' ability to discuss organic chemistry improves.

Plans for Improvement

After beginning experiments, we realize that more needs to be studied to determine the overall effectiveness of teaching organic chemistry through these methods. Various issues still need to be considered. A crucial area is evaluation of computer-based courses and the students' performance in them. How well do the students do on exams? Based on several years of experimenting, they seem to do as well as with the standard teaching format. However, we must confess that we are bewildered that they are not doing much better, for we have evidence of sophisticated knowledge from the transcripts of the conference sessions. We suspect that part of the problem is in the translation of knowledge first honed on the computer and then reproduced on paper for the exam. Some of the students who answer brilliantly in the conference miss similar questions on paper exams. Developing strategies for evaluation is a current focus of the faculty using the interdisciplinary laboratory, for all disciplines are trying to determine how effectively computers assist learning.

To understand this discrepancy between practice and testing, we are investigating the effect of state-dependent learning as it applies to computer learning. Gooden and Baddeley's classic study of memory tasks performed underwater found the following: If the memorization was done underwater, then the clients did better if the recall was also underwater (on waterproof slates). If the memorization was done above water, then recall was better if also performed above water.[24] We have all seen this difficulty. How many times have our students done a chemistry wet laboratory successfully, and then found themselves unable to recall the products on an exam? We would think having actually done a reaction would really seal the knowledge, but we believe it may get sealed in the laboratory—and not in the transfer area. We believe that we are seeing a similar effect with the knowledge gained in the computer laboratory. Written conversations with students are a true pleasure, for they provide evidence of their more sophisticated knowledge. However, we are trying to learn more about state-dependent learning. We plan to work toward methods of easing the knowledge transfer from one medium to another.

Other plans for the future with our method include analyzing our prior conferencing sessions. Using as a basis the work of T. Zielinski with modules in physical chemistry,[25] we want to develop nested questions that carefully build a framework of knowledge. Again, our emphasis will be on mechanisms and developing the ability to predict which mechanisms are responsible for a particular reaction. We also want to bring more visualization to the computer lab, using more molecular modeling.

We are also continually working to make the functioning of the laboratory as smooth as possible. Particularly as the laboratory supports more and more software packages, we find that one piece of software may negatively affect how another piece works. We've seen this especially with a "player" version of the program "Hypercard": It takes over how the computer runs any other Hypercard files, sometimes causing disruptions. Networks are also complex and difficult to keep running: Observant and trained laboratory personnel have been our best defense against all of the odd things that happen for seemingly no reason as more and more people use the laboratory.

At times, even the amount of software may be a problem, as hard disks become full and certain computers must be dedicated for certain uses. In many ways, the success of this program has led us to the opening of a dedicated computer laboratory for chemistry and the physical sciences in general.

We also plan to make use of the vast storehouse of information on the Internet. It is our great desire to use materials from the Internet to introduce our students to small research problems, using as a guide the work done by D. Craig with Brookhaven protein files.[26] We look forward to following the lead of M. Swift[27] in decreasing the dependence of the student on the textbook and letting the student experience functioning as a professional by using primary sources available on the computer. We do feel that our use of electronic conferencing has given our students, who do not have e-mail or listserv access through our college, some idea of how computers can allow people to communicate in new and potentially improved ways (an idea suggested by Rickley,[28] Scott,[29] and others).

Summary and Conclusions

Conducting an English lesson in the computer writing laboratory to improve the sophomore organic chemistry laboratory reports led to an expanded project for using writing for clarity as a learning tool in the organic lecture also. By substituting conferences on networked computers for chalkboard lecture time after students have developed sufficient vocabulary, the class engages in common written discussion about problems, visible to all participants, with ample time for student responses, questions, and instructor follow-up. Anonymous passwords for conferencing and group work with computer tutorials additionally encourage student participation. The result is elevation of the plane of organic instruction, from memorization to comprehension; the simplicity and cost-efficiency of the method is ideal for community colleges or small colleges.

Can you use this system at your campus? Yes, just talk your way into the writing laboratory. Usually, the writing laboratory will have conferencing software, something more and more writing teachers are experimenting with across the country. You do not have to devote your entire semester to this method, but you could try it for a few classes, as do many classes from various departments in our laboratory. At a school comfortable with network e-mail and the Internet, much of what we are doing could occur on a grander scale, with modifications for asynchronous messages extending over a longer time period and students contributing responses at various sites (see Oakley[19] for an example). The key idea is the focus on students sharing their written responses about chemistry with each other.

Large-enrollment classes could also possibly use this method, if the multiple group function of conferencing software were employed. Aiken and Hawley suggest how a large laboratory at the University of Mississippi's School of Business Administration uses conferencing to conduct large class discussions, with a follow-up use of the resulting transcript.[30] Teaching assistants could serve as moderators for each 20-member group, while the professor remains on line also, visiting each group to add comments. Especially if the teaching assistant and the professor were examining student messages from different but important perspectives, the benefit to the student might double. The teaching-assistant–led groups might have more significance for the students than do the normal recitation sessions. By watching their teachers exchange messages also, students might better understand that group work is not only for students, but is an important component of the delivery of chemical knowledge, whether at the laboratory bench or in the classroom. (See also Andriole, Lytle, Monsanto for additional ideas for asynchronous conferencing.[31])

Because of the abundance of excellent spectra databases and spectrum simulator programs (see Appendix I for the software list), we suggest that a good starting place for someone choosing to use conferencing along with other software in the laboratory is the analysis of NMR spectra. Ask a question you might use as a pop quiz; see how students can practice writing skills to help themselves learn what they need to learn in your course. Encourage students to use software or textbook tools to submit ideas as they develop responses, and then look at the results. As a tool to help you get started, we have provided a sample syllabus and a software list in the appendices.

Acknowledgment

We thank Yueh-Neu Lin, Ph.D., head of chemistry, Sandra Boyd, Ph.D., instructor of psychology, and Thomas Montgomery and Jyoti Shah, network and laboratory specialists, for their encouragement and assistance throughout this project. We also acknowledge our late dean of instruction, Anthony Chee, Ph.D., for providing the initial support to found the Macintosh Interdisciplinary Laboratory, and our current dean of instruction, Gail James, Ph.D., for her continuing support.

Appendix I: Sample Software List

Chemistry: Software/CD/VideoDisk

Beaker. John Werner, Joyce Brockwell, Steve Townsend, and Nim Tea; Brooks/Cole Publishing Company: Pacific Grove, CA, 1991.

Bonding. Ron Lemay; accessible via WWW at URL ftp://ftp.shef.ac.uk/pub/uni/academic/A-C/chem/chemistry-hyper-card/bonding-hc.sit.hqx.

Demonstrations in Organic Chemistry: VideoDisk. Gary Trammell; *J. Chem. Educ.: Software*. 1993, Special Issue #6.

Introduction to Organic Chemistry. Stanley Smith; Falcon Software Inc.: Wentworth, NH, 1991.

MacStereo. J. Phillip Bays; Falcon Software Inc.: Wentworth, NH, 1990.

Organic Reaction Mechanisms. Andrew F. Montana and Jeff R. Buell; Falcon Software Inc.: Wentworth, NH, 1993.

Proton NMR Basics: CD. Carolyn Sweeney Judd, Joel D. Morrisett, Mohan V. Chari, and Jeffrey L. Browning; *J. Chem. Educ.: Software*. 1996, Special Issue #11.

Proton NMR Spectrum Simulator. Kersey A. Black; *J. Chem. Educ.: Software*. 1990, IIc (1).

SpectraDeck. Paul F. Schatz; Falcon Software Inc.: Wentworth, NH, 1993.

Conferencing: Software

Daedalus Integrated Writing Environment. The Daedalus Group, Inc.: Austin, TX. http://www.daedalus.com/.

PacerForum. AGE Logic, Inc.: San Diego, CA. http://www.age.com/forum/foru-minf.html.

(For additional electronic conferencing, please see the Computer Conferencing Software Vendors List.[32])

Appendix II: Annotated Syllabus for Organic Chemistry, Special Section: Computer-Assisted Course

This is a partial syllabus for the purpose of explaining our method. Obviously, students will also complete numerous problems from the text, take examinations, and perform chemistry wet laboratory experiments. The text we are currently using is by John McMurry; *Organic Chemistry;* Brooks/Cole Publishing Company; Pacific Grove, CA; 3rd edition; 1992.

See Appendix I, Sample Software List for information about particular software used. For clarity, software titles are placed in parenthesis throughout the syllabus.

Decisions about topics to include are based on a 1994 report from *Chemical and Engineering News* of a national poll of organic chemistry faculty.[33]

New topics are either covered in the lecture or in the computer laboratory; the computer lab augments, but does not echo, the lecture.

Most (PacerForum) sessions address about three questions; however, only one example question is given in this annotated syllabus. Often, the three questions are nested questions, building on each other. While some students post one well-crafted response to each question, other students routinely divide their answers into parts, in a more conversational manner. The instructor posts a reply to each response, encouraging continuing effort, correct thinking, and deeper thought.

Semester I

Week 1: Chapter 1: Structure and Bonding
Lecture. Ionic and covalent bonding; Lewis structures of common molecules
Computer Strategy. Review of bonding
Introduction to the computer lab by R. Ford.
(Introduction to Organic Chemistry: Alkanes): Bonding review.
(Bonding): Short self-quiz.

Week 2: Chapter 2: Bonding and Molecular Properties
Lecture. Electronegativity; acid-base definitions; pK_a; organic compound analysis
Computer Strategy. Introduce conferencing and a drawing program; conduct a writing lesson
(PacerForum) Introduction conducted by R. Ford.
(PacerForum): "Writing the Chemistry Laboratory Report" conducted by R. Ford.
(Beaker): Practice drawing structural formulas.

Week 3: Chapter 3: Nature of Organic Compounds: Alkanes & Cycloalkanes
Lecture. Nomenclature of alkanes
Computer Strategy. Learning to recognize cis-trans isomers
(PacerForum): QUESTION: " Click on the icon. Give the IUPAC name of each compound. Are any of the structures of the same compound?" (GRAPHIC: three 1,2-dimethylcyclobutane structures, two of which are both cis.)

Week 4: Chapter 4: Stereochemistry of Alkanes and Cycloalkanes
Lecture. Molecular model kits; alkane conformations; ring strain; heats of combustions
Computer Strategy. Animation: alkane conformations and energy diagram; conformers of substituted cyclohexanes
(Organic Reaction Mechanisms: Conformation Analysis) Rotational energies of Newman structures of ethane, butane, and methylbutane are coordinated with energy profile diagrams.
(MacStereo) Introduce and examine axial and equatorial arrangements of substituted cyclohexanes.

Week 5: Chapter 5: Overview of Organic Reactions
Lecture. How to write mechanisms; general outline of radical, polar, and concerted reactions; bond dissociation energies
Computer Strategy. Animation: study in detail a radical and a polar reaction, emphasizing natures of the transition states and charge/shape of intermediates
(Organic Reaction Mechanisms: addition, Markovnikov, HBr + Propene) Introduction to this very important example of a polar reaction.
(Organic Reaction Mechanisms: Halogenation of Methane) Visualization of this classic radical reaction.

Week 6: Chapter 6: Alkenes: Structure and Reactivity
Lecture. Degree of unsaturation; nomenclature and sequence rules; alkene stability; Hammond postulate
Computer Strategy. Carbocation stability and electrophilic addition reactions (PacerForum): QUESTION: "Two products result from the addition of HBr to 3,3-dimethyl-1-butene. Describe the mechanism in complete sentences for the formation of one of the products, 3-chloro-2,2-dimethylbutane. Be sure to include whether the reaction is polar, radical, or pericyclic. What is the charge and the nature of the intermediate formed? Next, explain the formation of the second product, 2-chloro-2,3-dimethylbutane."

Week 7: Chapter 7: Alkenes: Reactions and Synthesis
Lecture. Alkene reactions and mechanisms: formation of halohydrins, alcohols, and HBr radical addition
Computer Strategy. Animation: halogenation and catalytic hydrogenation of alkenes; impact of small-membered ring transition states and intermediates
(Organic Reaction Mechanisms: bromine to a (Z) alkene) Three-membered bromonium ion and resultant stereospecificity.
(Organic Reaction Mechanisms: Catalytic Hydrogenation) Four-membered transition state and resultant cis addition.

Week 8: Chapter 7, continued: Alkenes: Additional Reactions
Lecture. Formation of cis-diols; oxidative cleavage of alkenes and diols, alkene preparation; carbenes
Computer Strategy. Syntheses involving several steps
(PacerForum): QUESTION: "Show how 2-bromopropane could be synthesized from 1-bromopropane. Write your answer in complete sentences, clearly indicating the number of steps, the reagents, the nature of the intermediates, and the reasons for your products."

Week 9: Chapter 8: Alkynes

Lecture. Nomenclature; reactivity rates of alkynes and alkenes toward electrophilic addition and reduction; preparation of alkynes

Computer Strategy. Animation: alkyne acidity; keto-enol tautomerism

(Organic Reaction Mechanisms: Alkylation of an acetylide) Abstraction of a proton from an alkyne, followed by alkylation with RX.

(PacerForum): QUESTION: "The hydration of an alkene leads to products that are quite different from the hydration of an alkyne. Illustrate this difference using 4-octene and 4-octyne. Explain in complete sentences the specific reagents used, intermediates formed, and the resultant products."

Week 10: Chapter 9: Stereochemistry

Lecture. Optical activity; enantiomers; chirality; sequence rules; R, S, meso, and racemic compounds

Computer Strategy. Diastereomers; Fischer projections; cyclic compounds (MacStereo): molecular model kits are used with this program; short self-quiz.

Week 11: Chapter 10: Alkyl Halides

Lecture. Preparation; allylic bromination of alkenes; radical stability; resonance; Grignard reagent; organolithium reagent

Computer Strategy. Relative C–H bond dissociation energies

(PacerForum): QUESTION: "There are three possible isomers of pentane. Choose one of the isomers and discuss the number and percentage of monochlorinated products. Explain your answer in complete sentences."

Week 12: Chapter 11: Reactions of Alkyl Halides: Substitutions

Lecture. SN2, SN1, and kinetics; importance of substrate, nucleophile, solvent, leaving group

Computer Strategy. Animation of SN2; relative nucleophilic strength

(Organic Reaction Mechanisms: Substitution: Nucleophilic, bimolecular) Substitution with inversion is correlated with an energy profile.

(PacerForum): QUESTION: "The formation of an alcohol by the reaction of t-butyl chloride with aqueous hydroxide and acetone is independent of the concentration of the hydroxide ion. Give the name of the alcohol formed, and explain the mechanism for the formation. Give details about any intermediates formed."

Week 13: Chapter 11, continued: Reactions of Alkyl Halides: Eliminations

Lecture. E2 and E1 and kinetics: importance of substrate, base, solvent, and leaving group; deterium isotope effect

Computer Strategy. animation of E2 reaction; synthesis

(Organic Reaction Mechanisms: Elimination, bimolecular) Animation shows elimination and major/minor products.

(PacerForum) QUESTION: "Treatment of *trans*-1-bromo-methylcyclohexane with KOH gives 3-methylcyclohexene. Is this the expected elimination product? Explain your answer, giving information about the mechanism for the actual product."

Week 14: Chapter 12: Mass Spectroscopy and Infrared Spectroscopy

Lecture. Chromatography: liquid and gas; mass spectroscopy; electromagnetic spectrum

Computer Strategy. Interpretation of IR and MS
(Introduction to Organic Chemistry: IR) Understanding IR; some important absorptions.
(PacerForum): QUESTION: "Please go to (SpectraDeck) and print out the MS and IR spectra for the three compounds listed by your name below. When you return to the computer laboratory next week, be prepared to enter your interpretations on (PacerForum).

Week 15: Nuclear Magnetic Spectroscopy
Lecture. Carbon NMR: theory and interpretation of spectra
Computer Strategy. Proton NMR and interpretation of spectra
(CD: Proton NMR Basics) Theory and Instrument portions are projected for class use; students use Spectral Interpretation section individually.
(PacerForum): QUESTION: "Click on the icon to see the proton NMR spectrum. The molecular formula is C13H12. Identify the compound, detailing in complete sentences the reasons for all peaks and splitting patterns." GRAPHIC: proton NMR spectrum of diphenylmethane.

Semester II

Week 1: Chapter 14: Conjugated Dienes and Ultraviolet Spectroscopy
Lecture. Conjugation; stability; MO description; 1,2 and 1,4 electrophilic addition; carbocations; kinetic vs. thermodynamic control
Computer Strategy. Animation: Diels–Alder reaction; addition of HX to a diene
(Organic Reaction Mechanisms: Diels–Alder Reaction) Animation showing pericyclic mechanism.
(PacerForum): QUESTION: "The addition of one equivalent of HCl to 1,3-pentadiene forms both 1,2 and 1,4 adducts. Using the computer program (Beaker), draw the possible intermediates, including resonance structures. Copy and Save the structures and Paste into your response on (PacerForum). Also include the names of the possible products, explaining how each product was formed."

Week 2: Chapter 15: Benzene and Aromaticity
Lecture. structure; stability; MO description; Huckel rule; naphthalene
Computer Strategy. Relative reactivity; heterocyclic aromatic compounds
(Videodisk: Demonstrations in Organic Chemistry) Reaction of bromine with cyclohexane, cyclohexene, and benzene.
(PacerForum): QUESTION: "Click on the icon to see the structure of furan. Compare the aromatic character of furan with tetrahydrofuran. Describe your conclusions in complete sentences."

Week 3: Chapter 16: Benzene: Electrophilic Aromatic Substitution
Lecture. Friedel–Crafts alkylation; acylation; nitration, sulfonation; oxidation; benzyne
Computer Strategy. Animation of bromination; substituent effects
(Organic Reaction Mechanisms: Substitution, electrophilic with benzene and toluene) Illustrated substituent effects using bromination of benzene, toluene, and nitrobenzene.

(PacerForum): QUESTION: "Click on the icon to see the compound, styrene. What will be the product of the reaction of HBr with styrene? Give reasons for your conclusions, describing the mechanism for the reaction? Would p-methoxystyrene react faster or slower than styrene with HBr. Justify your answer." GRAPHIC: structure of styrene drawn with (Beaker).

Week 4: Chapter 17: Alcohols
Lecture. Nomenclature; hydrogen bonding; boiling points; acidity; preparation; oxidation; reduction
Computer Strategy. Acid catalyzed dehydration; Grignard reagents
(Organic Reaction Mechanisms: elimination, dehydration of n-BuOH) Mechanism of this reaction, reinforcing steps common to other mechanisms.
(PacerForum): QUESTION: "List a possible combination of a Grignard reagent and carbonyl compound that could be used to make 2-pentanol. Describe the intermediate, including charge and shape. Detail the mechanism of the reaction in complete sentences."

Week 5: Chapter 18: Ethers and Epoxides
Lecture. Structure, Williamson ether synthesis, addition of an alcohol to an alkene; formation and ring opening of epoxides
Computer Strategy. Acid cleavage of ethers; IR/NMR
(PacerForum): QUESTION: "What are the products of the acidic cleavage of methyl isopropyl ether? Describe the mechanism in detail."
(PacerForum): QUESTION: "Click on the icon to see two spectra, both of molecular formula C3H8O. Which is the ether? Give reasons for your choice. What is the structure of the other compound, based on the spectrum?" GRAPHIC: (Proton NMR Spectrum Simulator) generated spectra of 1-propanol and methyl ethyl ether.

Week 6: Chapter 19: Aldehydes and Ketones: Additions & Oxidations
Lecture. Nomenclature; preparation; relative reactivity; nucleophilic additions; oxidations
Computer Strategy. Acetal formation; IR/NMR of aldehydes and ketones
(Organic Reaction Mechanisms: Carbonyl compound, acetal formation) Animated mechanism, including role of acid catalyst.
(PacerForum): QUESTION: "Click on the icon to see the NMR spectrum of a compound with the molecular formula C7H14O. The IR spectrum reveals a strong absorption at 1710 cm^{-1}. Using complete sentences, give reasons based on the spectra for your conclusions about the identity of this compound." GRAPHIC: (Proton NMR Spectrum Simulator) generated spectrum of 4,4-dimethyl-2-pentanone.

Week 7: Chapter 19, continued: Aldehydes and Ketones: Reductions; Additions to Unsaturated Carbonyls
Lecture. Review of reductions; direct and conjugate additions to alpha,beta-unsaturated carbonyls; organolithium and Grignard reagents additions
Computer Strategy. Animation: Wittig reaction; synthesis
(Organic Reaction Mechanisms: Wittig reaction) Complex mechanism is shown in motion.

(PacerForum): QUESTION: "How could you make 1-methylcyclohexene from cyclohexanone? Several synthetic pathways are possible. Answer in complete sentences, indicating all reagents and the mechanism in your answer."

Week 8: Chapter 20: Carboxylic Acids
Lecture. Nomenclature; acidity; preparation; oxidation; reduction; decarboxylation; IR/NMR
Computer Strategy. Substituent effects on acidity
(PacerForum): QUESTION: " Consider these three acids: p-nitrobenzoic acid, acetic acid, and benzoic acid. Place them in the order of increasing acid strength. Explain the reasons for your order."

Week 9: Chapter 21: Carboxylic Acid Derivatives: Nucleophilic Acyl Substitution Reactions
Lecture. Relative reactivity of acid derivatives; lack of basicity of amides; nucleophilic substitution reactions; nitriles; IR/NMR
Computer Strategy. Animation: amides; esters
(Organic Reaction Mechanisms: basic hydrolysis, amide) Animation shows this 3-step mechanism.
(PacerForum): QUESTION: "Describe the Fischer esterification of acetic acid and ethanol. Use complete sentences, taking care to remember the acid catalyst conditions."

Week 10: Chapter 22: Carbonyl Alpha-Substitution Reactions
Lecture. Enol formation; alpha halogenations of aldehydes and ketones; alpha bromination of carboxylic acids; acidity of alpha hydrogen atoms
Computer Strategy. Enolate ion formation; alkylation of enolate ions; haloform
(Organic Reaction Mechanism: Carbonyl Compound, alkylation)
(PacerForum): QUESTION: "What alkyl halide can be used in the acetoacetic ester synthesis to form 2-butanone? Comment on the strength and the purpose of the added base. Which carbon atoms of 2-butanone originate with the alkyl halide?"

Week 11: Chapter 23: Carbonyl Condensation Reactions
Lecture. Aldol, crossed aldol, Claisen, Dieckman and Michael reactions; comparison of alpha-substitution vs. condensation reactions
Computer Strategy. Predicting the product of a condensation reaction
(PacerForum): QUESTION: "The basic aldol condensation of benzaldehyde and methyl phenyl ketone is followed by dehydration. What will be the product? Explain the mechanism for the product formation, emphasizing intermediates that are compatible with the basic conditions."

Week 12: Chapter 26: Aliphatic Amines
Lecture. Nomenclature; structure; basicity; preparation; reactions; Hofmann elimination, IR/NMR
Computer Strategy. Synthesis
(PacerForum): QUESTION: "Starting with 1-butanol, show how pentylamine could be made. Several steps are involved. Describe each step in detail."

Week 13: Chapter 27: Arylamines and Phenols
Lecture. Aniline, basicity of arylamines; preparation and reactions of arylamines; diazonium salts; IR/NMR
Computer Strategy. Phenol, acidity, preparation and reactions of phenols
(PacerForum): QUESTION: "Starting with benzene, describe how carvacrol, 5-isopropyl-2-methylphenol, could be made. Several synthetic pathways are possible."

Week 14: Chapter 24: Carbohydrates
Lecture. Fisher and Haworth projections; D,L sugars; hemiacetal formation of monosaccharides; mutarotation; disaccharides
Computer Strategy. Oxidation of carbohydrates
(PacerForum): QUESTION: "Sugars which have aldehyde groups will reduce Benedict's solution, and are therefore called reducing sugars. However, fructose, a ketose, will also reduce Benedict's solution. Explain this observation."

Week 15: Chapter 27: Amino Acids, Peptides, and Proteins
Lecture. Amino acids; isoelectric point; zwitterion; peptides
Computer Strategy. Shape of proteins, electrophoresis
(CD: Proton NMR Basics) Research Laboratory part uses NMR to study the active site of an enzyme.
(PacerForum): QUESTION: "Which form of glutamic acid would be expected to predominate at pH 1.0? What would be the net charge? What would be the net charge if glutamic acid were at pH 11.0?"

References

1. Wilson, J. M.; Redish, E. F.; McDaniel, C.; "Building a Comprehensive Unified Physics Learning Environment"; *T.H.E. Journal* 1992, IBM Higher Education Supplement, pp 5–7.
2. Reif, R. J.; Morse, G. M.; "Restructuring the Science Classroom"; *T.H.E. Journal* 1992, *19*, 69–72.
3. Johnson, D. W.; Johnson, R. T.; Smith, K. A.; "College Teaching and Cooperative Learning"; Active Learning: Cooperation in the College Classroom; Interaction Book Company: Edina, MN, 1991; pp 1–26.
4. Twigg, C. A.; "The Changing Definition of Learning"; *Educom Review* 1994, July/August, 23–25.
5. Oakley, B.; "Electronic Interaction in ECE 270 Section & A Using PacerForum Software"; accessible via WWW at URL http://sloan.ece.uiuc.edu/Oakley/PacerForum-Manual/PacerForumManual.html.
6. Kirwan, W. E.; "Keynote Address: Role of Faculty in the Disciplines in Undergraduate Education of Future Teachers"; in *The Role of Faculty from the Scientific Disciplines in the Undergraduate Education of Future Science and Mathematics Teachers;* Haver, W. E.; Levitan, H., Eds.; National Science Foundation: Washington, DC, 1992, pp 17–20.
7. Hermann, C. K. F.; "Teaching Qualitative Organic Chemistry as a Writing-Intensive Class"; *J. Chem. Educ.* 1994, *71*, 861–862.
8. Beall, H.; "Report on WPI-NEACT Conference: General Chemistry: Approaching the 21st Century"; *J. Chem. Educ.* 1991, *68*, 835–837.
9. Hughes, E.; "Voices We Hear and Voices We Silence: Galloping Through InterChange"; *Wings* 1994, *2*, 10.

10. Condon, B.; "Renegotiating Empowerment: Moving a Collaborative Writing Assignment into Virtual Space"; *Wings* 2, p. 5.
11. Cooper, M. M.; Selfe, C. L.; "Computer Conferences and Learning: Authority, Resistance, and Internally Persuasive Discourse"; *College English* 1990, *52*, 847–869.
12. Madden, E.; "Pseudonyms and InterChange: The Case of the Disappearing Body"; *Wings* 1993, *1*.
13. "Computer Conferencing: Alternative to Lectures"; Institute for Academic Technology; Teleconference, November 13, 1995.
14. "Computer Conferencing: Alternative to Lectures Selected Bibliography"; accessible via WWW at URL http://ike.engr.washington.edu/iat/events/sb_bib95.html.
15. "The Daedalus Group, Inc. Bibliography"; accessible via WWW at URL http://www.daedalus.com/.
16. Wilson, J. W.; "Writing to Learn in an Organic Chemistry Course"; *J. Chem. Educ.* 1994, *71*, 1019–1020.
17. Black, K. A.; "What To Do When You Stop Lecturing"; *J. Chem. Educ.* 1993, *70*, 140–144.
18. Pressley, M.; McCormick, C. B.; "Science"; Cognition, Teaching, & Assessment; Harper Collins College Publishers, New York, 1995, 297–300.
19. Oakley, B.; "A Virtual Classroom Approach to Learning Circuit Analysis"; accessible via WWW at URL http://www.sloan.org/oakley/Maine/index.html.
20. Lipson, A.; Tobias, S.; "Why Do Some of Our Best College Students Leave Science?"; *Journal of College Science Teaching* 1991, *21*, 92–95.
21. Bodner, G. M.; "Why Changing the Curriculum May Not Be Enough"; *J. Chem. Educ.* 1992, *69*, 186.
22. Cooper, M. M.; "Cooperative Learning"; *J. Chem. Educ.* 1995, *72*, 162–164.
23. Wright, J. C.; "How Effective Are Our New Teaching Methods?"; Abstracts of Papers, 210th National Meeting of the American Chemical Society, Chicago, IL; American Chemical Society: Washington, DC, 1995; CHED 297.
24. Gooden, D. R.; Baddeley, A. D.; "Context-Dependent Memory in Two Natural Environments: On Land and Underwater"; *British Journal of Psychology* 1975, *66*, 325–331.
25. Zielinski, T. J.; "Promoting Higher-Order Thinking Skills: Uses of Mathcad and Classical Chemical Kinetics to Foster Student Development"; *J. Chem. Educ.* 1995, *72*, 631–638.
26. Craig, D. W.; "Accessing the Protein and Nucleic Acid Structural Literature in the Undergraduate Curriculum"; Abstracts of Papers, 210th National Meeting of the American Chemical Society, Chicago, IL; American Chemical Society: Washington, DC, 1995; CHED 256.
27. Swift, M. L.; "Teaching Basic Science Computer Skills Without a Textbook"; Abstracts of Papers, 210th National Meeting of the American Chemical Society, Chicago, IL; American Chemical Society: Washington, DC, 1995; CHED 289.
28. Rickly, B.; "Teacher Talk"; *Wings* 1993, *1*, 3–4.
29. Scott, G.; "E-Mail's Expanding Role in Education"; *T.H.E. Journal* 1991, *18*, 54–56.
30. Aiken, M. W.; Hawley, D. D.; "Designing an Electronic Classroom for Large College Courses"; *T.H.E. Journal* 1995, *23*, 76–77.
31. Andriole, S. J.; Lytle, R. H.; Monsanto, C. A.; "Asynchronous Learning Networks: Drexel's Experience"; *T.H.E. Journal* 1995, *23*, pp. 97–101.
32. "Computer Conferencing Software Vendors List"; accessible via WWW at URL http://ike.engr.washington.edu/iat/library/liblinks/ccvendor.html.
33. Illman, Deborah L.; "Chemists Ponder Reforms to Organic Chemistry Curriculum"; *Chemical & Engineering News* 1994, September 5, p 39.

Computer-Assisted Instruction in Chemistry

Stephen K. Lower

Introduction

Computer applications such as spreadsheets, visualization, molecular modeling, data acquisition, and most of the other topics covered in this monograph are now firmly within the mainstream of chemistry, and in these roles the computer is considered an essential tool for teaching the *practice* of chemistry. However, computers are much less commonly employed (and even less often *well*-employed) in teaching the much larger body of knowledge upon which chemical science is based.

Chemical education is by no means unique in this regard; educational practice in general has always been strongly buffered against change, especially in the academic world. Even when a new technology is finally accepted, it is more likely to become incorporated into existing practice than for these practices to be altered or their goals extended in ways that fully exploit the capabilities of the new technology.

> As an example of this phenomenon, consider the long time lag between the first appearance of the overhead projector (in bowling alleys in the 1940's) and their general entry into classrooms about twenty years later. Even now, however, many teachers still use them as if they were blackboards or slide projectors, rather than exploiting their ability to combine elements of both and thereby enhance the teacher's spoken presentation in a dynamic and visually interesting way.

The purpose of this article is to explore some of the reasons for this, and to suggest why this must change if chemical education is to meet the expectations that society is likely to place on it. We will also describe some special benefits of computer-assisted instruction (CAI) that are difficult to achieve by conventional methodologies, and how they must be utilized within the overall instructional scheme in order to yield good results. Finally, we will look briefly into the design and implementation of CAI lessons.

By CAI, we mean use of the computer as a major tool for the delivery of instruction, as opposed to its employment (even in an educational setting) for extending instruction into the areas mentioned in the opening sentence.

It is often pointed out that a number of other technologies that enthusiasts had once predicted would revolutionize education have largely dwindled into obscurity as far as their educational applications are concerned. Witness the Skinner-inspired programmed learning of the 1950s, instructional TV (which peaked around 1960) and audiotape-based learning, and Keller plan mastery learning, which enjoyed some popularity about a decade later. It is important to bear in mind, however, that all of these tools have been demonstrably successful in the contexts in which they were developed. That they never became "mainstream" is largely a reflection of the difficulty in implementing them and in getting others to accept unfamiliar methodology.

Although CAI may have been little more than just another newfangled candidate for failure in its early days,[1-3] it has probably been saved from the usual fate by the unprecedented degree to which computers have entered so many other aspects of our society, culture, and, above all, our private lives.

Traditional vs. Nontraditional Methods of Instruction

The use of the computer as a tool for instruction might at first appear to be too specialized a topic for a "what every chemist should know" collection such as this. On reflection, however, it becomes apparent that the health of chemistry, not only as a science but also as a tool of industry and as a basis for informed decisions on public policy, is critically dependent on technical training and education of the much broader range of the public that interacts with chemistry either directly or indirectly. Traditional methods of instruction that may have proven adequate for the smaller, more academically inclined, and more homogeneous groups of students we dealt with in the past are increasingly hard-pressed to serve the needs of the very diverse clientele of today, which extends well beyond those enrolled in formal physical sciences courses in high school and college. Even among the latter groups, a substantial number of students who begin as science majors eventually abandon these programs for other fields; according to a survey reported in *Chemical and Engineering News* about five years ago, the most common reason given was "poor quality of instruction".

How can this be? Most of us believe we are rather good teachers, always ready to provide assistance to the few who seek us out. With the aid of various degrees of force-feeding, we usually manage to get most students to the point where they can work out the empirical formula of a compound, and calculate the pH of a weak-acid solution and (with luck) of its titration end point. But is this what chemistry is all about? We know that it is not! It is nevertheless easier to teach and test these topics, than, for example, to ensure that students understand why electrons don't fall into nuclei, or why the sky is blue. Besides, getting through the "basics" consumes so much time that little is left for anything else. What we teach is limited to some degree by the capabilities of our methodology, which, if not still largely rooted in the 19th century, has advanced little beyond the 1950s and has certainly not kept up with the growth in what we *need* to teach.

There is not even universal agreement that any radical changes should be made to instructional methodology; some university instructors argue that an acceptable fraction of their classes are quite able to meet the standards they set (which are in many cases already well below the standards they would prefer to maintain) and that nothing more is needed than a bit of fine-tuning of the curriculum and perhaps some effort to improve skills in problem solving and mathematics at the precollege level. Some may even voice the suspicion that an aid to learning such as CAI amounts to a form of "spoon-feeding" that the "good" students don't need, and could even lead to a lowering of standards by improving the performance of the students who might otherwise be unable to survive on their own.

While this may be a reasonable attitude amongst those who teach honors classes intended to prepare students for careers in chemistry-related research, it fails to address several major aspects of instruction that are rapidly departing from the traditional.

The Students Themselves

In an era in which science is growing more and more a part of our society but seems to be increasingly alienated from the popular culture, it would be a folly of the worst kind to discount the need to capture the interest and respect of those who will not be continuing in science. In short, traditional instruction is increasingly unable to meet the needs of a student clientele that has long since ceased to be traditional in terms of background, language skills, motivation, and cultural values. Even in long-established and prestigious institutions, there is considerable pressure to recognize the special needs of students whose backgrounds would have made serious study of a science out of the question in an earlier era.

The Educational Setting

Perhaps in part because of these factors, education and training outside the context of the traditional school setting is increasing in importance. Distance education, originally conceived as a means of making instruction available to students in remote locations, has in many instances attracted a significant fraction of its clientele from urban areas already provided with adequate institutions. Many of these students find that employment or family responsibilities preclude regular attendance of thrice-weekly classes, but more than a few say they simply prefer nonclassroom instruction.

Industrial training is another important area of education that tends to be all but invisible to the outside community. In order to meet the demands of changing technology, more stringent environmental and safety regulations, and the pressure to improve workplace diversity, industry is devoting considerable resources to in-house training and upgrading. Some of this is classroom-oriented, but perhaps because its students must be paid a salary during training (which provides a strong incentive to make learning as efficient as possible—quite different from the situation in colleges and universities!), industry in general is far more open to nontraditional instructional methodology.

The Economics of Education

Anyone who reads newspapers knows that the growth of taxpayer support to education that occurred between 1960 and 1990 can no longer be sustained. During that time, the term "cost-effectiveness" was not a polite word in the higher education community. In recent years, however, traumatic events have begun to occur that would formerly have been unthinkable. Among these were the announced intention of several institutions to phase out majors programs in some subjects (including chemistry) whose costs could not be justified by the small numbers of graduates they produced. Also in this category are the draconian cutbacks in the Oregon higher education system that are the indirect result of the decision of voters to support public schools from state revenues rather than from local property taxes. Educational institutions both public and private are being challenged as never before to reduce their costs and to justify the continuing use of instructional methods that some outsiders would view as dated if not self-serving.

What We Teach

One aspect of departure from tradition of which we are all aware is the continually changing nature of the subject matter. There have been subtractions as well as additions: Chemistry majors no longer take the courses in phase rule and chemical microscopy that were common in the 1940s, nor are general chemistry students expected to know about the lead chamber and contact processes for the manufacture of sulfuric acid that were common fare in the 1950s. In balance, however, the additions always seem to outweigh the subtractions, and most teachers would agree that there is now more to teach than ever before.

There is, however, another facet to the question of what we teach that has not been very well explored. As teachers, we lay out a course of study and construct examinations whose contents reflect what experience has shown will present a challenge that a reasonable fraction of our students can attain. The point I wish to make here is that this experience is largely based on the traditional model of instruction centered on lectures and homework problems. If a method as powerful as interactive CAI is added to this recipe, it seems likely that we would be able to change our expectations and perhaps add topics that we could not have effectively taught in the past. The interaction between curriculum and methodology can have far-reaching consequences.

An Assessment of CAI

What are the benefits and drawbacks of CAI? It is not easy to answer such a question in the abstract because the methodology is so critically dependent on the nature and quality of the instructional material it delivers, on the capabilities and availability of the hardware required to deliver it, and on the context in which it is employed—that is, the manner in which it relates to the other aspects of instruction. Our initial experiences with situations in which any of these aspects were unsatisfactory (as they all have been in so many instances) can color our perceptions and lead us to discount CAI completely.

Similarly, those of us who are enthusiastic about CAI find it all too easy to overlook or explain away its well-demonstrated instances of failure, especially if we have devoted much of our careers to it and (like the author) have a vested interest in it. Perhaps the best that can be hoped for in the following discussion is that skeptics and enthusiasts alike will encounter a few points they have not previously given much thought to that will temper whatever prejudices they bring to either side of the argument.

At first glance, one might think that the efficacy of CAI as an instructional tool could be assessed by taking two groups of students, using CAI as a major instructional tool in one group only, and then comparing the test results. This was a favorite pastime of educational researchers in the 1970s, and it told us very little: There were probably more positive than negative correlations, but the confidence intervals were rarely very convincing, and the most common result was, in fact, "no significant difference". It is now generally realized that simplistic comparisons of this kind are of little use because they do not address the nature of the other elements of the course such as lectures, reading assignments, and tutorial sessions, nor do they take into account the many human factors connected with the instructional process. In the real world of the classroom, the most impressive results from CAI usually take some time to develop, as instructors discover how to modify the other components of their course to make maximum use of CAI's advantages; by the time this happens, the original course has typically undergone so much modification that meaningful comparisons become very difficult to make.

CAI and the Textbook

Any rational assessment of CAI must look not only at the medium itself, but also at its strengths and weaknesses in comparison with those of conventional modes of instructional delivery. A good starting point might be the textbook.

It has often been pointed out that the printed book is the most cost-effective learning tool that has ever been devised. It is no doubt true that CD-ROM technology can now deliver reading material at a lower cost than the conventional book, but at some considerable disadvantage in convenience, satisfaction, and flexibility of use, not to mention the cost of the viewing device. For most of us, the question of whether textbooks will continue to be needed is a nonissue, although many would stress the need for them to change from their present bloated forms.

In theory, a book (or something like it) is the only thing a student should need in order to acquire the information necessary to meet the requirements of a course. In practice, few students are likely to find this medium entirely satisfactory, even if they are able to read for understanding, which all too many cannot do. Although at least the fortunate among us can expect our students to possess basic literacy, many students lack the experience or perhaps even the desire to utilize it effectively. Those who lack an understanding of punctuation or are unable to recognize and interpret dependent clauses may be able to go through the motions of "reading the words" and little more. Even if these basic skills are present, a certain amount of discipline is required to coordinate the material contained in the body of the text with equations, examples, illustrations, and problems.

Reading is, of course, a major part of most CAI lessons: Even those that rise above the level of being "page turners" must present some information as displayed text. However, delivery of text by CAI offers some special advantages that, if successfully exploited, provide a nice complement to the textbook. Perhaps the major one is focus. To the student, of all the book is available at once, page after page, all interspersed (in most present-day texts) with interesting but distracting illustrations and sidebars. In CAI, however, the text can be presented in controlled increments, focusing attention on one idea at a time, with an opportunity for interaction and reinforcement before the next idea is introduced. In this respect, the computer as an automatic page turner need not be as pejorative as it might first appear. Along with the page turning, there is the finger of the instructor pointing at the significant items on each page, relating one to another, and forcing the student to make some of these connections before proceeding to the next. In many ways, textbooks and

CAI as reading sources serve complementary needs, and each can be of greater value in the presence of the other.

CAI and the Lecture

Much the same can be said for verbal communication through lectures and similar classroom activities. There is probably no better tool than the formal lecture for presenting the broad picture of a subject, laying out the ways in which its various subtopics developed and relate to one another, and generally communicating the instructor's enthusiasm. On the other hand, numerous studies have shown that lecturing is one of the least effective methods of bringing about learning. Other classroom techniques, such as directed problem-solving and "cooperative learning" groups, have met with some success, but they are all limited by the very large differences in the rates that individual students are able to assimilate the material. Another limitation is time; there is rarely enough of it to deal adequately with all of the topics in the course, so many are simply dropped by default.

CAI can be a big help here by taking on the task of presenting most of the "instruction" and bringing the student up to a level at which classroom discussion and other activities (such as demonstrations and laboratory work) can be more interesting and meaningful.

CAI and the Student

The interactive nature of CAI is without a doubt the key to its ability to mediate learning by the student. For "good" students, of course, reading a textbook and attending a lecture involve a degree of interaction with the medium, but these activities are likely to be fully exploited by only a minority of the typical class. More commonly, listening and reading tend to be largely passive activities that are separated in both context and time from the kind of active effort that is involved in problem solving, for example. Learning seems to be most effective when the passive and active aspects of the process are well integrated, and CAI is capable of bringing about this integration to a degree that is probably greater than that of any other methodology save one-on-one instruction.

CAI, more than any other methodology, is able to actively engage the student with the material, narrowing the focus of attention to a single item and forcing decisions as the lesson progresses. Modern learning theory holds that knowledge is not something like a fluid that can be poured into one's head, but must be *constructed* by the individual learner upon an existing framework, which becomes able to assimilate more and more as it grows. A well-designed CAI lesson can aid this process by building on what the student already understands, extending it incrementally, and always challenging the student to go a step farther—very much what an experienced teacher would attempt to do in a one-on-one setting.

One objection that is commonly heard about CAI (or any other system of self-paced or distance learning) is that students are isolated both from the instructor and from their peers. If CAI were the only tool used for instruction, then there might be some substance to this, although the popularity of "open university" programs suggests that many students are quite willing to decouple their education from their social life. Most applications of CAI in fact occur in the context of a "course" in which student–instructor contact includes both classroom activity and direct consultation. In terms of hours per week, both kinds of contact are usually very limited, especially in large-enrollment courses, which often amount to "distance education" by another name. It is my experience that well-designed CAI materials can markedly reduce the number of students showing up at my office door on the days preceding an examination. The students who do come tend to be the very bright ones (who always have questions) as well as some of the weaker ones who often require more help than the CAI lessons can provide, and to whom more time can now be devoted, thanks to CAI's reducing the demand of the "average" students on the instructor's time.

There is another social repercussion of CAI that is worth noting. Most institutions enroll substantial numbers of students who come from disadvantaged backgrounds or have problems with the language, or who come from cultures in which fear of "losing face" (through asking questions or being in danger of giving an incorrect answer in class) seriously inhibits participation in classroom activities. Students falling into these categories seem to benefit from CAI in two ways. First, they appreciate the freedom of being able to give the wrong answer "without anyone knowing", and of course they do so freely, thus participating in the activity to a greater extent than they would otherwise. Second, by progressing through the lesson in this way (and being able to repeat it as often as required), they are likely to build a degree of self-confidence. One would even hope that a lesson that challenges the user with interesting questions would serve as a model for the kind of questioning that should be going on in the student's own mind when tackling any subject, not just chemistry. All of us who are teachers certainly know that the worst students are those who never have any questions; perhaps they never learned to ask them!

CAI and the Teacher

A major impact of any methodology in which the student interacts directly with the medium is to make the teacher more a manager of instruction, rather than its primary deliverer. When this works as it should, the effect is to allow the teacher to accomplish more with a given input of time and effort. "More" in this sense refers not only to the total amount of material taught, but also to the ability to minister to the individual needs of the students. As was pointed out above, if CAI can meet most of the needs of students of average background and ability, then the instructor has more time to devote to those having special needs.

Classroom time is usually a limited commodity that most teachers would prefer to devote to activities such as discussions and demonstrations that build interest and provide motivation. Perhaps the major benefit of CAI to the teacher is to allow more flexibility in the use of class time by taking on some of the more routine aspects of instruction, such as exercises and drills, which are often not well adapted to class instruction anyway.

My own favorite example of this is the balancing of oxidation-reduction equations, a necessary but dreary topic that I have found can be taught entirely by CAI, usually very much to the relief of everyone concerned. A good use of the classroom time regained is to discuss biological oxidation reductions and the origin of atmospheric oxygen—something that I rarely had time for in the past.

Similarly, I now devote only about 15 minutes of lecture time to VSEPR theory, in which I discuss its origins, the theory behind it, and its limitations. The task of actually learning to apply the theory to the usual variety of structural types is handled quite adequately and far more effectively by CAI lessons on the topic.

Most teachers are well aware of the wide range of backgrounds and abilities among the students who come to us each semester. Even in a course such as the one I teach, which carries two years of high school chemistry as a prerequisite, we get students who can barely do stoichiometry problems or cannot tell me what properties a block of wood must possess in order to float in water. There is little time to devote to these and similar topics in the course, and to do a thorough review would be rather dull for the half of the class that is adequately prepared. A set of CAI lessons specifically designed to bring students up to speed on fundamental topics such as these during the first three weeks of the semester has been invaluable in reducing the functional heterogeneity of the class, and there is reason to believe that it has reduced the number of students who get discouraged and drop out during this time.

Even for more advanced courses, CAI can be helpful in overcoming differences in background. For example, second-year organic chemistry instructors usually expect their incoming students to have a good understanding of formal charge and hybrid orbitals. The amount of emphasis placed on these topics in the preceding general chemistry course varies considerably depending on both the textbook and the instructor. A set of CAI lessons covering these topics can efficiently bring the class up to the required level, with minimal diversion from the main subject of the course.

CAI and the Laboratory

Another important area of CAI is in connection with the laboratory.[4] One application is the need, obvious to almost anyone who has taught a laboratory course, to ensure that students come prepared with sufficient understanding of the procedure and the theory behind it. A prelaboratory CAI exercise can help by requiring the student to assemble the apparatus (by dragging their images to the appropriate locations on the display screen), to take readings (of volumes, for example, by using an image of a burette showing the liquid meniscus), and to work up the raw data into a numerical result.

Stanley Smith and his associates at the University of Illinois have taken this further by preparing simulations of entire experiments that can be used in place of "real" experiments.[5] Although this strikes some chemistry teachers as heretical, a number of good arguments can be made for incorporating at least some simulations into the laboratory course. The major benefit is that students can be exposed to a wider range of experiments, including those that would be too complicated, expensive, or dangerous to perform in the laboratory. Also, experiments that have already been performed can be repeated several times in simulation to observe the effect of changing temperatures, concentrations, or other parameters.

These materials and others like them are now available commercially. Some institutions find it practical to have students do "wet" and "dry" labs on alternate weeks, thus doubling the effective capacity of laboratory space, which is frequently in very short supply.

CAI and the Institution

Institutions, or at least the academic departments within them, can benefit from CAI in numerous ways. Use of CAI lessons that relate closely to the kinds of problems found at the ends of textbook chapters can materially reduce the need for collecting and marking homework problems. Although many would expect otherwise, there is no evidence in my experience that this has reduced either the fraction of the class who go through and understand the problems, or the overall class performance. On the other hand, it has allowed me to reduce the number of teaching assistants I need for the course and to redeploy them into the laboratory, where they can be more useful.

As CAI becomes more widely employed within a department, there can be more flexibility in terms of course prerequisites and offering low-enrollment courses in which regular lectures might not be justified. Taken to the extreme, this could bring some aspects of academic education closer to the industrial training model in which CAI allows students to begin the study of a specialized topic whenever it is convenient to do so, and to proceed at their own pace.

Making CAI Make a Difference

Many years of experience ranging back to the late 1960s have shown that the degree to which CAI can improve instruction is a sensitive function of the way in which it is integrated into the course. Perhaps the most common mistake is to treat CAI as an "add-on" to the existing activities of the course, which remain largely unchanged. CAI is a sufficiently powerful tool that it will exert a perturbing influence on the other components of the course; if CAI is used in isolation as an "instructional aid", very little of its potential is likely to be realized, and the economics of using it at all become questionable.

This implies the need for some alteration in the attitude of the instructor that is often quite difficult to bring about, especially in university setting, in which "The Lecture" is seen as the central pillar of the course. In all too many instances, the instructor makes only a one-time announcement that CAI lessons are available for those students who are "having difficulty with the course or who want additional practice". To the typical student, this sounds much like stating that "CAI is available for those who are dumb or want to do extra work" (the latter being taken as solid evidence of the former!)

If CAI is to achieve its potential, it must be integrated into the course in a comprehensive way in which the various components of instruction are utilized in a matter that capitalizes on their own particular strengths. This also implies altering the way in which some components are used, even to the point of reducing their contribution to the mix. This turns out to be much easier to talk about than to achieve: Established practice is very difficult to alter in academic education, particularly when "contact hours" are rigidly defined and enforced.

In my own experience at the college level, the traditional course component that CAI can most profoundly affect is the small-group recitation or tutorial session. These typically meet for one hour a week and are often used to discuss homework problems and to provide a forum for questions and discussion. The major difficulty with these is getting a majority of the students to participate in an active way. Whether out of shyness, lack of preparation, or simple indolence, a significant fraction will not ask questions and will strongly resist answering them. If the instructor or a teaching assistant can be made available on a regular drop-in basis to field the few questions that do come up (usually from the more capable students), most of the purposes of the tutorial can be better served by well-designed CAI lessons, which are more likely to engage the attention of all the students more intensively and for a larger fraction of the time.

As someone who enjoys lecturing, I find CAI to be especially liberating in allowing me to devote the allotted time to what is most interesting, rather than to what has to be learned. The previous reference to oxidation-reduction equations is again relevant here, but equally valid examples can be found within most of the topics covered in general chemistry.

In order to bring all these components together in a way that is meaningful for the student, it is very helpful to have a course guide that sets out a schedule showing the lecture topics, homework problems, CAI lessons, and assigned reading for every week in the semester. It should be particularly explicit about what topics will be covered mainly (or only) by CAI.

Implementing CAI

Clearly, the first thing one needs is the hardware. Although there still are a small number of holdouts from the old mainframe days, virtually all CAI is now delivered by personal computers. In larger institutions these may consist of hundreds of workstations networked to one or more servers, but stand-alone systems are not uncommon, particularly in high schools. There is a tremendous disparity in computer resources across institutions of a given type and size, and this is especially the case with high schools, not a few of which are little beyond the "one computer on a trolley" stage. Until accreditation bodies begin to understand that computer resources can be as crucial to a school's ability to serve its students as are traditional indicators such as teacher-to-student ratio and the number of volumes in the library, this disparity is likely to continue.

Another source of computer hardware is that owned by the students themselves or by their families. In a typical North American urban area, roughly one-third of households have computers, and this is about the ratio commonly found in surveys taken of students in the larger universities. When one looks at different regions or even at different economic groups within a single region, however, very large differences can arise. Nevertheless, home computers can take a significant load off a school-based facility if students can obtain access to the software, and CAI lessons delivered in this way are likely to be used more effectively.

The matter of hardware is complicated by the existence of the two major platforms, IBM-PC and Macintosh. Although Macintosh computers represent a very small (and, sadly, declining) fraction of the overall personal computer market, their outstanding capabilities in terms of graphics, built-in sound, and built-in networking gave them an early start in the educational market, in which they remain strong contenders.

Some institutions, especially the larger ones, have decided to make both kinds of computers available in order to allow their students access to the best software written for either system.

Software

As is usually the case, software (or, more exactly, "courseware") is the major problem: There is simply not enough good material around. This was true 20 years ago, and although it is less true today, most teachers who have looked into the matter would agree that there is far too little to choose from.

It is interesting to compare a commercially available CAI lesson with a textbook in this context. A textbook comes into being through the labors of a variety of editors, designers, and publishing resource persons and marketing specialists. They have no trouble selling this product to school districts and (at the college level) to individual students at prices that allow them to pay for this labor and to earn a substantial profit.

CAI, on the other hand, is still largely a cottage industry in which individual authors, working independently, develop materials according to their own interests and in their own styles. Given the small size of the market and the limited provisions of institutional budgets for instructional software, the prospective economic return is generally far less than what the same authors might expect to receive from a successful textbook.

As long as CAI is regarded as little more than an add-on to conventional instruction (thus guaranteeing its status as an expendable item in the institutional budget), this situation is likely to continue.

Although the major textbook publishers have been aware of CAI since its inception and have occasionally tested the waters, they have in the past regarded CAI more as a marketing tool for their highly lucrative major product than as a parallel product. Publishers now usually offer computer materials as incentives to potential adopters of their textbooks, but the quality of these offerings (typically a set of multiple-choice questions that is supposed to pass for CAI) usually does little to encourage wider use of CAI. Now that "multimedia" is the rage, one can see this cycle repeating itself, with publishers desperate to pack just about anything they can get their hands on into a shiny CD-ROM disk.

At the present time the best (if barely adequate) collections of CAI courseware come from small companies[6] that specialize in this area and sell mainly to institutions. The time may be coming, however, when this pattern will change, with large collections of materials being sold as blister-packaged CD-ROMs in college bookstores. At the college level, this will have the effect of shifting the funding of CAI from the institution to the students. However, this may not be all that onerous, given the relatively small cost of the courseware within this much broader market, in comparison to the high costs of tuition.

Developing CAI

In principle, instructional software can be written in any programming language, and most of the current languages and a number of dead or dying ones have been used for this purpose. For fairly narrowly defined applications, such as teaching students how to name organic compounds or interpret titration data, this approach can work quite well, as is evident from some of the products currently available.

Experience has shown, however, that general-purpose programming languages such as BASIC, Pascal, and C are not very well suited to the task of preparing interactive instructional software of a more general nature. It is fairly easy to assemble the pieces needed to construct a simple linear dialog of questions and answers, but much more than this is generally required for an effective instructional program in which sequencing and display page design play crucial roles; it is in these areas that such languages become awkward to use. This discourages the fine-tuning and updating that are essential steps in the development of any instructional program. Long-term maintenance is also a very difficult problem, particularly if the programming is done by someone other than the designer or author.

Languages designed specifically for instructional programming allow the author to focus more on the content and structure of the lesson, rather than on the details of how to implement them in code. The development of such languages was a significant step in allowing teachers having minimal interest or expertise in programming to write useful CAI programs. The most famous of these languages was TUTOR, the language of the PLATO CAI system developed at the University of Illinois and implemented on a Control Data Corporation mainframe system. As important a factor as the programming language was the PLATO terminal whose display unit was a plasma-discharge panel that allowed the effective use of graphics at least a decade before this became common practice. Several derivatives of TUTOR are still in use; TenCore (for IBM PC systems) is probably the most well known of these.

The rise of the personal computer inspired the development of another language, PILOT, which in its various dialects was implemented on a wide variety of small computers. The IBM-PC implementation by Washington Computer Services became very popular in the mid-1980s and spawned the development of a number of chemistry CAI programs that are still in wide use.

Although authoring languages make it relatively easy to specify the content and organization of a lesson, they are somewhat awkward to use at the screen-design level. For anything other than display of relatively linear text, the location at which a text or graphics item appears on the screen must generally be specified in terms of some kind of a coordinate system. Because the optimum location of a given item can only be determined by viewing it in relation to the other material on the screen, the lesson designer must frequently jump between viewing the screen and the code in order to achieve a desired visual effect.

By 1990, the older text-with-graphics approach to screen design had grown into "multimedia", in which text, still pictures, and motion pictures from a variety of sources must all be coordinated. The only practical way of achieving this is to compose the material on-screen, so that what the author sees is exactly what the user will see in the finished product. The most advanced of these authoring systems treat displayed items as objects that can be created on the screen as needed (text and simple drawings) or imported from other sources and pasted onto the screen. These items can then be dragged to their desired locations and made to appear or disappear individually or collectively. On-screen editing and resizing of text and graphics makes it easy to achieve a unified and pleasing screen design. Animation editors allow objects to move under program control. Student response modes also become more flexible: In addition to the traditional methods of keyboard input, the student can click the mouse on a button or on a specified object or screen area, or an object may be dragged to a required destination.

Of the growing variety of multimedia authoring systems now available, only a few are designed specifically for development of instructional software. At the time of this writing, the two major contenders are AuthorWare and IconAuthor. Both are suitable for multiplatform development (meaning that versions for both Macintosh and Windows are available, and lessons developed on one system can be ported to the other). The authoring process involves dragging icons representing operations such as text display, calculations, question-response sequences, etc., onto a "flow line" that establishes the presentation sequence. Multiple nesting levels permit the construction of an organizational structure that can be arbitrarily complex. The author inserts the instructional "content"—text, graphics, animated motions, etc.—into the individual icons either when they are initially placed on the flow line or when they are encountered while running the lesson.

Because very little if any program code need be written, these systems allow the author to concentrate on the program design and content, and thus make the prospect of writing lessons attractive to a far wider range of authors. At the same time, fine-tuning and minor updating of the lessons becomes far easier, since this can be done on the fly while running the lesson and without the need to understand the underlying structure of the program.

These strong advantages are not without a few drawbacks, however. Because these systems do not produce editable code, it is very difficult to make small alterations in the way some features work or to make wholesale major changes in such things as the appearance of a prompt for user input, for example. In a programming-based system this functionality might be localized in a specific subroutine or function, which could easily be changed. For a lesson constructed by an authoring system, every instance of the kind of interaction being altered must be found and changed individually. Similarly, production of a version in another language requires opening every icon containing text and translating each individually.

In general, however, authoring systems of this kind tend to stay quite close to being state-of-the-art, and are now beginning to incorporate database access and hypertext capabilities. Although these systems tend to be rather expensive to acquire, they are undoubtedly more economical than programming languages if the investment of time and effort required to produce a nontrivial program is calculated.

Some mention should be made of several other systems that have been used to develop instructional programs. Several programming-language-based systems are now available with icon-based visual editors that simplify the initial generation of program code. TenCore, mentioned in a previous section, is one of these. ToolBook, a Windows-based product, also allows for a wide range of interactive functions and has been used for instructional programs. HyperCard and SuperCard for the Macintosh have been quite popular for CAI programs that do not require complicated sequencing or student-response analysis. LinkWay is a Windows-based product from the developer of PC-Pilot that embodies similar capabilities. While none of these products allows multiplatform development or offers the sophistication and ease of use of AuthorWare or IconAuthor, all are quite capable of producing perfectly good instructional programs within the scope of their limitations.

Authors and Authoring

Given the wide selection of authoring tools available, there is no reason why any teacher who has the time and motivation should not be able to develop a CAI lesson that is reasonably simple but can still serve as a useful supplement for instruction. As the scope of the work expands, and particularly if this includes wider distribution of the materials, the level of sophistication required tends to go up, and the time needed to achieve this rises even more rapidly. The amount of detail that must go into an interactive tutorial-type lesson of commercial quality is far greater than that required for a textbook chapter on the same topic. Estimates of the ratio of authoring time required per unit of student time to go through a lesson vary widely, but generally range from 50:1 to 200:1.

It is partly for this reason that most of the major extended sets of chemistry courseware that address large segments of the general chemistry curriculum have been developed by perhaps a half-dozen authors who have been active in the field for some time (since the mainframe days, in many cases) and have devoted a major part of their careers to this activity.

Attempts to provide a wider range of CAI materials though government-funded courseware-development projects have been made since the earliest days of CAI. Few have come even close to meeting their sponsors' expectations, and most have quickly sunk into oblivion. The most obvious reason for this pattern of failure is the nature of the grant process: A "principal investigator" obtains the funds, which are mostly used to employ graduate students and technical personnel, to design and implement the instructional materials. Few if any of these people have the writing and expository skills of the experienced and talented teacher, a fact that is all-too-often abundantly evident in the final product. Eventually, the project ends, the money runs out, and the participants disperse, usually leaving the software in a state of limbo from which it cannot be resurrected.

CAI is very much the art of the teacher/communicator, not of the programmer and not of the "multimedia designer". It should be pointed out, however, that CAI is largely a graphical communications medium, and careful attention to typography, layout, and other elements of graphic design is an important consideration. These talents are not difficult to acquire; perusal of existing state-of-the-art materials is a very good way to begin, and advice from experienced designers can be invaluable.

One thing to avoid is excessive dependence on multimedia effects such as fancy ways of making text appear and disappear, overly ornate displays, inclusion of sounds without good reason, and similar pizzazz. These are stock-in-trade of the bland and vacuous multimedia presentations that are very much in vogue in the corporate communications world, where form must often reign over substance, but they tend to be distracting and intrusive in a serious instructional program.

Lesson Design Considerations

CAI lessons have traditionally been categorized as expository (teaching new material), simulation, and drill-and-practice. These three categories are by no means mutually exclusive, and in fact most good tutorial CAI materials contain elements of all of them. Lessons intended exclusively to supplement regular instruction might be entirely simulation or drill, while those designed to provide a major part of instruction in a topic will be largely expository, usually following a so-called Socratic model. The latter term implies a sequence of questions followed by responses that in turn elicit further questions, so that the dialog progresses as an idea or concept is developed.

The major difficulty in getting a computer to approximate such a dialog is that of detecting and responding to individual differences between students. To some extent this can be accomplished by careful structuring of the material and providing convenient means to allow students to navigate through the structure. More ambitious programs use the student's performance to determine the level and sequence of flow through the lesson; this is known as adaptive CAI. More recently, there have been efforts to employ artificial intelligence techniques to this end, but there is not a great deal of evidence that the added benefits have been worth the greatly increased complexity of the program, except perhaps in certain areas of elementary education.

At the level of a single topic, it is often important to cover points that are not usually treated explicitly in conventional instruction, or are only mentioned in passing. As an example, consider a lesson whose purpose is to teach "simplest formula" calculations. Assuming that our intention is also to relate this to chemical principles and not to simply train students to perform an algorithm, one would probably want to include sublessons on such topics as relating decimal numbers to integer fractions ($1.67 = 5/3$, etc.), converting a sequence such as ($1, 0.33, 2$) to integers, and expressing a true formula as mole ratios.

Questions involving calculations and algebraic manipulations can be approached in various ways. Besides the obvious method of having the student use a handheld calculator or a calculator function built into the program, it is often more expeditious to present a simple proportionality calculation as a multiple choice of a/b, a*b, and b/a. More complex operations, such as setting up an equilibrium constant expression, are most conveniently handled by having the student drag the various terms to the appropriate locations in the skeleton equation.

One of the most important lesson design considerations, and one that must be established at a fairly early stage, is the user interface. This refers to menus and other navigational aids, including provision for exiting a lesson, and a consistent system of prompts so that the student always knows when a response is expected and whether it involves entering a numeric value, typing a word, clicking the mouse on an object, etc. For any except simple multiple-choice responses, provision should be made for a sequence of "help" responses that provide increasingly detailed hints and, finally, the correct answer.

There is some disagreement among teachers (and certainly among CAI authors) on the proper way to deal with questions in which students must enter chemical formulas. Some believe that students should explicitly type in subscripts and superscripts; programs that allow this typically make use of special keyboard operations for this purpose. Others feel that these operations tend to get in the way of smooth and easy operation of the program, and point out that few if any students have ever been corrupted by typing "h2po4-", "so4-2", "so4-", or even "so42-" as long as the context is clear and the proper representation of the formula is displayed after the answer is accepted by the program.

A similar consideration applies to the inclusion of units in numeric answers; some argue that if we require them in homework assignments or exams, we should similarly insist on them in CAI lessons. The opposite opinion is that as long as the unit is apparent in the context of the problem and is displayed after the number is accepted, it is more important to minimize the drudgery of typing than to use the computer as an implacable arbiter of proper notational syntax.

Above all, CAI lessons should be interesting. They should come across as dialogs between student and teacher, challenging without being intimidating, respectful without being condescending. Lessons should employ a variety of response patterns and require a minimum amount of text reading. Unnecessary sounds and visual effects should be avoided.

Conclusion and Outlook

There is reason to believe that the predictions that were made in the early 1970s about the potential of the computer as a tool for the delivery of instruction may finally be realized over the next decade. Ultimately, the same economic forces that have made the computer an indispensable tool in so many other endeavors will force educators to adopt them and adapt to them, although perhaps not without being challenged by the growth of home schooling and alternative forms of schooling, and the general trend toward part-time and continuing education among the working public through open-university institutions.

Software will continue to be a major problem, however; there will be a very limited selection of well-crafted, comprehensive CAI materials for some time to come. In the short term, the traditional publishers will not be of much help, and will likely exacerbate the problem as they issue CD-ROMs packed with about anything they can get their hands on (including, of course, their textbooks) in order to effect an image of keeping up with technology. Eventually, however, as the traditional textbook market collapses under its own weight, to be replaced by custom-produced reading materials and perhaps "reference texts" specifically designed to support CAI, the publishing industry is likely to assume a major role.

As multimedia evolves from being a novelty to a necessity, we will see interactive video and the spoken word incorporated more and more into CAI. Over the longer term, students should be able to purchase a CD-ROM containing the CAI instructional materials, all required reading, a comprehensive reference text, and other reference materials for an entire course. Accompanying this will very likely be a list of institutions offering credit for completing the course in a distance-education format.

Finally, there is always the question: Will CAI replace the teacher? My answer to this has always been "Any teacher who *can* be replaced by a computer *should* be replaced."

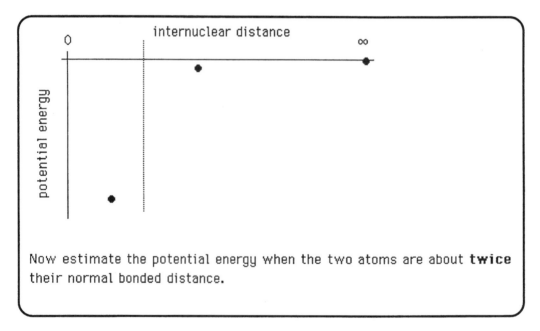

FIGURE I   Students see potential-energy diagrams in textbooks and on the blackboard, but often have little grasp of their form or meaning; all too frequently their first experience of sketching one is on a test. This exercise leads the student through the construction of such a diagram in four steps; only the position of the minimum is given. Once this operation has become a part of one's active personal experience, it is more likely to be remembered and understood.

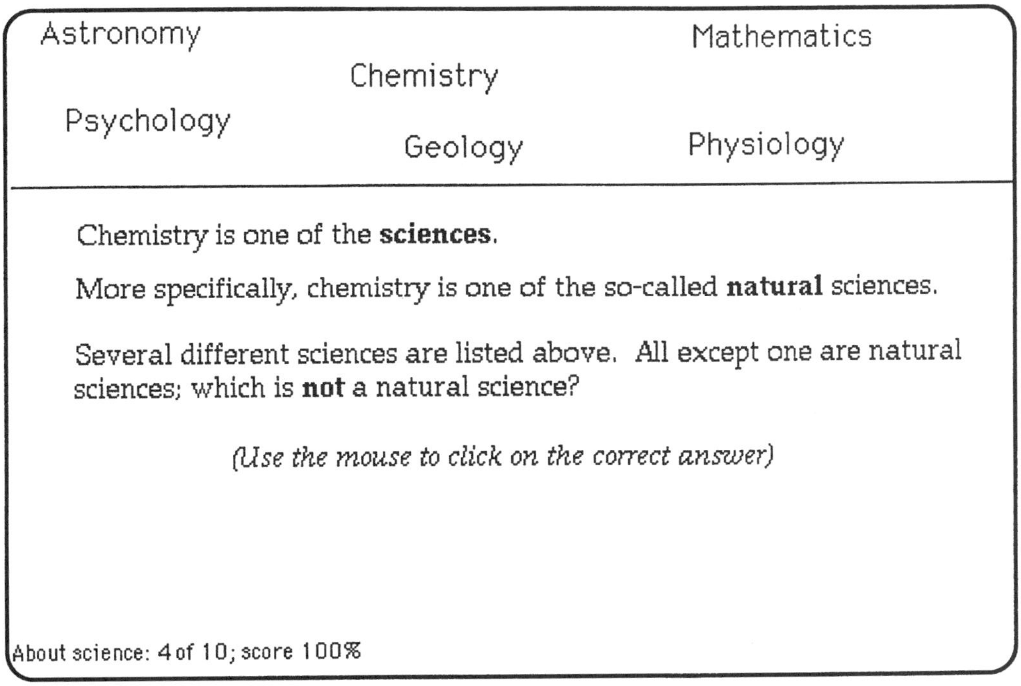

```
 Astronomy                                       Mathematics
                        Chemistry
    Psychology
                        Geology             Physiology
```

Chemistry is one of the **sciences**.

More specifically, chemistry is one of the so-called **natural** sciences.

Several different sciences are listed above. All except one are natural sciences; which is **not** a natural science?

(Use the mouse to click on the correct answer)

About science: 4 of 10; score 100%

FIGURE 2 CAI makes it much easier to implement a specific instructional strategy in a consistent way. In this example it is the well-known strategy of making the student distinguish between examples and nonexamples of a concept. In a well-designed lesson, even the "wrong" answers can yield informative responses.

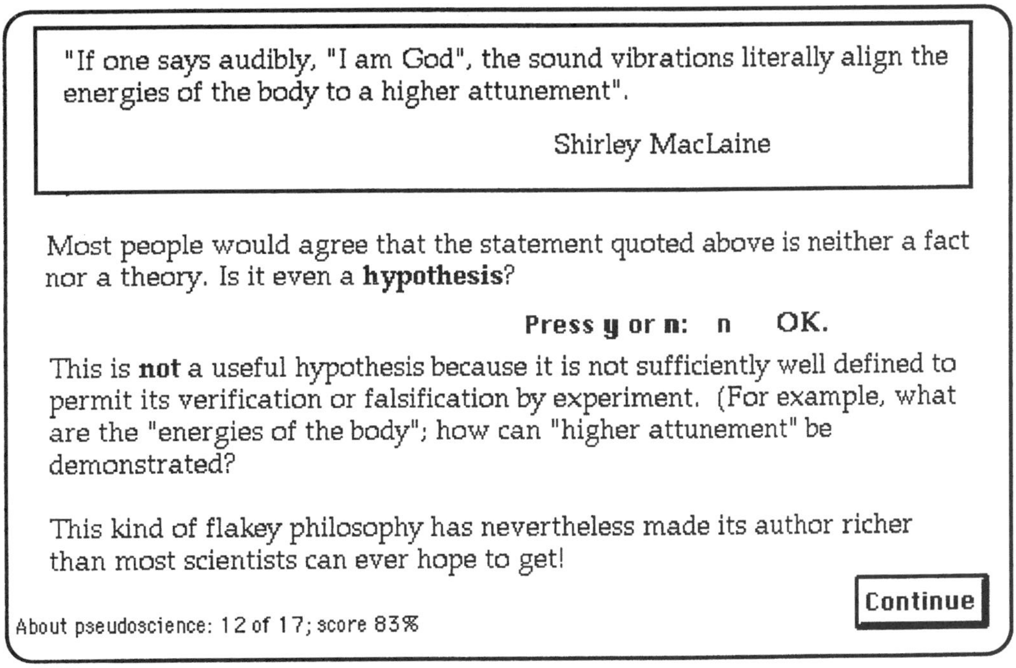

"If one says audibly, "I am God", the sound vibrations literally align the energies of the body to a higher attunement".

Shirley MacLaine

Most people would agree that the statement quoted above is neither a fact nor a theory. Is it even a **hypothesis**?

Press y or n: n   OK.

This is **not** a useful hypothesis because it is not sufficiently well defined to permit its verification or falsification by experiment. (For example, what are the "energies of the body"; how can "higher attunement" be demonstrated?

This kind of flakey philosophy has nevertheless made its author richer than most scientists can ever hope to get!

Continue

About pseudoscience: 12 of 17; score 83%

FIGURE 3 Even a simple multiple-choice or yes–no question can be a useful device for keeping the student focused on a topic and providing reinforcement. The lesson set from which this and the previous example are taken is in the public domain; see http://www.sfu.ca/chemed/ for more information.

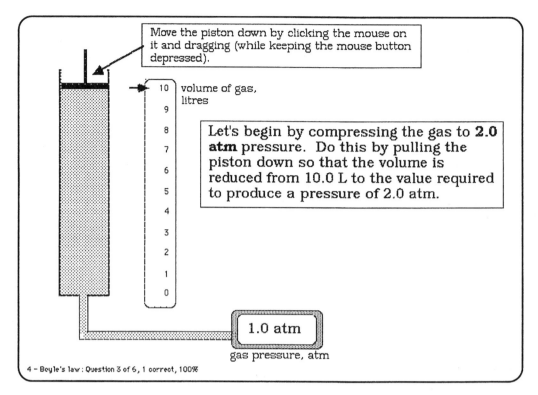

FIGURE 4 Even when students are able to manipulate algebraic expressions, they may have difficulty visualizing their meaning. This lesson allows the student to discover the nature of the relation, which is then cast into algebraic form.

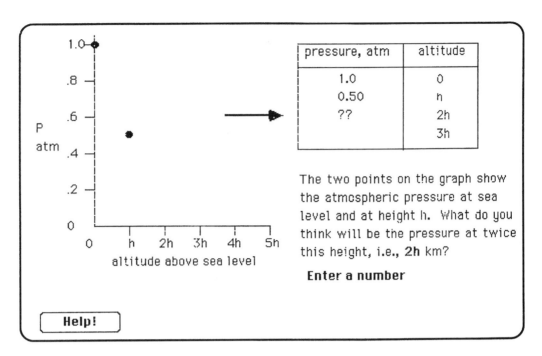

FIGURE 5   Although the barometric distribution law is not usually taught in general chemistry (its derivation requires a bit of calculus), there is no reason to completely shield students from it in a discussion of gases and the atmosphere. Given the general importance of exponential relations in the sciences, there is good reason to develop the concepts of constant fractional rate of change and of "half-values" as soon, and in as many different contexts, as possible.

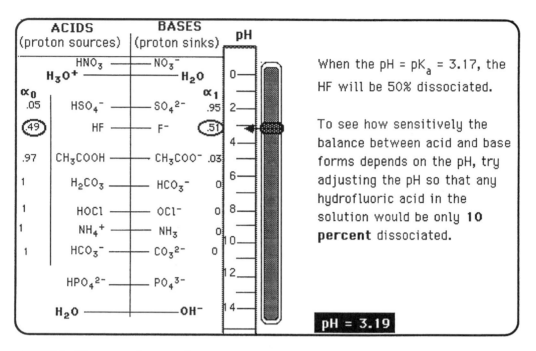

FIGURE 6   This screen, if presented in its completed form as shown here, would be difficult to comprehend without considerable study. In a CAI exercise, however, the various components can be developed and used individually to ensure that each is understood. Later, when several of these are combined, as in this lesson that relates the pH to average proton energy and dissociation fraction, the student can quickly focus on what is new.

This question is one of a sequence in which the student is required to adjust the pH so as to achieve a specific objective, thus building a "feeling" for the relationship and adding meaning to the detailed mathematical treatment that comes later. The purpose of this particular question is to point out a very simple relation that seems not to be widely known.

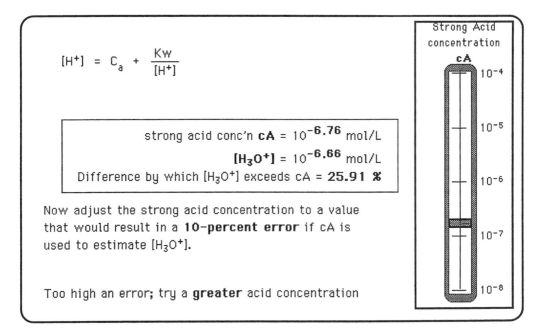

$$[H^+] = C_a + \frac{Kw}{[H^+]}$$

Strong Acid
concentration
cA

10^{-4}

10^{-5}

strong acid conc'n **cA** = $10^{-6.76}$ mol/L

$[H_3O^+]$ = $10^{-6.66}$ mol/L

Difference by which $[H_3O^+]$ exceeds cA = **25.91 %**

10^{-6}

Now adjust the strong acid concentration to a value
that would result in a **10-percent error** if cA is
used to estimate $[H_3O^+]$.

10^{-7}

Too high an error; try a **greater** acid concentration

10^{-8}

FIGURE 7   This screen is taken from a lesson in which the exact equation relating the pH of a strong acid solution to its concentration is developed. The point is not to "teach" this particular equation (which has little practical use), but rather to show that the trivial relation that says the pH of a 10^{-4} M HCl solution is 4 cannot be valid at 10^{-8} M. What *is* worth teaching is the idea that students ought to ask themselves questions such as the one posed here. A CAI lesson provides a convenient means of exploring relationships of different kinds in a context that can provide help, reinforcement, and connection to related topics.

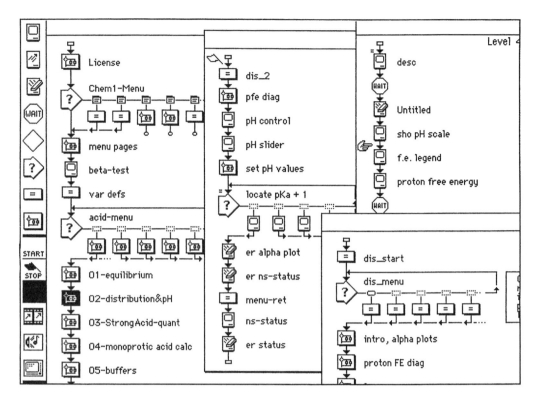

FIGURE 8 Modern CAI authoring systems such as AuthorWare (shown here) make only minimal use of traditional code. "Programming" consists largely of placing icons on the flow line and filling them with content, which may be text or graphics for display, animations, accepting various types of student responses, etc. Groups of related icons can be combined into group icons (the square boxes in the illustration) and nested at various levels. The ability of the author to open and modify icons while executing the course allows continual refinement of the content, appearance, and pacing of the lessons.

References

1. Castleberry, S.; Lagowski, J.; "Individualized instruction using computer techniques"; *J. Chem. Educ.* 1970, *47*, 91–96.
2. Lower, S.; "Audio-tutorial and CAI aids for problem solving in introductory chemistry"; *J. Chem. Educ.* 1970, *47*, 143–145.
3. Smith, S.; "The use of computers in the teaching of organic chemistry"; *J. Chem. Educ.* 1970, *47*, 608–611.
4. Cavin, C. S.; Cavin, E. D.; Lagowski, J. J.; "A study of the efficacy of computer-assisted laboratory experiments"; *J. Chem. Educ.* 1978, *55*, 601–604.
5. Smith, S.; Jones, L.; "Images, imagination and chemical reality"; *J. Chem. Educ.* 1989, *66*, 8–11.
6. "Publishers of commercially available instructional software", accessible via the WWW on the ChemCAI home page at URL http://www.sfu.ca/chemcai/.

Index

A

AAA, *See* Active analog approach
Ab initio methods
  current use of calculations, 267–268
  in molecular orbital calculations, 243–252
  limitations, 254–255
    to determine conformational energy differences in molecules, 250–251
Abstracts and indexes, electronic searching, 118–119
Accounting systems in universities, in support of computer services, 89
Active analog approach (AAA), use in molecular modeling of dopaminergic ligands, 319–323
Active compounds, molecular modeling, 319, 320–322
Active-site modeling, receptor structures, 328
Actuators and sensors, in computer interfacing, 197
Address and telephone directories, use by chemists, 47
Administrative systems, role of computers in universities, 77
Agonist pharmacophore development, and D1 receptor mapping, 320–323
Alerting/document delivery services, for electronic journals, 116
Algebraic manipulation and calculation, in computer-assisted instruction, 369
Aliasing, and analog-to-digital conversion, 193–194
Amdahl's law, for parallel software, 75
American Association of State Colleges and Universities (AASCU), survey of technology in university education, 77–78
Analog signal conditioning, in computer interfacing, 190, 197–199
Analog-to-digital conversion, in computer interfacing, 52, 191–194
Analysis of variance, integrating into curriculum, 217–219
Application tools, on the Internet, 99
Archiving of documents, 66–68
Assessment of computing knowledge in students, 14–15
Atomic orbitals, in Hartree–Fock models, 261
Authoring systems, and computer-assisted instruction, 366
Automated laboratories in universities, 86
Autonomous learners, characteristics, 12, 13t

B

Bacteriorhodopsin, and G-protein-coupled receptors, 327
Bandwidth, effect of low cost, 74
Basis sets, in Hartree–Fock models, 261
Beginners, computer courses, 10–12
Bimolecular reactions in solution, calculation of rate constants using Brownian dynamics, 301–303
Binary values, electronic representation, 190
Binding constants
  and ligand–receptor interactions, 313
  calculation, 325

Biochemistry
  current importance of computations, 283
  software, 13–14
  use of computers in an advanced course, 16–22
BioQUEST Library, adoption at Howard University, 14–15
Blood buffer system, modeling with spreadsheets, 153, 156–157
Bodner pedagogical model, 7
Bond lengths
  current use of Hartree–Fock models to calculate, 268
  in molecular mechanics calculations, 316–317
Bonds, carbon–carbon, in molecular mechanics calculations, 316–317
Born–Oppenheimer approximation
  in Hartree–Fock models, 260–261
  in molecular mechanics, 315
Brownian dynamics
  in calculation of rate constants for bimolecular reactions in solution, 301–303
  in molecular representation, 299–302
Browsing, in Internet searching, 100
Buffers
  blood, modeling with spreadsheets, 153, 156–157
  numerical differentiation with spreadsheets, 149–150
Bulletin board systems, on the Internet, 97

C

CAI, *See* Computer-assisted instruction
Calculations
  and algebraic manipulations, in computer-assisted instruction, 369
  important current uses to chemists, 267–274
  See also Computation, Numerical methods
Calculus, survey of numerical methods, 172–176
Calendar programs, use by chemists, 46
Campus services, developments related to computers, 82
Carter, Forrest L., work in chemistry of computers, 41
Case/Amdahl principle, for memory size and processor speed, 75
CD–ROMs
  chemistry journals, 115
  role in chemical information systems, 29
  searchable abstracts and indexes, 119
Centers for chemical computation, 37–38
Charge density, in molecular visualization, 231–235, 238f
Chemical Abstracts
  and other searchable sources, 118
  on-line database, 112–113
Chemical Abstracts Service (CAS), role in chemical information systems, 29
Chemical education listserv, 125
Chemical Educator, World Wide Web journal, 126
Chemical formulas, entering during computer-assisted instruction, 369

Production: Randall Frey, Susan D. Fisher

Copyediting: Matthew J. Hauber

Indexing: Jay C. Cherniak

Cover design: Amy O'Donnell

Typesetting: Roy A. Barnhill, Coconut Creek, FL

Printing/binding: Maple Press, York, PA